Evolutionary Genomics and Proteomics

Evolutionary Genomics and Proteomics

Mark Pagel
Andrew Pomiankowski

Editors

Sinauer Associates, Inc. Publishers
Sunderland, Massachusetts 01375

About the Cover:

The cover illustration shows part of an undirected phylogenetic tree for a short protein sequence. The proteins are represented as strings of colored spheres (amino acids), and proteins connected by a bar differ from each other by a single amino acid. The image alludes both to the evolutionary relationships revealed by genomics and proteomics and to the structure of biological networks. At the upper right, a ribosome translates a freshly transcribed mRNA into a protein. Art and cover layout by Joanne Delphia.

Evolutionary Genomics and Proteomics

Copyright 2008 by Sinauer Associates, Inc.

For information, address:
Sinauer Associates
23 Plumtree Road
Sunderland, MA 01375 U.S.A.

FAX: 413-549-1118
Email: publish@sinauer.com
Internet: www.sinauer.com

Library of Congress Cataloging-in-Publication Data

Evolutionary genomics and proteomics / edited by Mark Pagel,
Andrew Pomiankowski.
      p. ; cm.
  Includes bibliographical references and index.
  ISBN-13: 978-0-87893-654-0 (paperbound)
  ISBN-13: 978-0-87893-655-7 (hardcover)
 1. Evolutionary genetics. 2. Genomics. 3. Proteomics.
 [DNLM: 1. Evolution, Molecular—Review. 2.
 Genome—genetics—Review. 3. Proteome—genetics—Review. QU 475
 E956 2008] I. Pagel, Mark D. II. Pomiankowski, Andrew. QH390.E98
 2008 572.8′38—dc22
                2007035125

Printed in China

5 4 3 2 1

Table of Contents

CHAPTER 5   Evolution of Genomic Expression   81

Bernardo Lemos, Christian R. Landry, Pierre Fontanillas, Susan C. P. Renn,
Rob Kulathinal, Kyle M. Brown, and Daniel L. Hartl

CHAPTER 11   Molecular Signatures of Adaptive Evolution   241
Alan Filipski, Sonja Prohaska, Sudhir Kumar

Preface

Our idea to invite a group of leading researchers to contribute chapters to a volume on evolutionary genomics and proteomics grew out of the sense that a new kind of evolutionary analysis of organisms is emerging. This new kind of analysis is driven by technological developments that make genome sequencing possible at a scale not remotely possible only ten years ago, and equally make available large amounts of information on gene transcription, regulation and expression, protein structure, and protein interactions. Putting together these still-emerging technologies and the data they produce is the job of the fields of evolutionary genomics and proteomics. What makes the new science fresh is the perspective that, perhaps for the first time, the organism is in sight, but it is not going to arise as the sum of its genomic parts. Rather, if the organism resides in the genome, the trick is going to be to see how the instructions the genome encodes work together to make phenotypes, and how those instructions are written there by selection acting on individuals.

The need for an edited volume arises because the fields comprised by evolutionary genomics and proteomics are specialized and often rely on methods germane to particular kinds of questions. Equally, few broad generalizations have yet emerged. In some ways, evolutionary genomics and proteomics stands at about the same place that the neo-Darwinists did in the early part of the twentieth century—full of ideas with as yet enormous amounts of data still to come. And this—the production and interpretation of data—is likely to be the defining feature of genomic and proteomic science for some time. We are in for a period of intense empirical activity as researchers use new technologies to reveal how natural selection acting on phenotypes has molded the content and structure of genomes.

The chapters in this volume sample important pieces of the genomic and proteomic puzzle that will need to be assembled to produce a science of both the genome and the phenotype. These pieces include the determined work of understanding genes, where they come from, their regulation, the proteins they produce, and how those proteins interact in networks. The chapters include attempts to put diverse topics together and to study their links to phenotypes and to variation among them. For this sample, we are grateful to the authors of the chapters for providing summaries of their ideas.

Books, like phenotypes, don't simply arise as a sum of parts; they must be steered through a course of development. The kernel of the book was devised during a sabbatical Andrew Pomiankowski spent at the Collegium Budapest, Hungary, and during which Mark Pagel visited. We are extremely grateful for the invitation to visit from Eörs Szathmáry, and to the Rector and staff of the Collegium for providing such a pleasant and stimulating environment in which to work. Andy Sinauer was supportive of the project from its inception and forbearing when we encountered difficulties. Laura Green's flair at copyediting holds the chapters together as one, and Kathaleen Emerson's efficiency in coordinating the book's production has been a delight. David McIntyre lent his able assistance to the editorial effort and conceptualized the cover art. Joan Gemme produced an elegant and lively book design, and turned manuscripts, tables, and figures into high quality and colorful book chapters, Joanne Delphia created the beautiful cover illustration, and Chris Small oversaw the art production. Marie Scavotto has made important contributions to the promotion of the book.

Mark Pagel and Andrew Pomiankowski
Reading and London, UK
August, 2007

Contributors

Eric Bapteste Department of Biochemistry and Molecular Biology, Dalhousie University, Halifax, Nova Scotia, Canada B3H 1X5

Timothy H. Bestor Department of Genetics and Development, College of Physicians and Surgeons of Columbia University, New York, NY 10032, U.S.A.

Kyle M. Brown Department of Organismic and Evolutionary Biology, Harvard University, 16 Divinity Avenue, Cambridge, MA 02138, U.S.A.

Deborah Charlesworth Institute of Evolutionary Biology, University of Edinburgh, Edinburgh, EH9 3JT, U.K.

W. Ford Doolittle Department of Biochemistry and Molecular Biology, Dalhousie University, Halifax, Nova Scotia, Canada B3H 1X5

J. J. Emerson Department of Ecology and Evolution, University of Chicago, 1101 E. 57th St., Chicago, IL 60637, U.S.A.

Chuanzhu Fan Department of Ecology and Evolution, University of Chicago, 1101 E. 57th St., Chicago, IL 60637, U.S.A.; and Arizona Genomics Institute, Department of Plant Sciences, University of Arizona, Tucson, AZ 85721, U.S.A.

Alan Filipski Center for Evolutionary Functional Genomics, The Biodesign Institute, and School of Life Sciences, Arizona State University, Tempe, AZ 85287, U.S.A.

Pierre Fontanillas Department of Organismic and Evolutionary Biology, Harvard University, 16 Divinity Avenue, Cambridge, MA 02138, U.S.A.

Mary Grace Goll Department of Genetics and Development, College of Physicians and Surgeons of Columbia University, New York, NY 10032, U.S.A.

Daniel L. Hartl Department of Organismic and Evolutionary Biology, Harvard University, 16 Divinity Avenue, Cambridge, MA 02138, U.S.A.

Ines Hellmann Center for Comparative Genomics, Institute of Biology, University of Copenhagen, Copenhagen, Denmark

Laurence D. Hurst Department of Biology and Biochemistry, University of Bath, Bath, BA2 7AY, U.K.

Eugene V. Koonin National Center for Biotechnology Information, National Library of Medicine, National Institutes of Health, Bethesda, MD 20894, U.S.A.

Rob Kulathinal Department of Organismic and Evolutionary Biology, Harvard University, 16 Divinity Avenue, Cambridge, MA 02138, U.S.A.

Sudhir Kumar Center for Evolutionary Functional Genomics, The Biodesign Institute, and School of Life Sciences, Arizona State University, Tempe, AZ 85287, U.S.A.

Christian R. Landry Department of Organismic and Evolutionary Biology, Harvard University, 16 Divinity Avenue, Cambridge, MA 02138, U.S.A.

Bernardo Lemos Department of Organismic and Evolutionary Biology, Harvard University, 16 Divinity Avenue, Cambridge, MA 02138, U.S.A.

Manyuan Long Department of Ecology and Evolution, University of Chicago, 1101 E. 57th St., Chicago, IL 60637, U.S.A.

Camilla L. Nesbø Department of Biochemistry and Molecular Biology, Dalhousie University, Halifax, Nova Scotia, Canada B3H 1X5

Rasmus Nielsen Center for Comparative Genomics, Institute of Biology, University of Copenhagen, Copenhagen, Denmark

Brian Oliver Laboratory of Cellular and Developmental Biology, NIDDK, National Institutes of Health, Bethesda, MD 20892, U.S.A.

Mark Pagel School of Biological Sciences, University of Reading, Reading, RG6 6AJ, U.K.

Csaba Pál Department of Zoology, University of Oxford, Oxford, OX1 3PS, U.K.

László Patthy Institute of Enzymology, Biological Research Center, Hungarian Academy of Sciences, Budapest, H-1113, Hungary

Andrew Pomiankowski CoMPLEX and the Department of Biology, University College London, London, NW1 2HE, U.K.

Sonja Prohaska Center for Evolutionary Functional Genomics, The Biodesign Institute, and Department of Biomedical Informatics, Arizona State University, Tempe, AZ 85287, U.S.A.

Susan C. P. Renn Bauer Center for Genomics Research, Harvard University, 7 Divinity Avenue, Cambridge, MA 02138, U.S.A.

Christopher B. Schaefer Department of Genetics and Development, College of Physicians and Surgeons of Columbia University, New York, NY 10032, U.S.A.

Andreas Wagner Department of Biochemistry, University of Zurich, CH-8057 Zurich, Switzerland

Yuri I. Wolf National Center for Biotechnology Information, National Library of Medicine, National Institutes of Health, Bethesda, MD 20894, U.S.A.

Olga Zhaxybayeva Department of Biochemistry and Molecular Biology, Dalhousie University, Halifax, Nova Scotia, Canada B3H 1X5

CHAPTER **1**

The Organismal Prospect

Mark Pagel and Andrew Pomiankowski

Evolutionary Genetics in the Time of Genomics

THE MODERN GENOMIC ERA ARRIVED 28 JULY 1995 with the publication in *Science* of the complete sequence of the *Haemophilus influenzae* genome (Fleishman et al. 1995). The *H. influenzae* genome was 1.8 million base pairs long and the project took a year to complete. At that time the authors were able to announce proudly that "the *H. influenzae* genome sequence (Genome Sequence DataBase accession number L42023) represents the only complete genome sequence from a free-living organism" (p. 269). Progress in genomics was breathtaking. By October of the same year, the complete sequence of the *Mycoplasma genitalium* genome appeared—at 580,000 base pairs, the smallest known genome of any fully self-replicating organism. By 1996 the genomes for *Synechocystis* sp. PCC 6803, *Methanocaldococcus jannaschii*, and another *Mycoplasma* species were published, and by the middle of 1997 the complete sequence of *Saccharomyces cerevisiae*, the first eukaryotic organism to have its genome fully sequenced, was made available as the Yeast Genome Directory (1997). By the end of the century, 42 complete genome sequences had been published and, in 2001, rival genome sequencing consortia at Celera Genomics and the Sanger Centre simultaneously published the entire 3.2 billion-base-pair sequence of the human genome in *Science* and *Nature*. By the summer of 2007, 46 archaeal, 521 bacterial, and 65 eukaryotic genomes had been sequenced and annotated, with many more underway.

Fully sequenced genomes meant that researchers no longer had to estimate genome sizes by weighing their mass in picograms, and they no longer had to estimate characteristics of genomes, such as GC content, by extrapolating from shorter segments. They could count the number of genes in an organism, measure their average size, and calculate the proportion of genomes given over to coding and noncoding DNA, including repetitive

DNA. This yielded some unexpected findings. It had long been suspected that noncoding DNA accounted for most of the genome in many multicellular organisms, and fully sequenced genomes confirmed this. What wasn't expected was just how much of the noncoding DNA was attributable to mobile genetic elements. Transposons, retrotransposons, and other mobile elements alone make up as much as 50 percent of the noncoding DNA in the genomes of many plants and metazoa, and probably a good fraction of the rest is mobile element in origin but not identifiably so. Rampant mobile element activity can cause massive alterations to genomes, like the doubling of the maize genome in just 3 million years. Genomes seem to be playgrounds for transposable elements.

Gene counts (protein-coding sequences) also revealed some surprises. Some commentators had predicted that humans would have as many as 150,000 genes; others suggested 80,000, and most expected humans to have at least 40,000 genes. But the human sequence revealed a meager 35,000 genes, and even this number shrank on closer scrutiny; it is currently down to about 24,000 genes, and may drop to as low as 20,500 (Pennisi 2007). What makes this result doubly ironic is that the lowly nematode worm (*Caenorhabditis elegans*) and fruit fly (*Drosophila* spp.) require 18,000 and 14,000 genes, respectively, and neither has a backbone, a particularly large brain, or nearly as many tissue types as a human. Even a single-celled yeast needs about 5000 genes. What seems obvious in retrospect is that complexity arises from something other than large numbers of genes.

The availability of many fully sequenced genomes allowed comparative genomics to flourish, and researchers were quick to use these genomes to answer old, controversial, and fundamental questions of evolutionary genetics. In the 1960s, Motoo Kimura had proposed that the bulk of evolution at the molecular level is driven by mutation and genetic drift rather than by Darwinian selection. Kimura's "neutral theory" remained one of the most hotly debated topics of evolutionary biology throughout the final decades of the twentieth century, but little consensus was reached about the contribution of neutral substitutions to overall molecular divergence. Two papers in 2002, using data from *Drosophila* genomes, reported rates of adaptive evolution far higher than expected from the neutral theory (Fay et al. 2002; Smith and Eyre-Walker 2002). These reports were followed by work on the mouse and rat genomes (Bazykin et al. 2004) that also revealed an excess of adaptive substitutions. Increasingly, noncoding sequences are revealing themselves to be actively maintained and shaped by selection, rather than subject to just random turnover. The neutral theory was not dead, but its role in the shaping of genomes had been demoted. Still, Lynch and Conery (2003), noticing the tendency for organisms with small population sizes to have large genomes, suggested that so-called junk and other noncoding DNA could more readily accumulate in these species owing to the effects of drift. As a gross comparative generalization this may be true, but it seems an unlikely explanation of the huge variation in genomes seen among closely related species.

Another subject transformed by genomic data is the rate of deleterious mutation. This defines a species' genetic load, or how far the average genotype is from the optimal genotype. In the 1950s, H. J. Muller and J. B. S. Haldane realized that every deleterious mutation in a population must be eliminated by a selective "genetic death," lest a species inexorably decay and eventually go extinct. But as late as 1999 James Crow could write that "the deleterious mutation rate has been notoriously difficult to measure and no convincing estimates exist for any vertebrate" (Crow 1999, p. 293). Using sequence data from chimpanzees, gorillas, and humans, Eyre-Walker and Keightley (1999) calculated that humans may suffer up to three new deleterious mutations per person per generation. If each new deleterious mutation must be eliminated independently, this means that each person has to die a genetic death three times! How have humans avoided this fate or, more prosaically, a steady decline in their fitness? One way out is to propose some process by which deleterious mutations are eliminated in packets—that is, not one at a time. Two theoretical papers in 2001 suggested just this, arguing that if males carrying a greater number of deleterious mutations have lower mating success, then sexual selection can eliminate large numbers of deleterious mutations from populations, and even counter the two-fold cost of sex (Agrawal 2001; Siller 2001).

Novel genome-wide studies have cast new light on some old controversies. In the 1930s, Ronald Fisher claimed that deleterious mutations progressively become masked due to the accumulation of modifier alleles conferring dominance of the wild type (e.g., Fisher 1930). Sewall Wright famously dismissed this, arguing that the strength of selection on modifiers of dominance was exceedingly weak, and that dominance could simply be explained as a physiological safety margin inevitably present in metabolic pathways. The pendulum has swung back and forth on this argument about the strength of selection, with no particular resolution (Bourguet 1999). A parallel to the Fisher–Wright debate has emerged from considering the evolution of so-called redundancy. Using the fully sequenced yeast genome, it became possible to construct yeast knock-out lines, each of which has a different single gene inactivated. Measurements of the fitness of these knock-out lines showed that for a surprising 80 percent of genes inactivation had little or no phenotypic effect (Steinmetz et al. 2002). Similar results were reported for nematodes (Kamath et al. 2003), the plant Arabidopsis (Alonso et al. 2003), and mice (Wilson et al. 2005). The old arguments are back in an uncannily familiar form, but now on a genomic scale. Is this staggering level of redundancy in the genome a product of selection for robustness against mutation? Or is it an incidental outcome of the retention of gene duplicates for other reasons (Hurst and Pál, Chapter 7)?

The Organism in the Genome

Throughout the twentieth century, neo-Darwinism was largely dominated by questions about the properties of single genes, or of classes of genetic elements. Questions about genome size, the proportion of the genome given over to noncoding DNA, GC content, the size of fitness effects, and the proportion of molecular changes attributable to neutral and to selective forces—all treat the genome as something like a "parts list" in which the parts' combined effects on the organism are equal to the sum of their individual effects. If a new science of evolutionary genomics and proteomics is emerging, it is as a science of the evolution of phenotypes. Genomes need to be seen not just as collections of individual genes and other DNA sequences, but as complex sets of instructions and procedures for making a phenotype, written in a digital form. Genomes must be viewed from the perspective of the forces that shape organisms—thus, individual genes give way to biological pathways and networks that somehow produce higher-order features that in turn combine to make organisms. This means developing a science around where genes come from; how they are regulated; when, how much, and in what cells they are expressed; the topological properties of networks of proteins and metabolic elements; and how these produce the phenotypes on which natural selection acts (Pagel 2006; Koonin and Wolf, Chapter 2).

The genomic science of the phenotype is only just underway, but rapidly gaining momentum. The term "proteomics" does not appear in a Web of Science search of titles, keywords, and abstracts until 1997, when two papers used it. By the year 2000, the number had grown to 305, and by 2007 it is expected that proteomics will have been used about 2000 times. The term "transcriptomics" makes its first appearance in 1999, "metabolomics" in 2000, and "interactomics" in 2002. Researchers may sometimes regard the use of "omics" terms with suspicion—and the omics fad can get out of hand, as in the use of "mutome" to describe the set of single-nucleotide polymorphisms in a genome or the collection of somatic mutations in a body—but there is little doubt that a new perspective is emerging. If a biological system is thought of as a way to manage information, then the various omics categories can be arranged hierarchically: genomic information is stored in DNA, mRNA encodes the transcriptomic information, the proteome specifies proteins, and so on. Correlations among different classes of omics variables may reveal how patterns of between-species phenotypic diversity emerge, and how morphological variation arises within a species in response to varying environmental conditions. In effect, we wish to discover whether we can reveal a phenotypic biology in high-dimensional omics data (Koonin and Wolf, Chapter 2).

Complexity and Regulation

The most outstanding feature of the metazoa is the complexity of their genomes and their phenotypes. But the finding from whole-genome sequencing studies that large multicellular organisms have relatively few genes goads researchers out of the classical view of gene counting, and raises the question of how genomes encode the instructions that give rise to this phenotypic complexity. The human genome, at 3.2 billion bases of DNA, is 250 times larger than the yeast (*S. cerevisiae*) genome, yet we have only 5 times as many genes. Nearly all of this discrepancy arises from the non-protein-coding DNA that makes up about 99 percent of the human genome.

How do humans create and regulate their trillions of cells—divided among perhaps 300 different tissue types—and achieve their extraordinary complexity and flexibility of behavior from only five times as many genes as a yeast and not very many more than a nematode? The picture that is emerging is that large, multicellular organisms are virtuosos of gene regulation rather than champions of gene number (ENCODE Project Consortium 2007; Lemos et al., Chapter 5). Both epigenetic modifications and sequence-based features of the genome are involved (Baylin and Schuebel 2007; Lemos et al., Chapter 5). Changes in *cis*-regulatory regions are known to underlie evolutionary divergence in phenotype (e.g., the pigmentation patterns and sexual dimorphism of several *Drosophila* species), often with little or no accompanying change in structural coding genes and their attendant transcription factors. Borneman and colleagues (2007) documented a surprising degree of *cis*-regulatory sequence divergence between even closely related species. Wray (2007) investigated whether mutations in *cis*-regulatory and coding sequences have different phenotypic effects. Proving a disproportionate role for regulation will be hard, as there is no easy way to map mutations in *cis*-regulatory sequences to changes in regulation—unlike mutations in coding sequences, which accurately predict changes in amino acid composition.

If it is gene regulation that allows the eukaryotes to achieve their striking complexity, differences in the dominant modes of gene regulation between prokaryotic and eukaryotic organisms may help to explain how eukaryotes nevertheless get by with comparatively few genes. Prokaryote genomes are, to a first approximation, densely packed with protein-coding genes. Proteins are large, floppy, and three-dimensionally complex molecules. They are the principle mediators of gene regulation in prokaryotic cells, and are, in effect, analog regulatory devices. RNA, on the other hand, can have very precise sequence specificity in its binding and, therefore, in its regulation, leading to the suggestion (Mattick 2007) that it can act as a digital mechanism of regulation. The significance of this distinction for eukaryotes—and the human genome in particular—is that eukaryotic genomes can be viewed as vast and exquisitely regulated mRNA production machines. The ENCODE (Encyclopedia of DNA Elements) project (2007) reports surprising evidence that the entire human genome is "pervasively transcribed." It may be that eukaryotes have permitted their genomes to fill up with non-protein-coding DNA to support the digital regulation needed to achieve their extraordinary complexity (Mattick 2007); their genomes have, in effect, become information management systems that produce myriad, highly specific mRNA transcripts to carry out thousands of different regulatory tasks. Whether this regulation is required for real-time developmental flexibility or making a complex body (or both) is unclear.

However, unlike neutral theory–based explanations of genome size (Lynch and Conery 2003), this view of the genome gives a potentially adaptive role to every stretch of its DNA.

Further hints about how complexity is managed on a local scale may lie inside genes themselves. Average gene length (open reading frame only) is surprisingly similar in prokaryotes and eukaryotes (about 1000 bases; e.g., Xu et al. 2006). No one knows why. Could it reveal something about the average length of DNA needed to specify the basic architecture—α-helices and β-sheets—of most proteins? Despite the similarity in open reading frame length, an outstanding difference between prokaryotic and eukaryotic genes is that the latter have introns; eukaryotic genes tend to be very long, complex, and multidomain, whereas prokaryotic genes tend to have a single domain. Eukaryotic introns may act as built-in, RNA-based, digital regulatory sequences (Patthy, Chapter 6). From the organismal perspective, then, do multidomain genes reveal something about developmental mechanisms? Are they showing us how evolution has tended to clump together open reading frames that will often be expressed together, operon-like, either in time or space? If so, multi-domain genes may contain information about how development is organized, in effect spelling out what proteins need to be expressed together.

The Origins of Novelty

Darwin is sometimes criticized for never really explaining the origin of species, focusing instead on the origin of novelties by natural selection. Genomic scientists often ignore a similar question—how are we to explain the origin of new genes? It is difficult to think of them being built piecemeal. Even among closely related organisms, there can be many new genes (Fan, Emerson, and Long, Chapter 3; Doolittle et al., Chapter 4), and recent studies even document surprising within-species variation in the copy numbers of genes (Redon et al. 2006). These differences point to active and powerful evolutionary mechanisms for the origin of novelty. Classically, novelty was assumed to arise from beneficial mutations causing the gradual accumulation of positive effects at multiple loci. But beneficial mutations are held to be rare and some theories of development predict larger and more abrupt changes. Now both these assumptions are being challenged by two findings: (1) higher-than-expected rates of beneficial mutations of small effect in bacteria (Perfeito et al. 2007), and (2) an unusual example of multiple *cis*-regulatory substitutions governing changes to a single gene associated with morphological divergence between species of *Drosophila* (McGregor et al. 2007). Gene duplication has long been held to be the most common source of new genes, but many other processes—such as gene fusions and exon shuffling—are responsible for much gene diversity (Fan, Emerson, and Long, Chapter 3). The presence of so much gene diversity among closely related organisms raises the question of how genomes can withstand the disruption. Is there much more flexibility in the developmental programs that genomes define than we appreciate? Is this sort of flexibility an adaptive response to real-time variation in environmental conditions?

Lateral gene transfer is of course another source of evolutionary novelty, and perhaps the most pervasive one within the prokaryotes. Doolittle and coauthors' (Chapter 4) summary of its prevalence is salutary for questioning the extent to which strict, hierarchical descent with modification occurs among the prokaryotes and whether an evolutionary tree of life can meaningfully include these organisms. From an organismal perspective, why have eukaryotes become almost closed shops to outside genes, particularly as prokaryotes have used them to such great effect in achieving their diversity? The numerous endogenous retroviruses that can be found in metazoan genomes suggest that the germ line is not a barrier to lateral transfer. Equally, rates of lateral gene transfer, as Doolittle outlines, are low even in unicellular eukaryotes such as yeast. (And yet, there has been large-scale gene transfer from mitochondrial to nuclear genomes in nearly all eukaryotes.) The relative lack of lateral gene transfer in the plants and metazoa is an evolutionary puzzle that may ultimately tell us something about the best ways of making phenotypes. Perhaps the presence of complex gene regulation reduces the likelihood that a lateral transfer event will be successful.

Novelty can arise within a species and, when it does, may be a matter of gene regulation. Males and females with sex-chromosomal sex determination can be thought of as different species that just happen to share the same genome. They can have remarkably different morphologies, divergent life histories, and even occupy divergent ecological niches. Males often compete most directly with other males, and females with other females. Because of sexual reproduction, males and females have to cooperate to transmit genes successfully to the next generation. But, to the extent that males and females diverge ecologically, behaviorally, and morphologically, the demands they place on their genomes also diverge.

Genomic studies have revealed the high prevalence (~15%) of sex-biased gene expression in mice, chicken, flies, and worms (Oliver, Chapter 9). Naturally this affects genes expressed in the primary sexual organs, like the testis and ovary, but it also affects those expressed in other organs, like the liver, kidney, and embryonic brain. There also appears to be either a deficit or preponderance of these sex-biased genes on the X chromosome. Exactly why this occurs is open to a multitude of explanations, as the X chromosome has many distinctive properties (hemizygosity, dosage compensation, X-inactivation) and it is not yet clear how important these are in mediating the response to sexual and natural selection (Rogers et al. 2003). Genomic data are also revealing the Y chromosome to be more than just a pale imitation of the X chromosome (Charlesworth, Chapter 10). Many Y-linked genes in primates exist as palindromes (oppositely directed duplicates) that undergo within-chromosome recombination, and so save these genes from accumulating deleterious mutations. Y chromosomes also regularly recruit autosomal duplicates that are then turned into male-specialized genes.

Robustness and Evolvability

One view of robustness is that it is the ability of a genome to produce a successful phenotype under a range of unpredictable fluctuations in its environment. Evolvability, on the other hand, can be taken as the ability to adapt at a genetic level to new niches or environments. Organisms must be robust in their development and yet remain evolvable on an evolutionary timescale; this dilemma may have implications for the evolution of genomes as systems of instructions for development. For example, a rigidly specified developmental plan might be robust but unevolvable—that is, unable to acquire genetic variation without the program collapsing. Think of trying to add something at random to the motherboard of a computer.

Networks of interacting proteins can be thought of as the phenotypes of sets of functionally linked genes (Pagel et al. 2007). Interactions among networks and the processes and structures they define help to produce organismal phenotypes. In this way, protein interaction networks and metabolic networks—and network biology more generally—provide a methodology for developing a network-based view of development and phenotypes. Protein and metabolic networks often display what is called power-law scaling (Wagner, Chapter 12). This arises when a small number of so-called hub proteins are linked to many other proteins, but most proteins have only one or a very small number of links.

A network perspective may give insight into the twin problems of robustness and evolvability (Pagel et al. 2007). It may also help us understand the unexpected and puzzling findings to emerge from gene-knockout studies: that many of the genes an organism possesses are not necessary for survival, they are redundant or even "dispensable" (Hurst and Pál, Chapter 7). In designed systems, such as space rockets, redundancy is intentionally built-in, but a cardinal feature of evolved systems is that they are not designed—natural selection does not look ahead, anticipating what may come. Theoretical studies suggest that redundancy can, nevertheless, evolve if a gene must be expressed at a very high level or if mutations in it have a high probability of affecting somatic tissues whose maintenance throughout adult life is important for fitness. But this is not expected to be true of very many genes—certainly not the up to 80 percent identified as dispensable. What is going on?

Some studies suggest that a gene's contribution to a phenotype is proportional to its number of links in a network. Network power-law scaling implies that most of the genes in a network will have a small number of links. If we accept the premise that a gene's connectivity in the network is correlated with its phenotypic effect, then most gene knock-outs will have little effect. This in turn may mean that the resulting phenotype will not dramatically alter its form in response to variation in the inputs from these genes—variation that may arise from environmental fluctuations. At the same time, this scaling may confer evolvability on the organisms because most genes can be acquired or lost without a large change to the network phenotype. Wagner (Chapter 12) investigates whether a truly network-based biology is emerging and what it can tell us about the action of natural selection on organisms. A key, unanswered question is whether protein interaction networks evolve or merely emerge from simpler rules of assembly.

Genome Immunity

Most of the human genome is composed of DNA from repetitive sequences and mobile genetic elements. Retrotransposons—mobile DNA elements that move via an RNA intermediate—are the most abundant transposable elements and comprise approximately 40 to 50 percent of human genomic sequence. In the case of maize, they account for 90 percent of the genome. Studies estimate that among the hundreds of thousands of remnant sequences in the human genome that are derived from long interspersed nuclear repeats, or LINEs, up to 100 may be transcriptionally active. It may turn out that much of the genomic sequence that is associated with inactive remnants of past LINE transposition events is recruited to play some role in gene regulation; even so, the active elements can threaten the stability of transcriptional regulation of the genome, land inside active genes, and even change chromosome structure. For these reasons, all organisms have evolved diverse mechanisms to protect themselves against the harmful effects of transposons (Schaefer, Goll, and Bestor, Chapter 8); these include DNA methylation, RNA-mediated suppression, and enforced mutation. These mechanisms are expected to be found especially in sexually reproducing organisms, in which transposable elements can have up to twice the fitness of the host (Schaefer, Goll, and Bestor, Chapter 8).

Identifiying Selective Change

Humans and chimpanzees famously have at least 98 percent sequence identity in protein-coding genes, and yet the two species differ dramatically in size, shape, morphology, and behavior. The detection of positively selected genes is the most obvious way to identify the genes that make us human. Genome-wide screening for such genes is made possible by the availability of complete and well-annotated human and chimpanzee genome sequences. Genes such as *PRM1* (involved in sperm packaging), *FOXP2* (involved in language), and *HAR1* (related to brain growth) are among those showing strong selective divergence between apes and humans (Bustamante et al. 2005; Hellman and Nielsen, Chapter 13). At a whole-genome scale, additional signatures of evolution can also be used to identify the evolutionary forces that have given rise to differences within species, including disease-related variants (WTCCC 2007). These include chromosomal rearrangements, single nucleotide polymorphisms, and highly conserved regions. If found in non-protein-coding regions, the latter often identify transcription factor binding sites or other regulatory sites.

A reliable estimate of the underlying rate of neutral mutation is essential to accurately identify genes undergoing positive selection. Filipski, Prohaska, and Kumar (Chapter 11) outline some of the technical difficulties of estimating the true rate, and ways forward for detecting adaptive evolution. Hellmann and Nielsen (Chapter 13) describe the growing catalog of genetic sites that separate humans from the apes, and that vary within the human species. A picture is beginning to emerge of humans as a very young species that, nevertheless, shows surprising genetic differences among human subpopulations at some markers. These differences may indicate that humans have lived in relatively isolated groups that rapidly adapted to their particular environmental conditions.

What If Evolutionary Genomic and Proteomic Scientists Succeed?

If the organism really does reside in the genome, then the success of evolutionary genomic and proteomic science will be measured by progress in understanding, predicting, and manipulating phenotypes (e.g., Lartigue et al. 2007). A new evolutionary genomic theory, as a genomic analog to evolutionary genetics, needs to be built. This will take the form of a kind of network population genetics, in which the phenotypic consequences of a genetic mutation—whether regulatory or protein coding—will emerge from the network (population) of interactions in which it is embedded. This project's success will finally put some meat on the bones of essentially phenomenological notions like pleiotropy. This new network population genetics will also provide a theoretical basis for evolutionary developmental biologists to see how phenotypic changes—both ontogenetic and between individuals or species—emerge from a genomic foundation.

A fundamental and outstanding question for evolutionary developmental biology is this: Do abrupt phenotypic changes to body plans during evolution arise from large-scale and abrupt evolutionary changes to networks, or from the gradual accumulation of many changes of small effect? In other words, do they arise from changes to hubs or to satellite nodes of genomic–proteomic networks? In the simplest terms, the *Hox*-gene school of development links large-scale developmental changes to duplications of highly connected clusters of *Hox* genes. But is this sort of change the norm? We also need a hierarchical, developmental theory that somehow links together the separate developmental programs that make specific tissues, organs, metabolic pathways, and the like. Is there a homunculus overseeing things or does it all just somehow emerge? Physiologists, with their extensive knowledge of homeostatic processes within the body, will likely have much to offer to this hierarchical theory.

Finally, success will be measured by the ability of genomic and proteomic scientists to reliably encode in genomes the instructions for the production of phenotypes. The twenty-first century may be the century in which genomic and proteomic science turns science fiction into reality, a century in which it becomes possible to grow parts of complex phenotypes—and maybe even whole phenotypes—to order.

CHAPTER 2

Evolutionary Systems Biology

Eugene V. Koonin and Yuri I. Wolf

Systems Biology: The New Big Science of "Omes"

IF THE 1990S WAS THE UNDISPUTED DECADE OF GENOMICS, the beginning of the twenty-first century is equally obviously marked by the accelerating rise of the new field that usually goes under the rather inapt name *systems biology*. A precise definition of systems biology is elusive even as, in a general sense, its goals and principles seem to be quite transparent. The overarching aim is to achieve an integrated understanding of life forms, with all their characteristic complexity of interactions at multiple levels (Ge et al. 2003; Herbeck and Wall 2005; Pennisi 2005; Ivakhno 2007; Bruggeman and Westerhof 2007; Zhu et al. 2007). This superambitious undertaking decomposes into a variety of extremely diverse research directions that have in common certain crucial methodologies: they all measure and analyze integral characteristics of biological systems as opposed to focusing on an individual gene, protein, pathway, and so forth, as is characteristic of traditional, twentieth century biology. Above all, today's systems biology is about large-scale measurement—typically on the scale of a whole cell or even an entire multicellular organism. To put it in somewhat crude jargon, systems biology is a collection of "omics" that strive to investigate the relationships among various "omes." This terminology and, implicitly, the underlying train of thought have been vigorously attacked as a vacuous attempt to inflate some people's ambitions of "big science" (Petsko 2001). However, six years later, the omes have not disappeared and look stronger than ever, suggesting that they are here to stay as a manifestation of a new way to do biology and the new mentality that accompanies it (Koonin 2001). Omes are definitely becoming household names—so much so that the journal *OMICS*, subtitled "Journal of Integrative Biology," was established in 2002; almost symbolically, *OMICS* emerged through transfiguration of a pre-existing journal, *Microbial & Comparative Genomics* (Kolker 2002).

Table 2.1 Major and some minor omes and omics

Ome/omics terms[a]	Number in PubMed[b]	Definition of the "ome"	Principal means of evolutionary analysis
Genome/ genomics	576556	The entirety of the genetic material in an organism	Sequence comparison and phylogenetic analysis (traditional methods of molecular evolution)
Proteome/ proteomics	8015	Complete set of proteins in an organism, organ, tissue, or cell	Comparative analysis of protein abundance
Transcriptome/ transcriptomics	2408	Complete set of coding and non-coding transcripts in an organism, organ, tissue, or cell	Comparison of expression levels, expression profiles, coexpression networks
Metabolome/ metabolomics	258	Complete set of metabolites in an organism, organ, tissue, or cell	Comparison of metabolite sets and metabolomic networks
Interactome/ interactomics	160	Complete set of protein–protein interactions in an organism, organ, tissue, or cell	Comparative analysis of interaction networks
Phenome/ phenomics	73	The entirety of phenotypic features of an organism	Traditional evolutionary analysis (e.g., comparative morphology)
Kinome/kinomics[c]	59	Complete set of protein kinases of an organism	Sequence comparison and phylogenetic analysis (traditional methods of molecular evolution)
Glycome/ glycomics[d]	47	Set of all glycans (sugar chains) in an organism	Comparative analysis of glycan structures and phylogenetic analysis of the relevant enzymes
Lipidome/ lipidomics[e]	20	Complete set of lipid molecules in an organism	Comparative analysis of lipid structures and phylogenetic analysis of the relevant proteins
Regulome/ regulomics[f]	8	Complete set of genes with regulatory functions in an organism	Integrative analysis of the regulatory networks and pathways in an organism, and comparison of the regulation between organisms
Resistome/?[g]	6	The complete set of antibiotic or pathogen resistance genes in an organism or microbial community	Methods of comparative and evolutionary genomics
Phylome/ phylogenomics?[h]	2	Complete set of gene phylogenies for a given organism	Phylogenetic tree analysis and comparison
Mobilome/?[i]	1	Complete set of mobile elements in an organism	Methods of comparative and evolutionary genomics

[a]The omes are sorted by decreasing number of occurrences in PubMed; references are given for rare omes only. Question marks indicate that the omics version of the term has not been used in the literature or, in the case of phylogenomics, does not precisely correspond to the ome form.
[b]The PubMed database was searched for the number of occurrences of the respective ome term in any field.
[c]Johnson and Hunter 2005.
[d]Freeze 2006.
[e]Mutch et al. 2006.
[f]Howard-Ashby et al. 2006.
[g]Wright 2007.
[h]Sicheritz-Ponten and Andersson 2001.
[i]Frost et al. 2005.

In principle, the definition of an ome is clear and coherent: an ome is a compendium of all entities of a certain class found in an organism (or in a tissue, an organ, and so forth). Thus, the human transcriptome denotes the entire collection of transcripts made from the human genome, whereas the yeast proteome consists of all proteins produced in this organism. Of course, the devil is in the details, so, for example, defining a single human transcriptome or proteome might not be practical (see also Chapter 6). Instead, it makes much more sense to speak of a huge array of human transcriptomes (and the corresponding proteomes) that are characteristic of different cells, tissues, organs, tumors, and so forth. Comparative analysis of these diverse transcriptomes and proteomes is a major direction of systems biology (De Preter et al. 2006; Habermann et al. 2007; Mueller et al. 2007).

In Table 2.1, we list the major omes and the corresponding omics—fields of study that focus on integral characteristics of biological systems—as well as several (relatively) minor, specific ones. A quick look at the number of occurrences of each ome in the PubMed database reveals curious trends. Obviously, none of the new omes even starts to compete with *genome* in the popularity contest. Moreover, only two omes besides genome, namely *proteome* and *transcriptome*, can be considered commonly used terms at the time of writing (February 2007). Some of the omes do not seem to be catching up strongly, for example, *mobilome* occurs only in the single article where it was coined. Nevertheless, it is hard to deny that the concept of a mobilome—the entirety of mobile elements present in an organism—and the term itself (Frost et al. 2005; Ou et al. 2007; Sundin 2007) are coherent and useful. Hence we venture to predict that this and quite a few other omes will become familiar terms—and objects of systems biology research—in the next few years. A simple examination of the dynamics of the usage of the major omes in the scientific literature seems to bode well for this prediction by showing with perfect clarity that omics is, indeed, a science of the twenty-first century (Figure 2.1). While all other omes, which were more or less unheard of until the advent of the new millennium, are currently dwarfed by the term *genome*, the growth rate is very slow for genome but rapid for the rest of the omes, emphasizing their increasing prominence (see Figure 2.1A,B).

This short excursion into the world of omes clearly shows that systems biology (also known as omics) is a discipline on a rapid rise. In this chapter, we examine some recent research efforts that we believe are making considerable inroads toward integration of omics with fundamental biological research—in particular, in evolutionary biology.

Systems Biology in the Light of Evolution: The Quest for the Link between Genomic and Phenotypic Evolution

Over 30 years ago, Theodosius Dobzhansky famously professed that "nothing in biology makes sense except in the light of evolution" (Dobzhansky 1973). Biology has changed immensely since the time these unforgettable

Figure 2.1 The dynamics of the usage of "omes" terms in documents in the PubMed database. (A) Six omes, including genome; the *y* axis is in the logarithmic scale. (B) Five major omes of Systems Biology. The PubMed database was queried for the occurrence of each ome term together with the respective year in the "Publication Date" field (e.g., "transcriptome AND 2001[PD]").

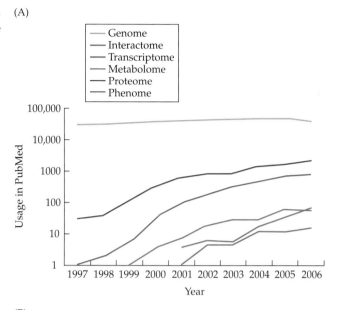

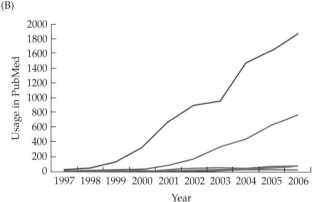

words were written, but never did Dobzhansky's motto ring more true than it does now.

Omics is often viewed as busy work—the collection of vast amounts of data without a clearly defined, biologically relevant perspective. Indeed, the omes are quantitatively overwhelming, especially as the data are inevitably noisy. Yet, in principle, the omes relate directly to the (arguably) most fundamental problem in all of biology: the nature of the connection between genotype and phenotype (Mahner and Kary 1997; Streelman and Kocher 2000). The omes mark the path from the genome to the phenome (the phenotype of the organism), which is specified by the genome through gene expression, protein–protein interactions, metabolic and signal transduction pathways, and so forth. The availability of increasingly reliable data on each of the omes allows us to address deep problems of evolution that used to be so far out of reach in the pre–systems biology era that they were not even usually articulated. Such basic questions include, but, of course, are not restricted to, the following:

- How much in the structure of each ome is functionally relevant and how much is intrinsic, biological noise (distinct from experimental noise) that is the product of unconstrained, neutral evolution?

- How fast do different omes change during evolution in relation to the genome?

- How are specific evolutionary changes in each ome coupled to (or uncoupled from) changes in the genome sequence?

Each of these questions relates to the central problem: How and to what extent does genome evolution determine organismal (phenome) evolution?

An important realization of the first years of the twenty-first century that far transcends biology is that any complex system can be presented in the form of a network, where elements of the system are the nodes, and the connections between them are the edges (Barabasi 2002). The omes are no exception; consequently, the analysis of gene coexpression, regulatory, interaction, and other networks is one of the key methodologies of systems biology (Barabasi and Oltvai 2004; Koonin et al. 2006; Mueller et al. 2007). Biological and nonbiological networks of very different natures show remarkable topological similarities. Again, evolutionary questions loom large: Which aspects of network topology are functionally relevant and, consequently, maintained by purifying selection; which ones have no functional underpinning and change neutrally; and which ones might be specific adaptations subject to positive selection?

At the time of this writing, evolutionary systems biology is a field in the making; it continues to evolve rapidly such that even its central tenets and range of fundamental problems are not yet firmly established (Media 2005; Koonin and Wolf 2006; Stumpf et al. 2007). Nevertheless, methodologies exist for evolutionary analysis of each of the omes (see Table 2.1), and the number of observations that clearly belong in this emerging area is already quite large. Attempts to introduce some order into the increasingly confusing maze of revealed connections are badly needed.

In this chapter, we discuss recent studies that attempt to achieve a degree of coherence in our understanding of the relationships between genomic and phenomic evolution as they can be gleaned from the accumulating omic data. Adequately covering all research directions that, arguably, fall under the umbrella of evolutionary systems biology has already become a daunting task that is not attainable within the limited space available to us here (and with our limited knowledge). Accordingly, we have chosen just one, albeit fairly broad and far-reaching, avenue of research: the study of correlations between various omic variables and the selective forces and biological phenomena that are thought to underlie these correlations. At the conceptual level, we address, primarily, the last of the fundamental questions listed above: the nature of the connections between the evolution of gene sequences and the evolution of various omes. On many occasions, we employ a network representation of omes (in particular, the transcriptome and the interactome) to address this evolutionary question. It is our hope that this selection of topics highlights the trends that are typical of the entire burgeoning field of evolutionary systems biology. Above all, our goal is to emphasize the *biology* in evolutionary systems biology—in other words, to show that this emerging discipline has the potential to provide new answers to old and fundamental biological questions.

Correlations between Variables Characterizing Genome Evolution and Functioning

Perhaps the most basic approach used to investigate the connections between genome evolution and phenotypic characteristics using omics data involves finding phenomic correlates of genome evolution rates. As soon as genome sequences (which allow for large-scale determination of gene evolution rates) and the requisite data on other omes became available, the hunt for links between the omic variables began in earnest. Table 2.2 summarizes the most notable correlations that have been reported to exist between variables that characterize the evolution, expression, and function of genes. The validity and, especially, biological relevance of most of these correlations

Table 2.2   Correlations between omic variables

	PPI[a]	GI	EL	CAI	PA	KE	NP	PGL	ER
PPI									
GI	+[b]								
EL	+++	–							
CAI	+++	ND	+++						
PA	+++	ND	+++	+++					
KE	+++	–	+++	+++	+++				
NP	++	++	+++	ND	ND	+			
PGL	– –	NS	– – –	ND	ND	– –	NS		
ER	– – –	–	– – –	– – –	– – –	– –	– –	+++	

[a]Abbreviations: PPI, number of physical protein–protein interaction partners; GI, number of genetic interactions (i.e., the number of genes with which the given gene forms a synthetic lethal pair such that simultaneous knockout of both genes is lethal, whereas knockout of any one is not); EL, expression level; CAI, codon adaptation index; PA, protein abundance; KE, lethal effect of gene knockout; NP, number of paralogs; ER, (sequence) evolutionary rate; PGL, propensity for gene loss (see text); ND, not determined; NS, not significant.
[b]Plus signs indicate positive correlations, and minus signs indicate negative correlations; the number of signs roughly reflects the strength of the reported correlations.
(Data from Krylov et al. 2003; Drummond et al. 2005; Lemos et al. 2005a; Wall et al. 2005; Zhang and He 2005; Wolf et al. 2006b.)

have been questioned at one point or another, emphasizing the unsettled conceptual issues that currently besiege evolutionary systems biology.

An excellent case history illustrating these problems involves the biologically fundamental relationship between a gene's dispensability and its rate of evolution. (A gene is dispensable if the organism remains viable when the given gene is knocked out and indispensable if such a knockout is lethal; of course, dispensability also can be expressed quantitatively, in terms of the decrease in fitness resulting from knockout.) The intuitively plausible prediction that genes making a large contribution to the organism's fitness would experience stronger purifying selection and, as a result, would evolve more slowly than less "important" genes is quite old and stems directly from Kimura's neutral theory of evolution (Kimura and Ota 1974; Zuckerkandl 1976; Wilson et al. 1977). Even small changes in important genes are likely to be deleterious and will be selected against, resulting in a slower rate of evolution. As soon as reasonable amounts of data allowing one to estimate evolutionary rates and genome-wide information on gene knockout effects became available, studies were launched to test this conjecture. Hurst and Smith were the first to perform such an analysis on a set of mammalian proteins (Hurst and Smith 1999). After excluding rapidly evolving immune system genes (which are thought to evolve under positive selection), they found no reliable correlation between gene sequence evolution rate and the severity of the knockout phenotype. However, subsequent studies in yeast (Hirsh and Fraser 2001) and in bacteria (Jordan et al. 2002) reversed this conclusion by demonstrating statistically significant, albeit relatively weak, negative correlations between the strength of a gene's knockout fitness effect and its evolution rate. The negative results of Hurst and Smith (Hurst and Smith 1999) have been attributed, primarily, to the smaller data set they used (for a more wide-ranging discussion see Chapter 7).

Obviously, studies on the connections between evolution rate and knockout phenotype did not develop in a vacuum—indeed, correlations between these and other genome-related variables were investigated simultaneously along several, partially overlapping paths. In particular, several independent studies have shown that a gene's sequence evolution rate is negatively correlated with its expression level—highly expressed genes, on average, evolve significantly more slowly than genes with low expression levels (Pál et al. 2001; Krylov et al. 2003; Jordan et al. 2004). A broader, systematic analysis of multiple variables has revealed significant negative correlations between sequence evolution rate or a gene's propensity to be lost (the relative frequency with which a gene is eliminated from a species' genome during evolution, another measure of long-term evolutionary conservation) and knockout effect, expression level, and the number of protein–protein interactions (Krylov et al. 2003).

In principle, these multiple correlations are mutually exclusive inasmuch as evolution rate could be independently correlated with several variables related to gene function and expression. However, partial correlation analysis of the yeast (Pál et al. 2003) and bacterial (Rocha and Danchin 2004) data has shown that, when controlled for expression level, the correlation between evolution rate and dispensability becomes insignificant. These observations have been taken to falsify the fitness–rate conjecture and have led to the notion that, apart from a gene's specific biological function, expression level is the principal determinant of its sequence evolution rate (Pál et al. 2003; Rocha and Danchin 2004).

However, this was not the end of the road for the intuitively attractive, more or less natural notion that important genes (i.e., those that cannot be knocked out without impairing the organism) should evolve slowly. Using more extensive and refined omic data, four independent studies have re-examined the connections between gene expression, dispensability, and evolution rate (Drummond et al. 2005; Lemos et al. 2005a; Wall et al. 2005; Zhang and He 2005). The first three studies analyzed substantially improved, genome-wide data sets on expression levels and dispensability of yeast genes. These studies also employed genome sequences from closely related yeast species to obtain accurate estimates of short-term evolution rates approximating the desired instantaneous rate. The fourth group performed similar analyses on the omic data from another model organism, Drosophila (Lemos et al. 2005a). Using these improved data sets and applying sophisticated statistical techniques to account for the experimental noise, all four groups concluded that there was a relatively weak but highly significant correlation between the evolution rate and dispensability, even after controlling for the expression level. Notably, the effect of dispensability on the evolution rate diminished rapidly with increasing evolutionary distance (Wall et al. 2005; Zhang and He 2005). The reverse test for the effect of dispensability yielded a significantly stronger dependence of evolution rate on expression level (Drummond et al. 2005; Wall et al. 2005; Zhang and He 2005). Three groups concluded that the effects of expression level and dispensability on gene evolution rate were independent, and that both were substantial (Lemos et al. 2005a; Wall et al. 2005; Zhang and He 2005). The contribution of expression level appeared to be greater than that of dispensability, but it has been posited that uncertainties regarding the noise level in the data made definitive conclusions on the relative importance of the two variables premature (Wall et al. 2005). By contrast, Drummond and coworkers held that the correlation between gene dispensability and evolution rate still might be spurious, depending on the noise levels for different variables (Drummond et al. 2005; Drummond et al. 2006). Thus, although the latest analyses with updated data sets generally seem to uphold the fitness–rate conjecture, a shadow of doubt persists.

We addressed the controversy around the connection between gene dispensability and evolution rate in some detail in order to demonstrate the complex state of affairs that is characteristic of current research in evolutionary systems biology. Very similar developments have taken place with regard to other correlations between omic variables (see Table 2.2). For example, the relationship between node degree (the number of nodes with which a given node is connected) in protein–protein interaction networks and evolution rate became the subject of an even more confounding debate following the original claim that highly connected genes evolved slowly (Fraser et al. 2002). A re-examination of the data suggested that the link between connectivity and evolution rate, even if real, was extremely weak; only the most prolific interactors (network hubs) were characterized by low evolution rate (Jordan et al. 2003). An analysis of an updated data set by the authors of the original observation appeared to vindicate the link (Fraser et al. 2003), but then further reanalysis showed that the correlation disappeared altogether after correcting for protein abundance (Bloom and Adami 2003; Bloom and Adami 2004; Drummond et al. 2006). Nevertheless, an independent, detailed study of protein–protein interaction networks from three species seemed to provide support for the notion that a gene's centrality in a network (defined as *betweenness*, i.e., the frequency with which a network node lies on the shortest path between pairs of other nodes, a quantity that strongly correlates with connectivity) constrains its evolution (Hahn and Kern 2005). A similar analysis of gene coexpression networks (see a detailed discussion of these networks later in this chapter) revealed a strong correlation between node degree and evolution rate (Jordan et al. 2004). The structure of the yeast metabolic network has been reported to similarly affect evolution of enzymes (Vitkup et al. 2006). Thus, despite the general attractiveness of the idea that network hubs are biologically important, slowly evolving genes, the validity of this connection remains an open issue.

A different, more biologically inspired way to address the relationships between genome-related variables is to step back from technical problems and consider the general structure of the correlations (see Table 2.2). The genome-related variables naturally fall into two classes that can be roughly defined as *phenomic* (expression level, protein abundance, dispensability, centrality in networks, and so forth) and *evolutionary* (sequence evolution rate and propensity for gene loss). Considering this simple classification, the pattern of correlations is perfectly clear: the correlations within a class are consistently positive whereas those between classes are typically negative (see Table 2.2 and Figure 2.2) (Wolf et al. 2006a,b). The phenomic variables show a "the bigger the better" pattern whereas, with the evolutionary variables, it is "the slower the better": Genes with high expression levels, a large number of interactions, and a drastic knockout effect typically evolve slowly and are not lost during evolution. In itself, this coherence does not exclude the possibility that some of the variables have no independent significance but are, instead, fully determined by one or more genuinely important quantities (e.g., gene expression level, which is most strongly correlated with the rate of evolution, might induce most, if not all, of the correlations between other phenomic and evolutionary variables). Alternatively, the pattern depicted in Figure 2.2 could reflect a genuine synergistic effect of the different variables within each class. In particular, it stands to reason that distinct characteristics of gene function might combine to determine the rate of evolution. This notion has been encapsulated in the so-called extended complexity concept, according to which the rate of a gene's evolution is, to a large extent, determined by the complexity of that gene's functional interactions: Genes that are involved in complex processes and multiple interactions tend to evolve slowly (Aris-Brosou 2005).

Figure 2.2 The two classes of variables in evolutionary systems biology. Red arrows show positive correlations, and blue arrows show negative correlations. The blue arrows are shown one-sided to emphasize that phenomic variables are thought to affect evolutionary variables.

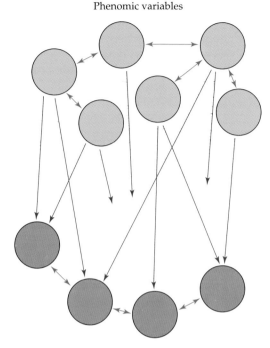

Phenomic variables

Evolutionary variables

The Multidimensional Space of Omics

Given the multiple, relatively weak correlations between individual phenomic and evolutionary variables and the coherent pattern of the connections within and between these two classes of variables (see Figure 2.2), multivariate statistics methods appear to be the approach of choice for dissecting the multi-dimensional space of evolutionary systems biology (Wolf 2006). In essence, these methods attempt to replace multiple, correlated variables like those shown in Table 2.2 and Figure 2.2 by a smaller number of composite independent variables. These are combinations of the original variables and account for a greater fraction of the variance in the data than any of the latter individually.

Principal component analysis (PCA) is the simplest of such methods, and two groups have employed it in attempts to make sense of the relationships between the omic variables (Drummond et al. 2006; Wolf et al. 2006b). The two studies differed considerably in the details of the statistical techniques employed and, probably more importantly, in the scope of the data analyzed. Thus, the conclusions came out substantially different as well, although there seems to be some agreement, in that the complexity of the omic data could not be reduced to the contribution of any single variable.

However, Drummond and coworkers came very close to claiming that there is, after all, a single determinant of protein evolution rate and that it can be identified by analysis of omic data (Drummond et al. 2006). They analyzed the data that has been amassed for a single model organism, the yeast *Saccharomyces cerevisiae*, and concluded that a single, dominant variable explained almost 50 percent of the variance in the rates of both protein and synonymous site evolution. In contrast, the contributions of other phenomic variables, including dispensability, were negligible. This dominant factor comprised variables related to protein translation rate, such as expression level, codon adaptation index, and protein abundance; as discussed below, this led to the suggestion that coding sequence evolution is primarily governed by selection at the level of translation.

Wolf and coworkers (Wolf et al. 2006b) took a broader approach by combining omic data from several model organisms using the KOG database of orthologous genes from diverse eukaryotes (Koonin et al. 2004) as the framework for the PCA of genomic variables. This study yielded more complex results than the single-organism analysis of Drummond and colleagues. The first three principal components (PCs) accounted for more than 60 percent of the variance in the data. The first PC received a natural biological interpretation as the *status* of a gene in the genomic community. Indeed, gene expression level, protein interactivity, and knockout effect make positive contributions to this PC, whereas sequence evolution rate and propensity for gene loss make negative contributions (Figure 2.3). Thus, high-status genes are those that are highly expressed, interact with many other proteins, often have multiple paralogs, are associated with a substantial fitness reduction upon knockout, are rarely lost during evolution, and evolve slowly at the sequence level. The correspondence of status with the intuitive notion of a gene's "biological importance" is obvious. The second PC is much harder to interpret in biologically sensible terms. The greatest positive contributions to this PC come from the number of paralogs and the number of genetic interactions, whereas the knockout effect makes a negative contribution (see Figure 2.3). This PC has

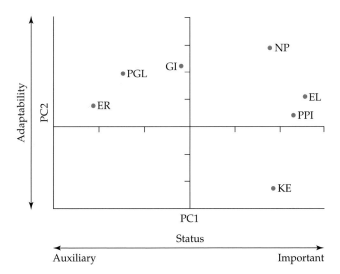

Figure 2.3 Contributions of seven phenomic and evolutionary variables to the first two principal components in PCA (loadings plot). Abbreviations: EL, expression level; ER, evolutionary rate; GI, number of genetic interactions; KE, lethal effect of gene knockout; NP, number of paralogs; PGL, propensity for gene loss; PPI, number of physical protein–protein interaction partners. (Data from Wolf et al. 2006b.)

been tentatively associated with *adaptability*, the capacity of a gene to respond to changes in the internal or external environment (Wolf et al. 2006b). The third PC has been interpreted as a gene's *reactivity*, another measure of functional and evolutionary plasticity (Wolf et al. 2006b). (The difference between adaptability and reactivity is that genes with high adaptability tend to interact with many other genes, whereas genes with high reactivity interact with few genes but tend to be highly expressed under certain conditions.) Interestingly, it has been found that genes with high adaptability and reactivity generally show a greater skew of expression profiles (an asymmetry of the distribution of expression values, e.g., those for human or mouse tissues) than genes with low values of these PCs, apparently in accord with the interpretation of these variables as measures of a gene's plasticity (Wolf et al. 2006b).

Regardless of differences in their specific conclusions, these studies reinforce the promise of multidimensional approaches in the analysis of omic data. In the best-case scenario, these approaches could lead to insights into the biological underpinnings of the connections between the omic variables. In the next section, we discuss some recent inroads in this direction.

Glimpses of Biology behind the Correlations

The links between the variables characterizing genome function and evolution are intriguing, and the observations that composite variables seem to better describe biological systems appear encouraging. However, these findings do not immediately reveal the biological underpinnings of the observed relationships or the evolutionary rationales for the emergence of these connections. Several recent studies have strived to address the biology behind the deluge of omic data in a more direct manner.

One notable example offers an unexpected twist on the previous observations on the connections between protein–protein interaction network topology and gene evolution rate (Fraser 2005). Fraser partitioned highly connected nodes in these networks (hubs) into two distinct classes: party hubs, which interact with numerous partners within a network module, and date hubs, which connect modules through interactions with partners from different modules (Han et al. 2004). Party hubs turn out to be strongly constrained and to evolve much more slowly than proteins that have no partners at all, whereas date hubs are only slightly more constrained than noninteractors. This seems to suggest that the organization and functions of network modules are conserved during evolution, whereas intermodule hubs are involved in network rewiring and contribute to innovation as they evolve.

The observations of Drummond and coworkers—that gene expression level or a composite variable that reflects the rate of protein production is a major determinant of the evolution rate of protein-coding sequences—led to a striking, novel hypothesis on the biological causes of evolution rate variation (Drummond et al. 2005; Drummond et al. 2006). The proposal, drawing on similar ideas of Akashi on the evolution of codon usage (Akashi 1994; Akashi 2003), is that genes for highly expressed proteins evolve slowly due to selection for translational robustness or, in other words, tolerance to errors of translation. Protein evolution is thought to be constrained by the requirement for the amino acid sequence to retain the ability to fold into the correct, functional structure despite relatively frequent translational errors. This constraint would most strongly affect proteins produced in large amounts, because of the burden that misfolded proteins impose on the cell (Goldberg 2003). A theoretical population genetics model supporting this proposal has been recently developed (Wilke and Drummond 2006). A dramatic corollary of the translational robustness hypothesis is that, under this view, the rate of gene evolution does not strongly depend on the specific function of the encoded protein—a direct challenge to the prevailing and intuitively obvious ideas about protein evolution.

The distribution of genes with different types of functions in the three-dimensional space defined by the three composite variables described above—status, adaptability, and reactivity—does not seem to be readily compatible with this startling conclusion (Wolf et al. 2006b). Indeed, the centroids (the centroid of a finite set of points is computed as the arithmetic mean of each coordinate of the points) of the sets of points corresponding to major functional classes of eukaryotic genes occupy markedly (and statistically significant) different positions in this space. For example, genes encoding components of information systems have, on average, a much higher status than other genes, and genes for metabolic enzymes and transporters have especially high reactivity (Figure 2.4). Moreover, substantial differences in status, adaptability, and reactivity have been noticed between genes at a finer level of functional classification (Wolf et al. 2006b). This seems to be most compatible with the more conventional notion that specific biological functions do have a major effect on the quantitative charac-

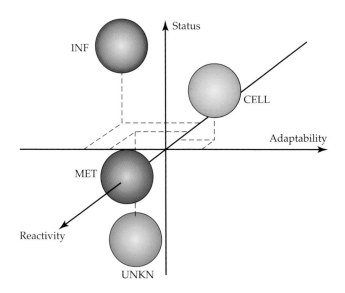

Figure 2.4 Distribution of the broad functional classes of eukaryotic genes in the three-dimensional space defined by status, adaptability, and reactivity. The spheres denote the centroids of the clouds of points for major classes of genes (defined as in Koonin et al. 2004): INF, information systems; CELL, cellular functions (e.g., compartmentalization and trafficking); MET, metabolic enzymes and transporters; and UNKN, genes with unknown function or general functional prediction only. (Data from Wolf et al. 2006b.)

teristics of genes, including their rate of evolution. However, it cannot be ruled out that the appearance of a strong connection between gene function and combinations of genomic variables is but a facade that conceals fine-tuning of genes with respect to maximum translational robustness. From a complementary standpoint, the contrast between the overarching translational robustness hypothesis and the more complex picture emerging from the analysis of the composite variables shown in Figure 2.4 might be determined, primarily, by genuine biological differences between the analyzed systems. Conceivably, the balance of forces that govern gene evolution is substantially different between yeast and more complex, multicellular eukaryotes.

Some Generalizations: Evolutionary Systems Biology— a Nascent Field in Turmoil

In this chapter, we covered only one research direction in evolutionary systems biology: the study of the correlations between omic variables and their biological context. We believe, however, that the state of affairs in this area is characteristic of today's evolutionary systems biology as a whole. Above all, this is still a nascent field where the methodology remains in constant flux, the data are noisy, the amount of experimental noise is variable and often unknown, and the effect of data quality on the conclusions is far from certain. Furthermore, the theory of evolutionary systems biology remains to be developed; at the very least, the systems approach must be integrated into the theoretical framework of modern evolutionary biology.

In this turbulent phase of development, evolutionary systems biology is overcome with intense debates on a variety of topics, some with extremely far-reaching implications. The dispute over whether protein evolution has a single, dominant determinant or is determined by a synergistic effect of multiple omic variables, as detailed above, is a case in point. It should be emphasized that at the heart of this debate is a truly fundamental biological question: Is the rate of protein evolution dictated primarily by the specific function of a given protein or by a generic principle, such as selection for translational robustness, that might be only indirectly correlated with function?

Another notable controversy centers around the nature of gene expression profile evolution—is it neutral or is it governed by selection? And, if the latter, what type of selection is prevalent? Observations of apparent clock-like evolution of expression profiles have been interpreted to indicate that this is predominantly, if not entirely, a neutral process (Enard et al. 2002a; Khaitovich et al. 2004; Khaitovich et al. 2005). However, the conservation of expression profiles between orthologous genes from different species suggests that evolution of gene expression is constrained by purifying selection (Jordan et al. 2004). The detection of local acceleration of evolution suggests a role for positive selection as well (Jordan et al. 2005). Clearly, further studies are required to characterize the evolution of gene expression in depth and, perhaps, to reveal fundamental differences between the modes of evolution of genomes and transcriptomes.

A debate of an even more general nature concerns the biological significance and evolutionary origins of the global topological properties of the so-called scale-free networks that are typical of biological and other evolving systems and are characterized by power law distributions of the node degree* (Barabasi and Albert 1999; Koonin et al. 2002; Luscombe et al. 2002; Barabasi and Oltvai 2004). The ubiquity of these distributions is often considered to reflect fundamental principles of evolution that are common to an enormous variety of systems (see Koonin et al. 2006 and references therein). However, an alternative view has been developed—namely, that the power laws are a manifestation of simple, generic physical properties of a system and tell us little, if anything, about evolution (Gisiger 2001; Keller 2005; Deeds et al. 2006; Siegal et al. 2007). A recent comparative analysis of human and mouse gene coexpression networks has shown that, although these networks are practically indistinguishable in their global properties, including the power-law node degree distribution, they have dramatically diverged at the local level (Tsaparas et al. 2006). While not resolving the dispute, this suggests that global topological properties might be of limited value for understanding network evolution.

All these controversies make it obvious that there are, at this point, many more questions than answers in evolutionary systems biology. However, it seems equally clear that evolutionary systems biology is not a trivial pursuit: Some of the most fundamental questions of biology cannot be meaningfully addressed without integrative analysis of the relationships between different omes in space and time.

Acknowledgments

The authors' research is supported by the Intramural Research Program of the NIH, NLM, NCBI.

---

*A power law distribution has the form $P(k) \sim k^{-\gamma}$ where k is the node degree, $P(k)$ is the frequency of nodes with the degree k, and γ is a positive coefficient; in double logarithmic coordinates, this distribution is a straight line with a negative slope.

CHAPTER 3

The Origin of New Genes

Chuanzhu Fan, J. J. Emerson, and Manyuan Long

Introduction

PARALLEL TO DARWIN'S CENTRAL QUESTION OF THE ORIGIN OF SPECIES is that of the origin of novel genetic elements. One of the most important roles of such new elements is to generate genes with new functions, which increases the biological diversity of organisms. Initial speculation claimed that gene origination must be accompanied by gene duplication (Ohno 1970), although recent studies provide evidence for other mechanisms (Long et al. 2003). In order to understand the molecular processes and mechanisms governing the evolution of novel genes and their functions, direct observation of newly originated gene copies is an indispensable approach. It is well established that many genes have persisted for long evolutionary times. Studying evolution over such timescales is difficult because the characteristic features that enable the elucidation of the evolutionary process erode with increasing time. Therefore, one productive strategy to learn more about gene origination is to investigate recently evolved genes. In this chapter, we will introduce the general features of new gene evolution, ranging from new genes that have been found in various organisms to molecular mechanisms and patterns of new gene origination. We will focus on the methods used to detect new genes and will describe a general genomic method adapted from microarray hybridization technology.

Recent Discoveries of New Genes

Experimental studies on new gene origination emerged in the early 1990s when a newly evolved gene, *Jingwei* (*Jgw*), was identified from two African species of *Drosophila* (Long and Langley 1993). Since then, by taking advantage of the availability of phylogenetic frameworks and rapidly expanding databases for the major model species, many examples of new genes

Table 3.1   Examples of new genes that have been fully described in mammals, flies, and plants

Organism	New gene	Parental gene(s)	Mechanism(s)	Age (my[a])
Primates[b]	PGAM3	Phosophoglycerate mutase (Pgam)	Retroposition	10
	RNASE1B	Pancreatic ribonuclease gene (RNAse1)	Gene duplication	4
	PMCHL1	Melanin-concentrating hormone (MCH)	Exon shuffling + retrotransposition	20
	PMCHL2	Melanin-concentrating hormone (MCH)	Segmental duplication	2.5–5
	Morpheus	Morpheus	Segmental duplication	12–25
	Tre2 (USP6)	USP32 (NY-REN-60) and TBC1D3	Segmental duplication	21–33
	CGβ	Luteinizing hormone β subunit gene (LHbeta)	Gene duplication	34–50
Rodents	4.5Si RNA	RNA gene	Gene duplication	25–55
	BC1 RNA	tRNA[Ala]	Gene duplication	60–100
	Insulin 1(Ins1)	Ins2	Retroposition	10–15
Flies	Jingwei	Alcohol dehydrogenase (Adh)+ yellow emperor	Retroposition and fusion	2.5
	Adh-Finnegan	Adh + unknown sequence	Gene duplication and fusion	30
	Adh-Twain	Adh + CG9010	Retroposition and fusion	5
	Sdic	AnnX + Cdic	Gene fusion	<3
	Exuperantia2	Exu1	Ectopic recombination + transposition	25
	Sphinx	ATP synthase chain F	Retroposition	2.5

have been fully described in eukaryotes from protozoa to *Drosophila* to primates. Here, we will briefly summarize our knowledge of novel genes identified in various organisms and the mechanisms thought to govern their origination and subsequent evolution (for a detailed review, see Long et al. 2003).

Drosophila species have served as the best model organisms for new gene research since the 1990s. Fourteen new genes, derived within the last 30 million years, have been fully described in several species of *Drosophila* (Table 3.1). For example, *Adh-Twain* was created by the fusion of a retroposed *Adh* sequence with a target gene in the common ancestor of *D. subobscura, D. madeirensis,* and *D. guanche* (Jones et al. 2005). *Siren,* in the *D. bipectinata* complex, is another chimeric gene involving *Adh,* also created by retroposition (Nozawa et al. 2005). New *Monkey King* genes were formed by duplication, followed by partial degeneration in complementary parts of the parent and copy gene sequences, and final fusion of these two adjacent genes in the

Table 3.1   *Continued*

Organism	New gene	Parental gene(s)	Mechanism(s)	Age (my[a])
Flies	*Dnth-2r*	*Nuclear transport factor*	Retroposition	5
	Monkey King	CG7163	Gene fusion and fission	1–2
	K81	CG14251	Retroposition	15
	siren	*Nanos+CG11779*	Retroposition	20
	Ifc-2h	*Infertile crescent (Ifc)*	Retroposition	1–2
	Hun Hunaphu	*Baochen*	Illegitimate recombination	1–2
	Ifc-2h	*Ifc*	Retroposition	2
	Quijote (CG13732)	*Cervantes* (CG15645)	Retroposition	5
Fish	Arctic *AFGP*	Polyprotein	Gene duplication	2.5
	Antarctic *AFGP*	Pancreatic trypsinogen	Gene conversion and duplication	5–14
Plants	*Sanguinaria rps1*	*rps1*	Gene transfer from mitochondrion to nucleus	45
	Plantago ap1	*Bartsia ap1*	Gene transfer from host to parasite plant	?
	Cytochrome *c1*	Cytochrome *c1*	Exon-shuffling	>110
	Nuclear *Cox2*	Mitochondrial *cox2*	Gene transfer from mitochondrion to nucleus in legume and exon-shuffling	50
	At1g71920	*At5g10330* (histidinol phosphate aminotransferase-like gene)	Gene duplication	0.5
	At1g05090	*At4g20720* (unknown function)	Gene duplication	0.6

[a]Abbreviation: my, million years.

[b] See Marques et al. (2005) for more primate-specific new retrogenes.

common ancestor of *D. simulans, D. mauritiana,* and *D. sechellia* (Wang et al. 2004). *Hun Hunaphu* was identified as a young gene created by illegitimate recombination at some time after *D. simulans, D. mauritiana,* and *D. sechellia* diverged from *D. melanogaster* (Arguello et al. 2006). *Quijote* is a recognizable retroposed copy of CG15645 present in the branch leading to *D. melanogaster* and *D. simulans* (Bertran et al. 2006). Sequence divergence and polymorphism analyses showed, in all these examples, that directional selection (positive selection or a recent selective sweep) played a crucial functional role in the early stage of their evolution.

Among vertebrates, a number of new genes have been found in fish, rodents, and primates (see Table 3.1). For example, *RNAse1B* is a new duplicate under strong positive selection because it is involved in the unique digestive system of leaf-eating colobine monkeys (Zhang et al. 2002b). *Pgam3* is a primate lineage–specific retroposed gene with testis-biased expression (Betran et al. 2002). *C1orf37-dup* is a human-specific retroposed gene, driven to evolve rapidly by positive Darwinian selection (Yu et al. 2006). *Ins1* is a rodent-specific insulin gene which was derived by retroposition 10–15 million years ago from *Ins2*, and is under positive Darwinian selection (M. S. Shiao, M. Long, and H. T. Yu, unpublished data). *Ubl4b* is a novel retroposed mouse ubiquitin-like protein with a testis-specific expression (Yang et al. 2007). In fish, antifreeze glycoproteins (*AFGPs*) help in adaptation to cold environments. Sequence comparisons indicate that two *AFGPs* arose independently in Arctic and Antarctic fishes through extensive gene duplication and gene conversion (Chen et al. 1997a,b). Finally, *TRIM5-CypA* in owl monkeys has been shown to be a novel chimeric gene, which evolved to resist HIV-1 virus infection (Nisole et al. 2004; Sayah et al. 2004).

Compared to other taxa, fewer young genes have been described in plants. The reason may be that whole genome duplications through hybridization are a common process contributing to plant species diversity (Otto and Whitton 2000). This may lessen the selective need for new gene origination via single gene duplication. Alternatively, retroposition may be relatively less common in plants because they lack active L1 retrotransposons. It is hard to evaluate these possibilities as the scarcity of new young genes may simply reflect the relative lack of attention this subject has received from plant biologists.

Nevertheless, several insights have been reached since genomic sequence data were completed for model plant species. First, horizontal gene transfer, which involves gene movements between species or between cytoplasm and nucleus, is far more frequent in the plant kingdom than in other organisms (see Table 3.1) (Bergthorsson et al. 2003; Park et al. 2007; Richardson and Palmer 2007). Second, a number of retroposed genes have been identified in the Arabidopsis genome by computational and experimental approaches, and some of them are newly derived genes that only occur within the *Arabidopsis* genus or are unique to *A. thaliana* (Zhang et al. 2005). For example, *At1g61410* and *At5g52090* are two retroposed genes derived from mRNA. Sequence analyses of these two genes indicate that they are only present in *Arabidopsis* species derived from Mediterranean Pleistocene refugia (Zhang et al. 2005). Third, a recent study using rice genome data demonstrates that extensive retroposition has resulted in thousands of functional retrogenes during grass genome evolution (Wang et al. 2006). A large proportion of these retrosequences are chimeric, having recruited new exons, introns, and coding regions from the sites in which they have inserted. Finally, Moore and Purugganan (2003) showed reduced nucleotide polymorphism in two new gene duplicates in *A. thaliana*, suggesting again a possible role for positive selection in the evolution of these genes.

Mechanisms to Generate New Genes

Molecular mechanisms involved in creating novel gene structures are now well understood and are described in detail elsewhere (Long et al. 2003).

Here, we will briefly describe the main forces at play. Many new genes have been created through a combination of two or more of the following mechanisms (see also Table 3.1):

- **Gene duplication** can be achieved at several cellular levels: whole genome duplication, segmental duplication, or single tandem gene duplication. Duplication allows the original functions to be maintained by one copy, while the second copy provides a substrate for other mechanisms leading to new gene origination (see below).

- **Exon shuffling** is achieved by illegitimate recombination of exons or retroposed exon insertions that create a new exon-intron gene structure. Such processes can lead to a new gene with novel function. The shuffling mechanism has been found to generate numerous chimeric proteins (e.g., Wang et al. 2005; Li et al. 2007). It can also create chimeric genes with previously unrelated regulatory sequences, for example, the *Siren* (Nozawa et al. 2005) and *Ste* genes in *D. melanogaster* (Usakin et al. 2005).

- **Retroposition** is a mechanism that generates new intronless gene copies (retrogenes) by reverse transcription of mRNA derived from parental genes. Three hallmarks can be used to identify retrogenes: (1) one member of the pair is intronless while the other contains introns in the coding regions; (2) the new (intronless) copy contains a poly(A) tail; and (3) the new copy may still have short duplicate flanking sequences. Experimental and computational genomics studies have both found large numbers of retroposed genes in eukaryotes, including yeasts, plants, and animals.

- **Mobile elements** can pick up host sequences and integrate them into new genomic positions. If the site of integration is near or within existing coding sequences, a new, chimeric gene structure can be generated. Many cases of new genes created by mobile elements have been described. For example, Pack-MULEs can recruit small chromosome fragments and combine with other genomic regions when they transpose to form chimeric gene structures (Jiang et al. 2004). Helitrons, which are helicase-bearing transposable elements, are likewise capable of shuffling genomic regions (Bennetzen 2005).

- **Horizontal gene transfer** is the movement of genes from one species to another or between organelles and the nucleus. This event occurs frequently in bacteria and yeasts. There is also some evidence for it occurring in plants (see Chapter 4; Bergthorsson et al. 2003).

- **Gene fusion/fission** occurs when two adjacent genes fuse together to form a single gene, or when a single gene splits into two genes that then evolve different functions. One example of gene fusion/fission is the formation of the *Monkey King* gene described above.

■ **De novo origination** is a final possible source of new genes which cannot be ruled out, even though no clear evidence has been reported for such an origin of a complete protein-coding gene. There is some evidence that frame-shift mutations in a number of duplicated vertebrate genes created quasi-random sequences from which new genes subsequently evolved (Raes and Van de Peer 2005). This supports the possibility of de novo origins of protein-coding genes. More recently, Begun and colleagues (2007) found a significant number of X-linked testis-biased de novo noncoding RNA genes in the *D. yakuba/D. erecta* clade.

Evolutionary Forces for New Gene Retention

Positive Darwinian selection is likely to be the most important force acting for the retention and evolution of novel genes (Long et al. 2003). Particularly, most new genes that originated through exon shuffling and gene duplication have undergone significantly accelerated rates of evolution compared to their parental copies. For example, *Jingwei* has a significantly higher rate of substitution in its protein sequences and gene structure, and sequence divergence analysis suggests a high rate of protein adaptive evolution (Long and Langley 1993).

Two methods have been used to test whether positive selection acted on novel genes during their evolution. The first is to estimate the K_a/K_s ratio in new gene lineages (K_a = the nonsynonymous substitution rate, K_s = the synonymous substitution rate). For example, *RNASE1B* is a new, duplicate ribonuclease gene that arose 4 million years ago in the leaf-eating colobine monkey. The K_a/K_s ratio (4.03) of *RNASE1B* is significantly higher than unity. In contrast, its paralog, *RNASE1*, has accumulated no amino acid substitutions over the same time period (Zhang et al. 2002b).

The second method to test for positive selection is to compare sequence divergence between species and sequence polymorphism within species, formalized as the McDonald–Kreitman (1991) test of neutral molecular evolution. Positive selection is indicated when there is an excess of amino acid replacement substitutions between species compared to the neutral prediction that the variation of replacement substitutions and synonymous substitutions should be positively correlated. For example, McDonald–Kreitman tests of *Hun Hunaphu*, a recently evolved (within the last 2–3 million years) chimeric gene found in *D. simulans*, *D. sechellia*, and *D. mauritiana*, reveal it to have been subject to positive selection in the *D. simulans* branch (Arguello et al. 2006).

However, we cannot rely on analyses of individual cases to determine whether positive selection has a general role in driving new gene evolution and retention. An approach that uses genomic data derived from different chromosome regions might be a feasible way to detect general forces that drive the evolution of duplicated genes. Thornton and Long (2002) compared the K_a/K_s ratios of more than 100 paralogous gene pairs on the X chromosome with 1743 paralogs on the autosomes in *D. melanogaster*. They found that X-linked duplicates have higher K_a/K_s ratios than autosomal duplicates. They further estimated the K_a/K_s ratios of single-copy genes and found no accelerated rate of amino acid substitutions of X-linked genes. Such an inconsistency suggests that different forces might act on single-copy and newly duplicated genes, which likely acquire new functions under positive selection.

The Location and Movement of New Genes

A new gene can be located adjacent to (tandem duplication) or far away from its parental copy. A random distribution of new gene movement might be expected. However, studies of the pattern of gene movements show a surprising asymmetry. Betran and colleagues (2002) computationally screened the genome sequence data of *D. melanogaster* to check the location of retroposed genes and their parental copies. They found that there was a significant excess of retrogenes originating from the X chromosome and retroposed to autosomes, and relatively few new genes retroposed in the opposite direction. This result was further supported by a recent study that investigated retrogene movement between and within chromosomes in the *D. melanogaster* genome (Dai et al. 2006). Emerson and coworkers (2004), extending this approach to human and mouse, showed that the mammalian X chromosome also generated a significantly higher number of functional retroposed genes than autosomes. In contrast to *D. melanogaster*, mammalian X chromosomes also recruited an excess of new retrogenes. Further experiments in *D. melanogaster* demonstrated that most new autosomal retroposed copies exhibited testis-biased expression, unlike the parental X-linked genes (Betran et al. 2004; Emerson et al. 2004). These observations provide strong evidence that genome position also plays a very important role in the recruitment of new gene copies (Betran et al. 2004).

Positional effects of new gene locations have also been observed in plants. In a study of retroposed genes in the rice genome, Wang and colleagues (2006) observed that functional retrogenes tend to "avoid" centromeric regions and prefer to insert into the middle of chromosomal arms. Compared to the random distribution of processed pseudogenes, the biased distribution of functional retrogenes probably reflects natural selection.

Inferring the Functionality of New Genes

In principle, the functionality of a new gene can be inferred in both direct and indirect ways. Direct experimental tests in various functional analyses provide explicit information about a new gene's functions, but these are costly and time-consuming. Indirect approaches, using bioinformatic techniques, are therefore valuable, as they are easily accomplished and provide candidates for further direct functional analysis. The simplest way is to detect evolutionary constraints that are associated with functional genes.

We can examine evolutionary constraints by calculating the K_a/K_s ratio in a new gene lineage. $K_a/K_s < 1$ indicates strong functional constraint under purifying selection. $K_a/K_s = 1$ indicates no functional constraints and neutral evolution—typical of pseudogenes. $K_a/K_s > 1$ indicates an accelerated amino acid evolution rate under positive selection. A more appropriate model may be to assume that the parental gene is subject to strong purifying selection with $K_a/K_s = 0$, whereas the new gene is a functionless pseudogene with $K_a/K_s = 1$. Therefore, the substitution rates calculated by comparing the parental copy and new pseudogene copy yield $K_a/K_s = 0.5$. Thus, a ratio of $K_a/K_s < 0.5$ suggests the new gene copy is likely to be functional, $K_a/K_s = 0.5$ suggests it may be a functionless pseudogene, and $K_a/K_s > 0.5$ suggests it may have experienced positive selection (Thornton and Long 2002). It should be noted that the criterion of $K_a/K_s = 0.5$ is very conservative, because it was derived based on the specific assumptions of evolutionary stagnation of the parental gene and equal synonymous substitution rates for the new and parental genes. The functional specificity of new genes can also be explored by analyzing their expression profiles. By comparing a new gene's transcription pattern (with respect to tissue or developmental stage) to that of its parental gene, we can tell whether the new gene has acquired new functions. For example, *Dnth-2r* was identified as a new retroposed gene that is only transcribed in testis, while its parental gene is ubiquitously expressed in both sexes of *D. melanogaster* (Betran et al. 2003).

There are two general direct approaches to detecting function in new genes. First, we can use biochemistry and immunological technology to obtain protein products, and then test their functions in vitro. For example, Jones and colleagues (2005) used western blotting to analyze the protein synthesized from a new chimeric retroposed fusion gene. Zhang and colleagues (2004) investigated the function of the new *Jingwei* gene in *Drosophila* by studying the enzymatic properties of JGW proteins collected from a microbial expression system. They found JWG was a novel dehydrogenase with alternative substrate specificity compared with the ancestral ADH protein. JGW protein also uniquely prefers to catalyze reactions involving the long-chain primary alcohols found in insect pheromone metabolism.

As a second direct approach, the functions of new genes can be tested by genetic silencing in vivo, for example by gene knockout or RNAi. In addition, transgenic lines carrying *gene::GFP* fusions can be produced to observe the location of a new gene's expression, which can provide insights into its biological functions (Loppin et al. 2005). Competition experiments between a strain that carries the new gene and a strain in which it is silenced can be used to measure the effect of the new gene on fitness. Observation of changes at the phenotypic, physiological, behavioral, and population genetic levels would provide further understanding of gene function, but as yet this approach has not been taken.

General Methods to Detect New Genes

Early findings

New genes were initially found by serendipity rather than intentional searches. For instance, the first novel gene, *Jingwei*, was identified in 1993 based on previous studies which had considered it to be a processed pseudogene in *D. yakuba* (Long and Langley 1993). Several more young *Drosophila* genes, for example *Adh-Finnegan*, *Sdic*, and *Exu2*, were accidentally found four to five years later. *Adh-Finnegan* was initially claimed to be an *Adh* pseudogene, but later analyses showed it was a functional gene recently descended from an *Adh* duplication (Begun 1997). The discovery of *Sdic* was built solely on an earlier observation that the genetic organization of the 19DE region on the *D. melanogaster* X chromosome differs from that of other species in the subgroup (Nurminsky et al. 1998). The gene *Exuperantia2* was detected because of a complete linkage disequilibrium between two single nucleotide polymorphisms in *D. pseudoobscura* (Yi and Charlesworth 2003).

Comparative molecular cytogenetic analyses

Phylogenetic comparison of genetic signals (e.g., fluorescence in situ hybridization [FISH] and genomic Southern blotting), has proved to be an efficient and reliable way of identifying young protein-coding genes in *Drosophila* and mammals. Wang and colleagues (2002 and 2004) used FISH to systematically search for new genes in *Drosophila* species, taking advantage of publicly available cDNA collections (Berkeley Drosophila Genome Project, www.fruitfly.org). They amplified and labeled the cDNA inserts and then hybridized them to polytene chromosomes of each member of the *D. melanogaster* subgroup. By counting hybridization signals on the polytene chromosomes of theses species, new homologs translocated to different cytological loci were detected. This approach detected about 100 new duplicates across eight species of the *D. melanogaster* subgroup, and three of these have been fully described: *Sphinx, synathase chain F*, and *Monkey King* (see Table 3.1).

Despite this success, the limitations of FISH screening are also obvious. First, FISH (or genomic Southern blotting) is technically demanding and laborious. Second, many organisms do not have polytene chromosomes. Third, even in the case of the *Drosophila* genus, which has polytene chromosomes, FISH cannot be used to detect new genes in heterochromatic regions, because polytene chromosomes do not include heterochromatin. Finally, FISH cannot resolve tandem duplications, where the duplicates are adjacent.

Computational genomic analysis

Extensive comparative sequencing and expression studies, coupled with evolutionary analyses and simulations, have been applied in several organisms to identify gene duplication events at the whole genome level. Completed genome sequences in model organisms provide opportunities to search for duplicated genes and further examine the pattern of gene origination.

Betran and coworkers (2002) surveyed the whole *D. melanogaster* genome to search for new retroposed genes. They inferred parental and derived copies by examining potential retroposed genes for hallmarks of the retroposition process (see previous discussion). At a threshold of more than 70 percent protein sequence identity, they identified 24 retroposition events, all within the last 30 million years. They further reported a new gene in the *D. melanogaster* subgroup, *Drosophila nuclear transport factor-2-related* (*Dntf-2r*). Its sequence and phylogenetic distribution indicate that *Dntf-2r* is a functional retroposed gene that originated in the common ancestor of *D. melanogaster, D. simulans, D. sechellia,* and *D. mauritiana* within the past three to five million years and is under positive Darwinian selection.

Marques and colleagues (2005) systematically screened the human genome for retrogenes by comparing the genome sequences of human, chimpanzee, and mouse. They identified 57 retrogenes in the human genome, estimating that one retrogene per million years has emerged on the primate lineage leading to humans. Comparative sequence and gene expression analyses suggest that a significant proportion of recent retrocopies represent human-specific genes. They concluded that retroposition significantly contributed to the formation of recent human genes and that most new retrogenes were progressively recruited during primate evolution by natural and/or sexual selection to enhance male germ line function.

Zhang and coworkers (2005) identified 69 retroposons in the *Arabidopsis thaliana* genome. Most of them were derivatives of mature mRNAs. Of them, 22 are processed pseudogenes and 52 genes are likely to be actively transcribed, especially in tissues from roots and flower apical meristems. This study estimated the rate of new gene creation by retroposition as 0.6 genes per million years. Forty-five of the parental genes were highly expressed in the germ line cells, which presumably predisposes them to be templates for retroposition.

Genomic computational searching also has noticeable restrictions. First, this method is limited to model organisms whose genomes have been sequenced. Second, although retrogenes can be found using this method, it is less useful for finding other types of gene duplication, such as exon shuffling and gene fusion. Finally, such a screening method will often miss new duplicates in genomes sequenced by the whole genome shotgun approach (e.g., International Chicken Genome Sequencing Consortium 2004). The time of gene duplication events can be estimated by sequence divergence analyses (e.g., sequence identity or the synonymous substitution rate K_s) or from the phylogenetic distribution of the duplication.

Comparative Genomic Hybridization to Detect New Genes

Microarrays are a developing technology used to study gene expression at the whole genome level. Single-stranded DNA (ssDNA), referred to as *probe,*

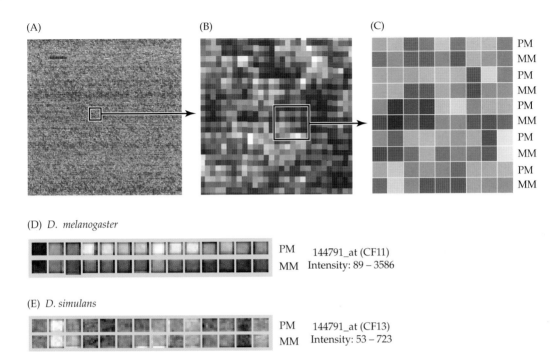

(A) (B) (C)

(D) *D. melanogaster*

PM 144791_at (CF11)
MM Intensity: 89 – 3586

(E) *D. simulans*

PM 144791_at (CF13)
MM Intensity: 53 – 723

Signal intensity comparison between *D. melanogaster* and *D. simulans*

Figure 3.1 An example of images from an Affymetrix GeneChip hybridization experiment with *D. melanogaster* genomic DNA. The intensity of hybridization signal is characterized by the ratio of black and white: the darker the color, the more intense the hybridization signal (the more labeled DNA fragment was bound). (A–C) Images at three different resolutions of a GeneChip after hybridization with labeled genomic DNA. (D) Hybridization intensity of 14 probe pairs from feature 144791 hybridized with *D. melanogaster* genomic DNA. (E) Hybridization intensity of 14 probe pairs from feature 144791 hybridized with *D. simulans* genomic DNA. Abbreviations: PM, perfect match; MM, mismatch. (Fan and Long, unpublished data.)

is printed in a regular grid-like pattern. The target RNA or DNA from a particular biological sample is fluorescently labeled and allowed to hybridize to the array. Depending on the specific experimental design, the intensity of each spot or the average intensity difference between matches and mismatches can be related to variation in gene expression (mRNA abundance), DNA polymorphisms, or mutations caused by changes in copy number in whole genome samples (Figure 3.1; Pinkel et al. 1998; Barrett et al. 2004; Greshock et al. 2004; Toruner et al. 2007). In an effort to develop a more generally useful method to detect new gene candidates, we have adapted this technology to detect the variation in duplicate copies (gene gain or loss) in closely related species by hybridization.

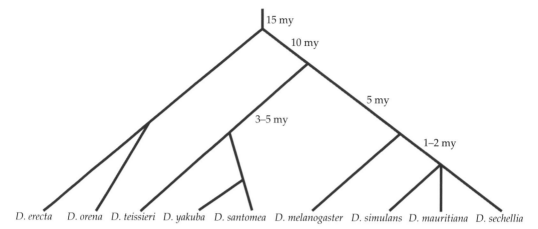

Figure 3.2 Phylogeny of the *D. melanogaster* subgroup with estimates of divergence times.

The availability of genomic sequences and GeneChip arrays for *D. melanogaster* has allowed us to systematically search for new genes using array-based comparative genomic hybridization (CGH). The subgroup of *D. melanogaster* includes nine species with a divergence time of less than 15 million years (Figure 3.2; Lachaise et al. 1988; Lachaise et al. 2000). The relatively small genome size of *D. melanogaster* and its well-defined phylogeny provide a convenient context to find young genes. Because the GeneChip arrays were made using genomic sequence data from *D. melanogaster*, the initial data analyses were conducted using *D. melanogaster* as a baseline to calculate the ratio of hybridization intensity for each gene between *D. melanogaster* and the other species. Considering the sequence divergence between different species, we took a ratio of 1.5 or higher as the threshold for gene duplication. This was based on an initial calibration using a known duplicated region (with 31 genes) in *D. melanogaster*, which yielded a ratio distribution for duplicates of about 1.3–1.5.

Examination of the microarray intensity ratios of pairwise comparisons allowed us to identify candidate duplicates with ratios of 1.5 or higher in these species. Next, we applied genomic Southern hybridization and BLAST searching of genomic sequence data of *D. simulans* and *D. yakuba* to confirm the duplicate copies and survey their phylogenetic distribution. Examples of candidate duplicates identified by CHG are given for *D. simulans* and related sibling species in Table 3.2.

A specific example of a new gene identified by array-based CHG is *infertile crescent-2h* (*Ifc-2h*), derived from its parental gene *infertile crescent* (*Ifc*). The hybridization ratios of *D. simulans* (1.74), *D. mauritiana* (1.73), and *D. sechellia* (1.56) to *D. melanogaster* were higher than those of other species (1.15–1.49), suggesting that the former group of species had a new gene copy. This was supported by a BLAST comparison of the *D. melanogaster Ifc*

Table 3.2 Examples of new duplicates identified by array-based CGH in the three sibling species *D. simulans*, *D. mauritiana*, and *D. sechellia*

GeneChip ID	Parental copy	Location of parental copy[a]	Copy number in *D. simulans*	Copy number in *D. melanogaster* and *D. yakuba*	Location of new duplicate[a]
153965	CG12081	X	2	1	2h
143838	*Rpl9*	2L	2	1	2R
153842	*Ifc*	2L	2	1	2h
153010	*Chmp1*	3L	2	1	X
145353	CG7914	X	2	1	2L

[a]Chromosomal location is given for arms (L = left, R = right) and heterochromatin (h).

sequence to genomic data of *D. simulans*, which revealed two sequences homologous to *Ifc*. One copy had a single intron and is therefore the parental copy; the other lacked an intron and is likely to have been derived via retrotransposition (Figure 3.3A). A Southern hybridization using *Hind*III digested DNA confirmed the BLAST result. Only single bands exist in *D. melanogaster*, *D. teissieri*, *D. santomea*, and *D. erecta*. Two bands are found in *D. simulans*, *D. mauritiana*, and *D. sechellia*. Two bands are also found in *D. yakuba* but are generated by a unique *Hind*III digestion site in the intron of *Ifc*, so *D. yakuba* actually has only one copy (Figure 3.3B).

Sequence analysis of *Ifc-2h* shows 12 indels—four in coding and eight in noncoding flanking and UTR regions (Figure 3.4). The lengths of all four coding region indels are in multiples of three. This pattern is significantly different ($P = 0.0123$) from the random distribution that occurs in noncoding regions, revealing evolutionary constraint to maintain a nondisrupted reading frame in the coding region of the new gene copy. However, the *D. sechellia* copy is probably degenerating to a pseudogene, because a premature nonsense mutation in its coding region drastically shortens the reading frame.

The expression profile of *Ifc-2h* was examined by RT-PCR in *D. simulans* (Figure 3.3C). The results show that *Ifc-2h* is highly transcribed in eggs, second larvae, and adults, but the transcription in third larvae and pupae is relatively low. This expression pattern differs from the parental copy, *Ifc*, which is ubiquitously transcribed at all developmental stages. All the analyses described above indicate that *Ifc-2h* is a functional protein-coding gene (Fan and Long 2007).

Challenges in using array-based CGH in new gene studies

Array-based CGH is a potentially powerful tool to detect duplication events in different species. However, there are a number of challenges to this approach. In particular, sequence divergence between species limits hybridization of heterospecific DNA. We noticed that a one percent sequence

(A)

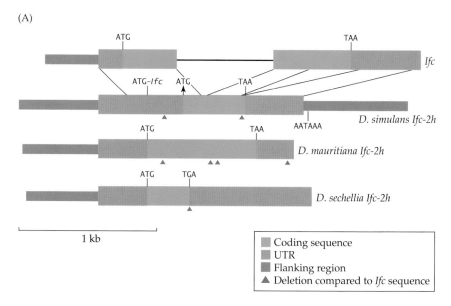

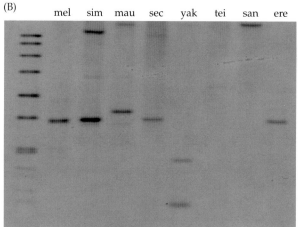

(B)

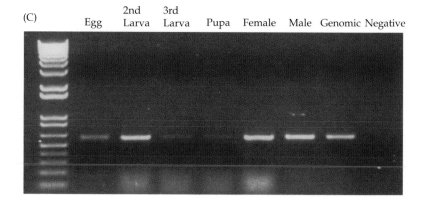

◀ **Figure 3.3** Schematic diagram of the gene structure of *Ifc* and *Ifc-2h*. (A) Start and stop codons and an adenylation signal are shown. (B) Genomic Southern blotting using a probe for *Ifc-2h* against *Hind*III digested DNA. Species names are shown at top of each lane (mel, *D. melanogaster*; sim, *D. simulans*; mau, *D. mauritiana*; sech, *D. sechellia*; yak, *D. yakuba*; tei, *D. teissieri*; san, *D. santomea*; ere, *D. erecta*). Note that two bands are seen in *D. simulans*, *D. sechellia*, *D. mauritiana* and *D. yakuba*. (C) RT-PCR for *Ifc-2h* transcripts in *D. simulans* at various developmental stages; similar patterns are seen in *D. sechellia* and *D. mauritiana*.

divergence usually accounts for a five to eight percent signal reduction in *Drosophila* genomic hybridization (Table 3.3). If the species is too distantly related (sequence divergence > ten percent), then many probes do not generate adequate signals from hybridization.

A related problem is sequence divergence between paralogs. Signals from duplicate copies are subject to a considerable range of variation. Furthermore, paralogous duplicates in the related species are associated with sequence divergence correlated both with the time of the duplication event and the degree of functional constraint on the new gene. New genes tend to evolve at an accelerated rate early after origination, thus lessening the signal intensity considerably and potentially limiting the ability to detect new genes.

Finally, we note that microarrays are a complementary technology to other tools in the arsenal of the molecular biologist. In the search for new genes, microarrays provide candidates that need to be verified by other molecular techniques (e.g., FISH and genomic Southern blotting, detailed sequences analyses, and expression profiles).

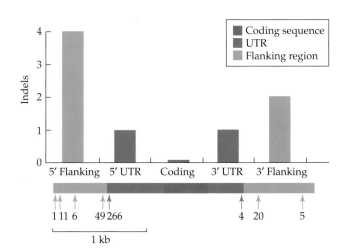

Figure 3.4 *Ifc-2h* polymorphic insertions and deletions (indels) found in *D. simulans* population sequences. Arrows indicate the relative positions of indels. The number below each arrow indicates indel size (bp).

Table 3.3 Sequence divergence between species and ability to detect related sequences by microarray hybridization

Species	Sequence divergence (%) from *D. melanogaster* (Canton-S)	Replicates	Detection by microarray (%)	Average intensity (three replicates)
D. melanogaster (Oregon-R)	0.5	3	96.6	522.4
D. simulans	4	3	77.9	532.2
D. mauritiana	4	3	79.0	536.0
D. sechellia	4	3	73.5	544.0
D. yakuba	7	3	47.0	562.0
D. teissieri	7	3	47.5	571.3
D. santomea	7	3	45.9	563.0
D. erecta	10	3	47.9	546.7

Outlooks and Perspectives

The power of searching for new gene candidates by comparative genomic screening is increasing with the expansion of genome sequence data from different species. For example, genome sequences have now been completed for twelve closely related *Drosophila* species; comparisons among these entire genomes will provide solid evidence of young gene candidates for each species and/or clade within this group. Such comparisons will also provide whole genome evidence for how often novel genes have arisen. We will also be able to see whether novel genes are associated with speciation and adaptation after new species evolved. Moreover, the role of selection in the retention of new gene copies can be fully characterized.

One of the major challenges for novel gene studies is to determine whether novel genes produce functional proteins. Fortunately, advances in global-scale analysis of proteins are expected to allow direct observation of protein function and regulation (Sauer et al. 2005). Endogenous proteins can be identified by two-dimensional gel electrophoresis and characterized using mass spectrometry. Further, we can examine the function of uncharacterized proteins using protein–protein interaction data produced by affinity-based proteomics (protein arrays) and other proteomic methods, such as yeast two-hybrid analyses (Y2H).

The survival value of new genes is determined by their networks of interactions with other genes, which give rise to biological processes. To understand how natural selection leads to the retention or loss of new genes requires placing the new genes' activities in this context, rather than simply inferring properties from the rates of gene sequence evolution (e.g., K_a/K_s ratios). Interesting questions to investigate include: (1) how gene–gene interactions evolve in a new gene copy, and (2) how these interactions affect biological functions.

One of the ultimate goals of gene origination studies is to uncover general patterns for gene origination and evolution, and to further understand how this contributes to organismal evolution. Using applied, array-based CGH, combined with molecular biological tools and computational comparisons, we can identify young genes with known functions in sufficient numbers to learn about new gene evolution at the genomic level, and, further, to precisely measure the origination rate of new gene functions. This achievement will show how quickly organisms adapt by changes in gene diversity, and to what degree this is correlated with environmental change.

CHAPTER **4**

Lateral Gene Transfer

*W. Ford Doolittle, Camilla L. Nesbø, Eric Bapteste,
and Olga Zhaxybayeva*

Introduction

THE EXTENT TO WHICH ORGANISMS ENGAGE IN LATERAL GENE TRANSFER, the general importance of the process in adaptation and divergence, and the threat it poses to natural classification have been widely appreciated only since the mid-1990s. Now, the science of *lateral genomics* is an independent discipline, with dozens of laboratories focused on documenting recent and ancient LGT events, undertaking global estimates of its prevalence, and searching for the constraints under which it operates. In this chapter we discuss the definition, detection, and quantitation of LGT, its role in microbial adaptation, and its implications for the phylogenetic basis of microbial systematics. Our focus is on prokaryotes, but eukaryotes (even multicellular animals and plants) are not immune to LGT, as we note in the concluding section.

Defining terms

The terms *lateral gene transfer* (LGT) and *horizontal gene transfer* (HGT) are interchangeable, but equivocal in their meaning. Sometimes these terms are used to describe evolutionary *patterns*, such as patchy distribution of a gene or incongruent phylogenies, and sometimes to refer to the genetic *processes* or mechanisms thought to be responsible for the patterns, such as conjugative DNA transfer, or uptake and chromosomal integration of foreign DNA by transformation.

In general, LGT is said to have occurred whenever genetic information has crossed boundaries between species, regardless of the genetic mechanism responsible. Even homologous recombination (HR) between orthologous genes is LGT, if the donor and recipient are of different species. But within a prokaryotic species, HR between orthologs is not usually (and we think should never be) considered LGT: if it were, then for consistency we would have to say that every event of human reproduction entails innumerable LGTs. Within prokaryotic species, only the introduction of donor genetic information for which orthologous sequences were previously lacking in the recipient should be considered LGT. Such introduction is most often effected by transfer of plasmids or mobile genetic elements from a donor strain into a recipient strain which previously lacked them, or by HR between homologous sequences adjacent to the novel region.

A Very Brief History of Lateral Genomics

The potential evolutionary and ecological impact of between-species gene exchange was already apparent by the early 1960s, when Watanabe and coworkers showed that antibiotic resistance among epidemic *Shigella* strains could be explained by transfer of plasmids from *Escherichia coli* (see Brock 1990). This sort of evolution was clearly distinct from that generally contemplated by neo-Darwinist microbiologists: a more gradualistic process of selection operating on mutant alleles arising within asexual populations. In the 1970s, Sorin Sonea and Darryl Reanney proposed that all prokaryotes comprise one global metapopulation, sharing a common universal gene pool and evolving as does a single "species" of multicellular organisms, albeit with significant differences in rate. Such proposals did not enjoy wide acceptance, however, perhaps because the spread of antibiotic resistance was seen as a special (and recent) phenomenon, and perhaps because pervasive LGT was incompatible with the molecular phylogenetic approach, then gaining popularity (see essays collected in Sapp 2005).

It was the appearance of complete genome sequences that first earned widespread attention for LGT as a potentially major evolutionary force. In the genome sequence of the hypothermophilic bacterium *Thermotoga maritima* published in 1999 (Nelson et al. 1999), nearly a quarter of the genes appeared to be products of LGT from phylogenetically distant sources; indeed, even from another domain—the Archaea. *Thermotoga* is no longer the between-domain LGT record holder: in the genomes of the archaeans *Methanosarcina mazei* (Deppenmeier et al. 2002) and *Haloarcula marismortui*, 30 and 50 percent of the genes, respectively, have their closest homologs in bacterial genomes, rather than other archaeal genomes. Similarly, a recent report estimates that three-quarters of the yeast genome is bacterial in origin, rather than archaeal, as would be expected if the widely accepted sister relationship of Archaea and Eukarya holds (Esser et al. 2004). There are complex issues of interpretation hidden within such broad phyletic comparisons, but there is no missing the message of these genomes: LGT, by whatever mechanisms it may occur, is a massively important evolutionary reality.

Mechanisms of Lateral Gene Transfer

Nevertheless, individual claims about LGT events, especially between phylogenetically remote participants, often provoke the question "how could this (possibly) have happened?" Such skepticism (for instance Kurland et al. 2003) often takes the form of insistence that a detailed transfer scenario should accompany the phylogenetic or gene distribution data supporting any claim for LGT. Such a scenario might include reasoned guesses as to biogeographic, ecological, cellular, genetic, and population genetic processes involved. We suggest that this attitude—the insistence that unless LGT is solidly proven, vertical descent should be our default assumption or null hypothesis—is no longer appropriate. Vertical descent through simple DNA replication is, of course, the means by which the overwhelming majority of heritable information is transmitted in the short term, but we have old, rich, and solid knowledge of a trio of mechanisms for LGT, which may be the dominant evolutionary processes in the long term. Furthermore, we now find that the effectiveness of each of these mechanisms—transduction, conjugation, and transformation—can be enhanced under natural conditions where transfer might be beneficial (for recent reviews, see Frost et al. 2005; Thomas and Nielsen 2005).

Transduction

Transduction is the transfer of DNA via viral particles. Generalized transduction (errors in packaging that replace phage DNA with host DNA in active particles) and specialized transduction (errors in prophage excision that produce chimeric host–phage genomes) have been understood for half a century: they made fine scale genetic mapping possible long before DNA could be sequenced. Renewed appreciation for the role of viruses (phages) in prokaryote ecology and evolution, and in particular as agents for LGT, comes from genomics at three levels (Brüssow et al. 2004).

First, complete bacterial genome sequences show that prophages (active and moribund) are present in most bacteria, and can make up 20 percent of a genome. Much of the variation in genome size among strains of a species can reflect variation in prophage content. Diversity in the numbers, locations, and types of prophages encoding Shiga toxins, for instance, characterize the rapidly diverging strains of the notorious pathogen *E. coli* O157:H7.

Second, comparative genomics of phages themselves demonstrates that LGT may be the dominant mode of viral evolution. Particularly well-studied are lambdoid phages and other double-stranded DNA phages with tails. Between close relatives, there is *quantitative mosaicism*—patchy variation in the extent of sequence similarity, indicative of HR among moderately diverged lineages. There is also *qualitative mosaicism*, in which syntenic modules are not recognizably homologous by sequence, and yet presumably produce analogous proteins. If the latter phenomenon is to be attributed to recombination, it will be "illegitimate recombination," not requiring extensive (or even any) sequence similarity between recombining genomes. Indeed, Hendrix (2002) has proposed that illegitimate recombination between nonhomologous sites is the basis of much bacteriophage evolution. It is of necessity coupled with a fiercely selective winnowing of the vast majority of recombinants, which would have lost function at the level of gene (reading frame), protein (structure/function), or virus (replication). This radical idea—entailing an enormous loss of failed recombinants to generate just a few that accrue advantage—is borne out by recent analyses (e.g., phages infecting mycobacteria; Pedulla et al. 2003). As well, it may explain the fact that extra, host-derived, genes are common in the genomes of the better-characterized lambdoid phages, interrupting, for instance, genes of the head and tail region. Often sporting their own promoters and terminators, such added genes may be expressible even in the prophage state. Hendrix calls them "morons" (because they add *more* DNA), and speculates that random illegitimate recombinational integration followed by selection for function is the creative force here. Lytic phages as well might find advantage in incorporating the occasional host gene, the better to reproduce themselves. Recent work with phages infecting marine cyanobacteria (*Prochlorococcus* and *Synechococcus*) illustrates the potential complexity of such phage–host interactions in an especially dramatic way. Many of these phages carry "photosynthesis genes" homologous to their hosts', such as *psbA*, which encodes a component of the photosystem II core reaction center (Lindell et al. 2004). Phage-borne genes can undergo HR with host copies, and cyanophages and host may together form a single population, as far as the forces operating on the evolution of such genes is concerned (Zeidner et al. 2005). Similarly, Tobe and colleagues (2007) concluded from a study of type III secretion effectors (likely virulence determinants) in pathogenic *E. coli*, that "the type III secretion system is connected to a vast phage 'metagenome', which acts as a crucible for the evolution of pathogenesis in this species."

The third element in our renewed appreciation for the evolutionary potential of transduction, broadly defined, comes from metagenomic studies (see Edwards and Rohwer 2005), which make it clear that viruses (mostly phages) are everywhere, astonishingly numerous (perhaps 10^{30}–10^{31} planetwide), and in many environments incredibly diverse (perhaps a million different viral genotypes in a kilogram of marine sediment). They control host abundance (killing up to half of the bacteria produced every day in the ocean) and diversity, by preventing prolonged dominance of any single bacterial type ("killing the winner"). With such enormous numbers and rapid reproductive rate, the impact of phage-mediated LGTs, even if few confer selective advantage to their hosts, can be considerable: estimates range up to 10^{15} transduction events globally, every year. Figure 4.1 attempts to summarize the view of interacting metagenomes that is encouraged by such studies and others discussed below.

Conjugation

Conjugation mediates the transfer of single-stranded DNA from donor cells to recipients. Excellent recent reviews of this process are those of Chen and

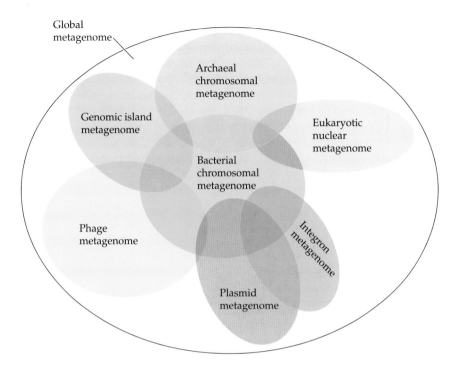

Global metagenome

Archaeal chromosomal metagenome

Genomic island metagenome

Eukaryotic nuclear metagenome

Bacterial chromosomal metagenome

Phage metagenome

Integron metagenome

Plasmid metagenome

Figure 4.1 Intersecting metagenomes. The global metagenome is the collection of all genes currently on the planet. The various intersecting metagenomes (not an exhaustive representation) each have genes which are more likely (or, in some cases, exclusively) to be found in other genomes in that assemblage. The intersecting regions (not to scale) indicate the exchange of genes between metagenomes.

colleagues (2005) and Sørensen and colleagues (2005). Often, conjugation functions are encoded by "selfish" mobile genetic elements (MGEs): self-transmissible and mobilizable plasmids, conjugative transposons (MGEs encoding necessary intracellular transposition and intercellular conjugation functions), and integrative conjugative elements (chromosomal gene clusters bearing integrases and conjugation genes). Gram-positive and gram-negative bacterial conjugation systems have shared components, and differ most obviously in the nature of the apparatus mediating cell contact—a pilus for the former and surface-localized adhesins for the latter. Archaeal conjugation functions, thus far characterized only for the self-transmissible plasmids of *Sulfolobus*, show scant similarity to their bacterial counterparts; conjugation of these plasmids may involve the transfer of double-stranded DNA (Greve et al. 2004).

Although conjugation mediated by the F plasmid was discovered almost a decade before the structure of DNA, the realization that bacterial "sex" (transfer of chromosomal genes) is frequent in natural environments is quite recent. Some plasmids have enormous host ranges: an experimental study of a complex sludge-derived microbial community (Bathe et al. 2004) found that the conjugative plasmid pJP4 could be transferred to species of α-, β-, γβ-, and γ-proteobacteria. Community complexity itself can facilitate plasmid spread. Biofilms are often especially hot spots: the dense and structured population facilitates conjugation, while the conjugation machinery and the release of DNA may themselves stimulate biofilm formation and maintenance (Molin and Tolker-Nielsen 2003). It is clear that there are many interactive and coevolved systems, so far largely undetected in laboratory situations, whose net effect is to enhance the evolutionary role of LGT via conjugation in natural environments.

Transformation

Transformation, or the direct uptake of DNA from the environment, was the first route of LGT discovered in prokaryotes, and seemingly the simplest and least selective in terms of DNA source. It is also an evolved process, with inducible competence for DNA uptake requiring dozens of genes (Chen et al. 2005; Thomas and Nielsen 2005). Whether or not the benefit of being of transformable is primarily evolutionary (the acquisition of novel genes and/or the repair of double-strand breaks) or nutritional (catabolism of DNA and recycling of its constituents) remains hotly debated. A tantalizing recent report (Meibom et al. 2005) shows that for *Vibrio cholerae*, competence can be activated by chitin from the copepods on whose surface it grows. No doubt laboratory studies in model systems have given us only a very limited appreciation for the evolutionary and ecological importance of transformation in natural settings, where DNA released by bacteria or host eukaryotic cells can reach high levels (1 µg per g of soil or sediment).

Since uptake specificity in transformation can be minimal, stable transformation (and LGT) is limited by the ability of the incoming DNA to be integrated by HR (or at much lower frequency illegitimate recombination) into the recipient's genome. Current studies suggest that HR in transformation generally requires stretches of very similar sequence from 25 to 200 bp long to initiate DNA pairing and strand exchange (Thomas and Nielsen 2005). Thus genomes of more than 25–30 percent sequence divergence (corresponding roughly to the genus level in bacteria) should not recombine by HR, a result confirmed for several bacteria by Fred Cohan (2002) and others.

Some consider that this HR barrier to between-taxon information exchange circumscribes prokaryote species in Nature, but there are circumstances under which the requirement for sequence similarity between incoming genes and their chromosomal homologs can be abrogated. Sequence-independent end-joining at double-stranded breaks does occur, albeit infrequently, in *E. coli*, and recombination initiated at a single region of similarity can extend into adjacent regions of little or no similarity. Of more general effect, it has long been known that in mutator strains (defi-

cient for instance in mismatch repair) and environmentally induced mutator states, the requirement for sequence similarity in HR is dramatically
reduced. This makes possible the integration by HR of less similar (and thus
possibly more useful and various) sequences in such mutator strains or
states. The notion that much evolution occurs through this mechanism is
very appealing, and confirmed both at the level of evidence and theory (see
Giraud et al. 2001; Townsend et al. 2003).

Detecting LGT

A variety of methods for detecting or confirming LGT events are available;
these identify LGT events at different phylogenetic distances and of different ages, and often return nonoverlapping sets of candidates for a given
genome (Ragan et al. 2006). All the methods are acknowledged to be inac-
curate, but in general they concur in detecting much more LGT than microbiologists would have predicted a decade or two ago.

Compositional methods

Genomes differ in G + C content, codon usage, and other compositional
characteristics. Laterally transferred genes may retain signatures of their
previous "home" genome for some time after incorporation into a new one.
LGT detection methods using either atypical nucleotide composition and/or
atypical codon usage patterns to infer which genes in a genome are instances
of LGT do not rely on any phylogenetic information or between-genome
comparisons. Thus they avoid some pitfalls. But transferred genes ultimately
ameliorate—adopt the nucleotide composition and codon usage typical of
surrounding genes—when subjected to the same mutational biases and
selection pressures. Thus compositional methods, first described by Antoine
Danchin and developed more recently by Howard Ochman and Jeffrey
Lawrence (Ochman et al. 2000; Ragan et al. 2006), are applicable to detection of relatively recent transfers only (at most hundreds of millions, not bil-
lions, of years old). These methods will, of course, also miss transfers
between genomes with similar composition, and thus within-species LGT
events. While easily applicable to sequenced genomes, compositional methods have been criticized for returning high rates of false positives and false
negatives.

Unusual phyletic similarity patterns

Perhaps the simplest way to assess whether a gene's presence can be
explained by LGT is through a BLAST search of a sequence database, such
as the National Center for Biotechnology Information (NCBI) nonredundant database, seeking homologs of the query gene. LGT can be suspected
when the top-scoring BLAST hit is from a taxon thought to be distinct from

that of the query gene. Eugene Koonin and his collaborators at NCBI used this approach extensively in the late 1990s (Koonin et al. 2001), estimating extents of transfer between bacterial phyla from none (*Buchnera* or *Mycoplasma genitalium*) to 33 percent (*Treponema pallidum*), and finding amounts of between-domain (Bacteria to Archaea) transfer as high as 16 percent (*Halobacterium* sp.). Such methods make the assumption that a gene's closest BLAST hit would be its closest neighbor in a true phylogeny, which is unfortunately often not so (Koski and Golding, 2001). Still, individual candidates identified by BLAST often do prove to be LGTs by other tests, and refinements are also possible (e.g., Clarke et al. 2002).

Phylogenetic incongruence

The gold standard in LGT detection is generally considered to be phylogenetic reconstruction, with the assumption that trees with unexpected phylogenetic histories—that is, topologically incongruent to some assumed "true" tree—identify laterally transferred genes. This implies that we know the expected phylogeny (the organismal tree) for reference, but, of course, we only have assumed proxies for it. These proxies include rRNA trees, genome trees, trees derived from concatenation of selected data sets, and trees supported by a plurality of sets of orthologous genes.

One drawback of phylogenetic approaches is that LGT events between neighboring taxa on the reference tree are invisible. Another is that weak phylogenetic signal in a set of orthologous genes often results in unresolved tree topology. Unresolved topologies cannot be used to identify or confirm LGT events, but by the same token should not be used as evidence for *absence* of LGT (Doolittle and Bapteste 2007). A third drawback is that often the choice of reference tree may bias the results of LGT detection. Still, phylogenetic analyses can be convincingly unambiguous, and illuminating in their detail (Figure 4.2).

Many LGT reports based on phylogenetic incongruence are anecdotal; that is, they result from the accidental discovery of genes whose phylogeny was not the one expected. However, there have been several systematic attempts to automatically detect, display, or quantitate all LGT events among genes in a collection of genomes (e.g., Sicheritz-Ponten and Andersson 2001, Beiko et al. 2005; Zhaxybayeva et al. 2006). Recently we described two tools that resolve incongruence among trees for genes shared among a collection of genomes, in terms of the most parsimonious inferences concerning LGT against a reference tree (MacLeod et al. 2005). These tools can independently display the number and direction (identity of donor and recipient) of inferred LGT events and the support for vertical descent, in a weblike diagram we call a *synthesis*.

Patchy (anomalous) gene distribution among phyla and species

Most genes are patchily distributed among genomes. Mirkin and colleagues (2003) surveyed the distribution among 26 genomes of genes in the COGs

(Cluster of Orthologous Groups) database and concluded that "the phyletic patterns (patterns of presence-absence in completely sequenced genomes) of almost 90 percent of COGs are inconsistent with the hypothetical species tree." [p. 2] More recently, Dagan and Martin (2007) concluded from a much more extensive distribution analysis that "among 57,670 gene families distributed across 190 sequenced genomes, at least two-thirds and probably all, have been affected by LGT at some time in their evolutionary past." [p. 870] As with phylogenetic incongruence, differential loss can always account for this (see discussion later in this chapter), but often seems a far less parsimonious explanation.

There are many good examples of patchy distribution. For instance, the recently completed genome sequence of the halophilic bacterium *Salinibacter ruber*, which lives in salterns and shares many physiological features of halophilic archaea, shows it to carry—in addition to a gene for a proteorhodopsin-like protein called xanthorhopsin—three rhodopsins of the haloarchaeal type (Mongodin et al. 2005). No other bacterium has ever been found to carry such archaeal rhodopsin genes, and no archaea other than haloarchaea (which are thought to be derived from methanogens) carry them either. It is not yet clear whether this is to be explained by an LGT event from some haloarchaean to an ancestor of *S. ruber* or vice versa, but one or the other must be the case. The alternative to LGT—that these genes were present in the last common ancestor of Bacteria and Archaea and then independently lost in the dozens of lineages that lack them—seems untenable in terms of parsimony or biology.

Not only individual genes but gene clusters, operons, and indeed whole suites of physiologically coordinated activity can show patchy distribution, and likely have LGT in their histories. Some pathogenicity islands, for instance, occur broadly throughout major groups, exhibiting significant conservation of gene content and synteny (Mohd-Zain et al. 2004). Other islands may themselves be mosaics, assembled in an *ad hoc* fashion from individually transferred genes. An example, again from the recently completed *S. ruber* genome sequence, is the "hypersalinity island" shown in Figure 4.3.

The *selfish operon theory* (Lawrence and Roth 1996) posits that for multistep pathways in which no single step (or its gene) is of use without the others, selection and LGT together create and maintain operons (or tightly linked clusters). Step-by-step pathway acquisition would require multiple unlikely events, so only genes that are tightly linked in donor DNA are likely to spread laterally—thus constituting a selection pressure for operon assembly. Although there appear to be many good cases of operons transferred as units, whether this transfer (rather than coregulation) is the pressure that gave rise to operons remains debatable (Omelchenko et al. 2003; Price et al. 2005).

Strain-specific genes (patchy distribution within species)

Even within a single species, strains characteristically differ in their possession of certain genes and gene clusters; some are unique to one strain and

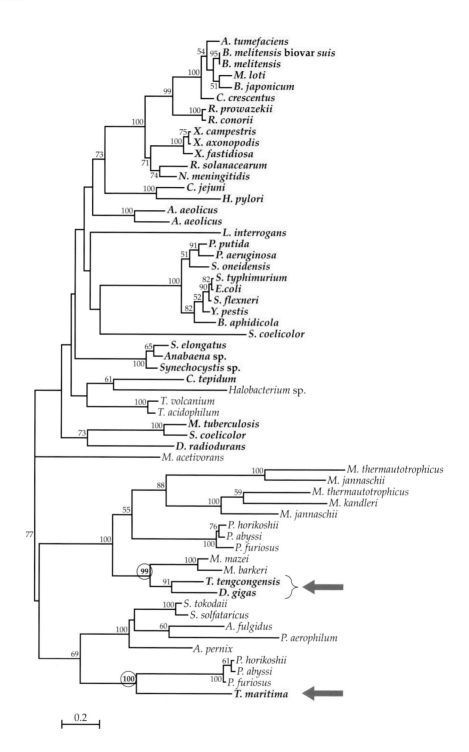

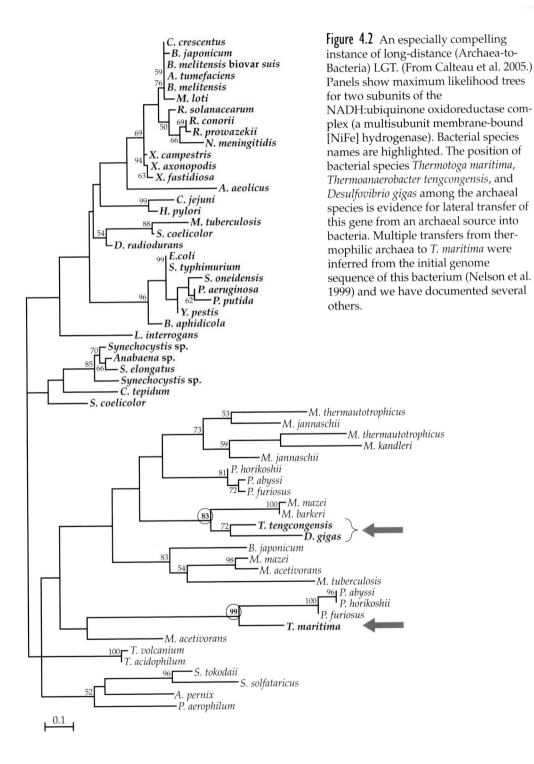

Figure 4.2 An especially compelling instance of long-distance (Archaea-to-Bacteria) LGT. (From Calteau et al. 2005.) Panels show maximum likelihood trees for two subunits of the NADH:ubiquinone oxidoreductase complex (a multisubunit membrane-bound [NiFe] hydrogenase). Bacterial species names are highlighted. The position of bacterial species *Thermotoga maritima*, *Thermoanaerobacter tengcongensis*, and *Desulfovibrio gigas* among the archaeal species is evidence for lateral transfer of this gene from an archaeal source into bacteria. Multiple transfers from thermophilic archaea to *T. maritima* were inferred from the initial genome sequence of this bacterium (Nelson et al. 1999) and we have documented several others.

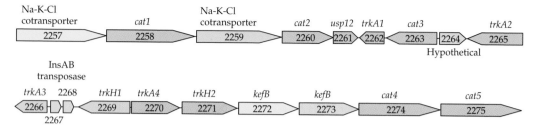

Figure 4.3 A mosaic "hypersalinity island" in the genome of *Salinibacter ruber*. This cluster of 19 genes from the genome of *S. ruber* includes K⁺ uptake/efflux systems and cationic amino acid transporters. It is mosaic in nature, apparently pieced together from a variety of bacterial and archaeal sources (as indicated by color). More than half the genes—the cationic amino acid transporters, one *trkH* gene, and three *trkA* genes—most closely resemble haloarchaeal homologs. The Trk system is responsible for the uptake of K⁺, where TrkH is the membrane-bound translocating subunit and TrkA is a cytoplasmic membrane surface protein that binds NAD⁺. The existence of multiple *trkA* genes in *S. ruber* suggests complex regulation of the Trk system, a feature likely shared with haloarchaea. An additional TrkAH system (ORFs 2266 and 2269) is most similar to that found in Firmicutes.

some are shared with their conspecifics in a patchy fashion. When the second (O157:H7) and third (CFT073) *E. coli* genome sequences appeared and could be compared to the first (K12), there was considerable surprise that only 2996 (39.2%) of the 7638 genes present in at least one genome could be found in all three genomes (Welch et al. 2002). This *core* of shared genes gradually gets smaller as more genome sequences appear. In this and many other species, the number of genes unique to a strain can comprise up to 20 percent of its total gene complement, and the collection of all genes present in the collection of strains (once called the *species genome* but now called the *pangenome*) may greatly exceed the number found in any one genome.

There will soon be many bacterial species for which multiple, closely related genome sequences are available. Claire Fraser-Liggett and her colleagues (Tettelin et al. 2005) consider that there are two types of pangenome—*closed* (e.g., *Bacillus anthracis*), for which each new genome sequence adds relatively (and increasingly) few new genes, and *open* (e.g., *Streptococcus agalactiae*), for which no limits to the number of genes can yet be discerned (Figure 4.4). In many cases, strain-specific genes are vital to strain-specific biology—pathogenicity and other genomic islands, integrons, and prophage-encoded "morons" may contribute to the ecological success of a strain. But we cannot exclude the possibility that many strain-specific genes are neutral, and that their detection simply speaks to the amazing tempo of gene gain through LGT.

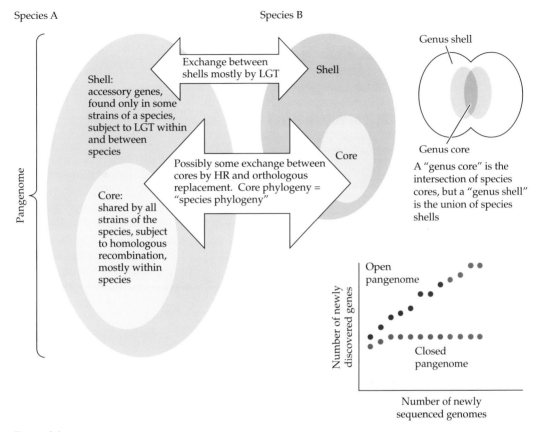

Figure 4.4 Exploration of some concepts concerning genomes of species. Although for prokaryotes the species concept is problematic, there is general agreement that strains of assemblages usually recognized as "species" will differ by as much as 20 percent in gene content. All strains share the *core*, while the genes found only in some strains make up a *shell* of accessory genes. The sum of core and all accessory genes was first called the *species genome* by Ruting Lan and Peter Reeves, but is now commonly referred to as the species *pangenome* (Tettelin et al. 2005). Pangenomes are thought to be of two types: open, for which the sequencing of more genomes adds still more accessory genes; and closed, for which additional genome sequences soon reveal no new accessory genes.

Physical methods

Complete genome sequences are the high road to LGT detection, but genes that are present in some strains of a species and absent from others (and thus candidates for LGT) can be identified by two complementary physical methods: subtractive hybridization and comparative genomic hybridization (DNA microarrays). The former identifies genes present in an unsequenced genome that are absent from a sequenced genome, while the latter tells us what genes in the sequenced genome the unsequenced genome lacks. Both methods have been used to good advantage in characterizing groups of closely related pathogens (such as strains of *E. coli* and *Shigella* or *Helicobacter* [Gressman et al. 2005]), and "environmental microbes," such as *Thermotoga maritima* and its relatives (Nesbø et al. 2002; Mongodin et al. 2005b).

Ever-Present Alternatives

Differential loss of genes present in a common ancestor is always a possible alternative to LGT (Figure 4.5). Often, many independent loss events must be invoked to produce a tree or distribution pattern that could result from a single LGT. But we cannot rule out loss on grounds of parsimony without some knowledge of the relative probabilities of LGT, gene creation, and gene loss as genomic events. Each case must be decided on its merits: these will often be quite persuasive and based on several independent lines of argument. Moreover, there is a compelling argument against assuming differential loss in *all* cases of doubt, in a well-intentioned effort to appear judicious and conservative about invoking LGT. Each case attributed to differential

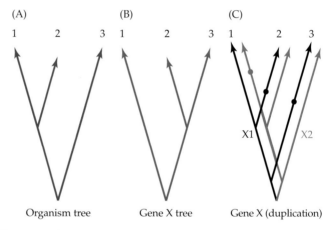

Figure 4.5 How differential loss of duplicated genes can be mistaken for LGT. Consider a simple, three-taxon, rooted organismal tree (A), which is known to be the true organismal phylogeny, and a phylogeny for gene X (B). The incongruence between the organismal and gene trees is consistent with an orthologous replacement, in which an LGT of X from taxon 2 to taxon 3 (or vice versa) was coupled to the loss of the original copy of X from the recipient. The gene tree is also consistent with a scenario of gene duplication and loss (C), in which gene X was duplicated in the lineage leading to the last common ancestor of taxa 1, 2, and 3, producing paralogs X1 and X2. X1 was subsequently lost in taxa 2 and 3 (black dots) and X2 was lost from taxon 1 (red dot).

loss logically requires the presence of yet another gene in the genome of an ancestor. Thus the inferred last common ancestor of a group will have had a larger genome and more physiological flexibility than any of its descendants. Ultimately, in the case of the last universal common ancestor (LUCA), which lies at the presumed root of the Tree of Life (Figure 4.6), we would have to imagine a totipotent entity, with an enormous genome. We call this the *Genome of Eden hypothesis*, and consider it unappealing.

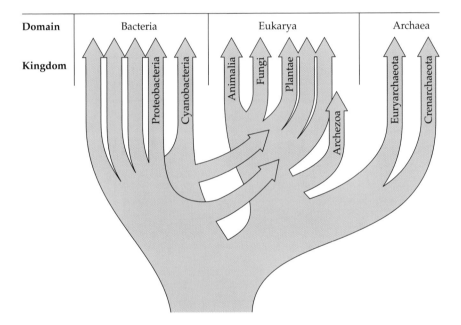

Figure 4.6 The three-domain universal Tree of Life. The division of all life into three domains was initially based on sequences of small-subunit ribosomal RNAs and the rooting was inferred from (1) the generally greater similarity of archaeal and eukaryotic informational genes; and (2) trees relating ancient duplications of such genes. Lateral branches from proteobacteria and cyanobacteria to Eukarya indicate endosymbiotic events that gave rise to mitochondria and plastids, respectively. It may be that more eukaryotic nuclear genes derive from these sources than from archaeal ancestors, in which case the indicated closer relationship of Eukarya and Archaea privileges the latter (largely informational) genes. The root of this tree is termed LUCA (last universal common ancestor), sometimes called the *most recent* universal common ancestor. LUCA (or MRUCA) marks the point of deepest divergence between any lineages that have left survivors. Any other lineages that may have existed at the time of LUCA have gone extinct, just as have all the other primate lineages that diverged from the last (most recent) common ancestor of humans, chimps, and bonobos.

(A)

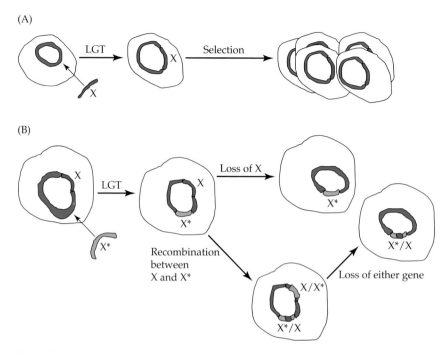

(B)

Figure 4.7 Two kinds of LGT. (A) In the simplest cases of LGT, a novel gene is added to a genome and is fixed in a population because of a selective advantage conferred by its possession. (B) In orthologous replacement (sometimes called xenologous displacement), a new (foreign-origin, or xenologous) copy is added to a genome. In a subsequent step, either the resident copy is lost or recombination between resident and xenologous copy is followed by loss of one recombinant copy. Orthologous replacement (with or without recombination between duplicate genes) can occur without specific selection for new genetic information.

In some cases, between-species LGT can be mistaken for within-species HR, even though the latter is not involved, or only involved at a subsequent step (Figure 4.7). In the phenomenon called *orthologous replacement* (or *xenologous displacement*), the integration of a foreign (xenologous) gene into a genome in which an ortholog already resides is followed by loss of the original resident. If all we have is gene sequence (no further information on gene position, say), we cannot tell what has occurred, unless the new sequence is so different from the original as to firmly rule out the possibility that it is just another allele within the species' gene pool. If intragenomic HR occurred between the introduced xenologous copy and the original, followed by loss of one of the resulting genes, this could further confound our ability to reconstruct the genetic past.

Quantifying LGT

Many participants in the LGT field think it is important to know how much LGT there is, or has been, from a global perspective. The detection methods previously described can be harnessed to this end, but will give different answers, as will the two ways of asking about the extent of LGT, *genome-centric* and *gene-centric*.

Genome-centric approaches

Genome-centric inquiries seek to know how many genes in a particular genome have arisen by LGT, rather than by vertical descent or duplication within a lineage. A recent compositional analysis of 116 genomes revealed that the number of transferred genes ranges from 0.5 percent in the aphid intracellular symbiont *Buchnera* sp. APS, to 25 percent in the anaerobic methane-producing archaean *Methanosarcina acetivorans* C2A (Nakamura et al. 2004). As noted earlier, compositional methods are imprecise and cannot detect events that occurred in the distant past: they will inevitably underestimate the extent of LGT.

The use of a sequence-similarity search tool, such as BLAST, to detect genes not present in relatives but found in more distant groups (unusual phyletic similarity patterns) is perhaps the most common genome-centric approach. An example from our own work is shown in Figure 4.8. Many publications quantify LGT as the fraction of genes whose best BLAST hits are found outside the taxonomic group (species, genus, phylum) to which the organism belongs. For instance, at the time of first publication, 1043 (31%) of the 3371 ORFs in the genome of the archaean *Methanosarcina mazei* had stronger hits to a gene found in the Bacteria than to any found among Archaea; of these, 544 (52%) had no significant hits to other Archaea (Deppenmeier et al. 2002). But not surprisingly the number of potential LGTs has been reduced with the appearance of other sequences of large archaeal genomes, in particular, that of the congeneric *Methanosarcina acetivorans*. Inclusion of more closely related genomes will inevitably reduce the number of apparent LGTs detected by any genome-centric method, but where we draw the line for exclusion (species, genus, phylum, domain?) is unavoidably arbitrary. We cannot address theses issues without knowing the phylogenetic history for each gene—ultimately all the way back to LUCA. We need gene-centric methods.

Gene-centric methods

Gene-centric methods attempt to quantify incongruence (conflicting topology) in a collection of different gene phylogenies. Their limitations arise from difficulty in distinguishing LGT from paralogy and differential loss, weakness of phylogenetic signal, and inadequacies of phylogenetic recon-

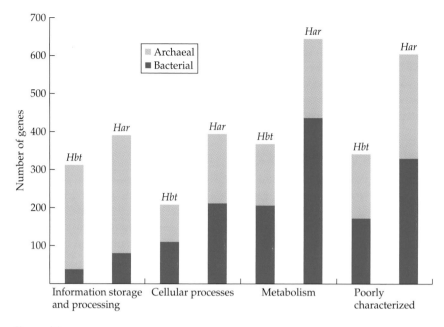

Figure 4.8 BLAST-based estimation of lateral transfer events of bacterial genes into the genomes of halophilic archaea. Each gene in the haloarchaeal genomes was used in a BLAST query against a database of about 200 bacterial and 20 archaeal genomes, and scored as to whether its best BLAST score was to a bacterial (blue) or to another archaeal (pink) genome. Informational genes seem the least frequently transferred. Abbreviations: *Hbt, Halobacterium* sp. NRC-1; *Har, Haloarcula marismortui.* (Analysis courtesy of David Walsh.)

struction algorithms or the assumptions (evolutionary models) on which they are based. Often, different gene data sets are analyzed statistically with respect to a reference tree, which can be variously derived and is never certain to be the "true" organism tree.

Gene-centric approaches were used in two recent, large-scale genome analyses to estimate LGT across many available sequenced genomes. Ge and colleagues (2005) analyzed 297 gene families from gene clusters of 40 microbial genomes. They compared gene family phylogenies with a genome tree reference to find incongruities, then increased the stringency for accepting genes as LGT candidates by checking if any observed topological differences between two trees could be explained by LGT scenarios (that is, mere incongruence of gene tree and reference tree was not considered sufficient to declare LGT). This approach identified 33 LGT events (11.1%).

In a very thorough study, Beiko and colleagues (2005) analyzed gene families from 144 prokaryotic genomes. They compared 95,194 strongly supported *bipartitions* from 22,432 gene family trees. In such a bipartition analysis, each internal branch of a given tree topology is seen as dividing all taxa into two groups. Different tree topologies will divide taxa differently at some internal branches, a difference that can be evaluated statistically through the bootstrap support of each tree at that branch. Beiko and colleagues found that 82,473 bipartitions (86.6%) were in agreement with bipartitions of a reference supertree constructed from all highly supported bipartitions. This might at first be taken to mean that LGT is not so important after all (only 13.4% occurrence). But that would be an overhasty conclusion, for several reasons—including that the genes most likely to have been transferred (those with patchy distributions) are the least well-represented in bipartitions, and that the majority of bipartitions address recent branchings. Most relevant here is the fact that many LGT events will leave many bipartitions undisturbed. Early LGT events of profound impact—such as transfer from a newly diverged bacterial phylum into the euryarchaeal subdomain of the Archaea just after its divergence from the other, crenarchaeal subdomain— might affect very few (in this case only one) of the many bipartitions of a many-taxon tree. Thus, relatively infrequent events of LGT can result in the "wrong" phylogenetic placement of many descendant genes, in many contemporary genomes.

Another global gene-centric assessment method, providing a relative rate estimate for LGT, is based on combined phylogenetic and phyletic approaches. Presence or absence of individual genes (patchy distribution) across a reference phylogeny is used to enumerate events of gain and loss, on the assumption that these events must be in equilibrium overall, if genomes are not to expand indefinitely or disappear altogether. Kunin and coworkers have applied this type of approach to 165 diverse microbial genomes and inferred about 40,000 LGTs, about 90,000 gene losses, and more than about 600,000 vertical descents in all analyzed gene families (Kunin and Ouzounis, 2003; Kunin et al. 2005). While such numbers might be interpreted as indicating limited LGT, these estimates do not consider LGTs resulting in orthologous replacement, which could constitute a substantial part of a genome. Dagan and Martin (2007) use ancestral genome size as a constraint to quantify LGTs from the time of LUCA. They "estimate the minimum lower bound for the average LGT rate across all genes as 1.1 LGT events per gene family and gene family lifespan." [p. 870] Again, we should remember that an "event" will result in the "improper" phylogenetic placement of all descendant genes, so that a very large fraction (even all) of the gene family could be "misplaced" in the global phylogeny. For instance, although eukaryotes are generally considered sisters to Archaea, genes that were imported with the mitochondrial endosymbiont show all eukaryotes to be sisters to alpha-proteobacteria.

An LGT-Resistant Core

Those who see LGT as an interesting but quantitatively *minor phenomenon*, and those who argue that it is the *dominant force* for adaptation and diversification of prokaryotes can both find comfort in the global analyses performed to date (Figure 4.9). More progress might be made by recasting our

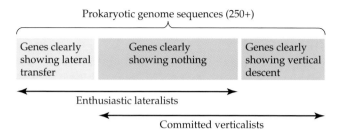

Figure 4.9 State of the debate over LGT. There are currently more than 250 complete prokaryotic genome sequences and many times as much data from incomplete genomes or isolates sampled for a few genes. Most of these data have too little phylogenetic signal or are represented in too small or too biased a taxonomic sampling to allow us to answer the question, How much lateral transfer has there been globally? Or, more specifically, How many genes in modern genomes have a history of uninterrupted vertical descent back to the time of LUCA? Enthusiastic lateralists imagine that the answer to this second question could be "none." Committed verticalists think the answer is "many." Neither side really knows, and both sides would like to claim the middle ground of phylogenetically uncommitted genes for their own.

questions about the magnitude of LGT and focusing on genes that appear to have been immune to it. It is commonly believed that there are such genes, making up what is often called the genomic *core*.

The complexity hypothesis

At the very heart of the genomic core, many believe, is the gene on which most prokaryotic phylogenies are based. This is the gene encoding 16S (small subunit) ribosomal RNA (SSU rRNA), of which databases now hold more than 200,000 examples. The initial choice of this molecule as a "universal molecular chronometer" was made by Carl Woese in the late 1960s. His reasoning, in part, was that rRNA is so tightly integrated with other coevolved structural and functional components of the translational apparatus, and this apparatus is itself so expensive and so necessary a piece of the cellular economy, that transfer of the rRNA gene (even over a short phylogenetic distance) would reduce the efficiency or accuracy of translation, and thus the viability of the cell (Woese 1987).

An elaboration of this notion by James Lake and his colleagues has been called the *complexity hypothesis* (Jain et al. 1999). Although entailing that *any* gene whose product interacts physically with those of many other genes would be recalcitrant to LGT, this notion has usually been invoked to explain or predict nontransferability of what have come to be called *informational genes*—those involved in DNA replication, transcription, and translation. There are other reasons, too, why *operational genes*—those determining metabolic and regulatory activities, transport, cell structural features, or interactions with hosts—might be more easily transferred. An evolving lineage might have, at different times in its evolutionary history, used one sugar and not another, manufactured an amino acid or imported it, respired or fermented, or been free-living or parasitic—thus needing (or not) certain genes which could be acquired (or reacquired) by LGT. It would never, however, have been free of the need to replicate and express DNA.

While genome-wide analyses indicate that all types of genes (including rRNA genes) can be among inferred LGTs, certain functional categories are indeed over- or underrepresented among apparent transfers, with prophages perhaps being the most frequently gained and lost. The early belief that genes subject to LGT are special was not altogether wrong, although the differences are matters of degree, not kind. Among LGTs detected by a compositional method, Nakamura and colleagues (2004) report biased representation of LGTs in the functional categories of cell surface, DNA binding, and pathogenicity-related genes. Beiko and colleagues (2005) found overrepresentation of "energy metabolism" and "mobile and extrachromosomal element functions" among genes with discordant bipartitions (that is, among LGT candidates). However, a transfer distance may need to be taken into account when types of transferred genes are evaluated. In genome-wide analyses of gene transfer within cyanobacteria and between cyanobacteria and other phyla, no bias is found for LGTs within cyanobacteria, but an excess of metabolic genes and deficiency of informational genes is observed in interphylum transfers (Zhaxybayeva et al. 2006).

The tiny universal core

If some genes are always needed and have never been transferred, they should comprise a core present in all genomes and exhibiting the same phylogeny. Is there such a core? How many genes does it include? Do they show congruent phylogenies? Most attempts to identify a set of orthologs shared by all sequenced genomes come up with similar results. There are less than one hundred such genes, and most of these are indeed informational in function. In a recent survey, Charlebois and Doolittle (2004) found only 34 genes to be present in all of the 147 sequenced bacterial and archaeal genomes compared. Universal presence is an onerous requirement, however: sequencing errors and misannotation, as well as such nonartifactual phenomena as gene fusion in a single genome, will eliminate a gene. A fairer comparison might be these authors' adjustable *phylogenetically balanced core*, defined as those genes present in at least X percent of the genomes sequenced from each of 12 bacterial phyla, plus Euryarchaeota and Crenarchaeota (the two divisions of Archaea). If X is set at 80 percent, the core amounts to 71 genes, of which 22 are ribosomal proteins and 13 are aminoacyl tRNA synthetases.

For taxonomic assemblages that do not include all prokaryotes (as defined above), core genomes are of course larger, not only because fewer genomes are involved, but because those that are included are more closely related. Values for all bacteria or all archaea are two to three times higher than for all prokaryotes. A recent estimate for the single-copy orthologous genes shared by 13 γ-proteobacterial genomes is 205 (Lerat et al. 2003), while among 11 sequenced cyanobacterial genomes, there are 663 shared genes (Zhaxybayeva et al. 2006).

Do core genes have a common history?

For the seemingly tiny universal (three-domain) core, the conclusions drawn after the earliest serious attempt to examine it phylogenetically seem still to hold. From an analysis of 32 gene families and nine genomes (at that time quite a comprehensive study), Teichmann and Mitchison (1999) concluded that "it is clear that the strong phylogenetic signal comes from three families, at least two of which are good candidates for horizontal transfer. The tree from the remaining 29 families consists almost entirely of noise at the level of bacterial phylum divisions, indicating that, even with large amounts of data, it may not be possible to reconstruct the prokaryote phylogeny using standard sequence-based methods." [p. 98] More recently, Creevey and colleagues (2004), using a large amount of data (61 genomes), concluded that "congruence among gene trees spanning deep relationships is not better than random" and that "our results do not preclude levels of HGT that would be inconsistent with the notion of a prokaryote phylogeny." [p. 2551]

The situation for a shallower core may not be so dire. An analysis by Lerat and colleagues (2003) of the 205-gene γ-proteobacterial core found only two genes that had robustly incongruent phylogenies, while 203 genes could not reject (at a 95% confidence level) the topology obtained with a data set comprising all 205 gene sequences strung together (concatenated) as a single gene. However, almost as many of the 205 genes could not reject at the 95% confidence level a number of alternative phylogenetic trees (including the rRNA topology) examined in the study. Indeed, a subsequent statistical study by us (Bapteste et al. 2004) showed that more than half of the genes would accept (not reject) more than half of a set of 105 topologies constructed to represent a range of possibilities many would consider biologically reasonable for these taxa. So, to claim that these core genes are "resistant to transfer" (as Lerat et al. in fact did) is to assume that which was to be proven.

In more recent work, some of us (Bapteste et al. 2005) have introduced a new method that more finely discriminates between noise and genuine conflicting signal in such core gene sets. The likelihood (p-value) of each gene data set given one of a large collection of reference trees is displayed in a two-dimensional array, the p-value represented as a color. The resultant heat map looks like the expression microarray patterns familiar to functional genomicists, in which expression responses of genes to experimental conditions are represented by colors. Like such microarray data, the p-value arrays can be subjected to two-way clustering: grouping genes that have the most similar responses (p-values) to topologies along one axis, and topologies that are the most similar in their responses to genes along the other axis. Within the 205-gene γ-proteobacterial core, or collections of ribosomal informational genes for archaea or bacteria, there are few clear cases of recent LGT. But, although most genes do show evidence of phylogenetic signal (since they can reject many trees), most cannot distinguish between several likely but distinct topologies, and there is some evidence of support for several different topologies by small cohorts of genes. We concluded (Baptiste et al. 2005) that "we simply cannot determine if a large portion of the genes have a common history." [p. 33]

Nevertheless, attempts are still being made to resolve the Tree of Life using the core genes. Recently, Ciccarelli and colleagues (2006) analyzed 191 genomes to find genes with "indisputable orthology" and ended up with only 31 such genes. The joint analysis of these genes results in a phylogeny that has some resolution; however, the tree's backbone is poorly resolved (<80%), perhaps due to insufficient phylogenetic information. More to the point, Dagan and Martin (2006) rightly refer to this tree as "the tree of 1 percent," and question its value as a history of genomes or the organisms they produce.

So, as with whole genome analyses, studies of shared genomic cores leave room for both *committed verticalists*, who see LGT as a nuisance for phylogenetics, and *enthusiastic lateralists*, who believe it to be the very essence of phylogenesis (terminology of Charlebois et al. 2003). Many genes have little phylogenetic signal, and even if all genes have the same true phylogeny, it would not be surprising that at any reasonable phylogenetic depth many cannot distinguish between similar topologies, differing by a few branches. Indeed, even with genes for which we have good biological reasons to assume a common evolutionary history (such as those of higher plant chloroplasts), incongruence is often observed (Delsuc et al. 2005). Much depends, of course, on how far back in evolution we want to look. If the aim is to go back to LUCA (see Figure 4.6), it seems possible that most genes outside the universal genomic core will have a history dominated by transfer. The core itself is a very small fraction of any genome's genes, and the genes in it do not necessarily have the same evolutionary histories that far down the Tree of Life.

LGT and Microevolution

Taking LGT fully on board requires a rethinking of models for microbial adaptation and divergence at the species level. Here we discuss just a few of the implications.

HR and the biological species concept (for bacteria)

Until recently, population genetic models and phylogenetic models for prokaryotes reinforced a belief in the clonal nature of bacterial species, as might be expected from their asexual mode of reproduction. Fitter types arising by mutation within populations outcompete their wild-type sisters, carrying not only the selected mutation, but all polymorphisms already accumulated in the same genome to fixation: genomic cohesion is thus maintained through periodic selective sweeps. Separated populations diverge genomically thanks to separate selective sweeps of this sort. Fred Cohan (2002) considers such diverging clonal populations, which he calls *ecotypes*, to be the prokaryotic equivalent of eukaryotic species. But since the beginning of the 1990s, more and more analyses, based on the rich data sets provided by MLST (multilocus sequence typing), indicate that HR is a significant and (in some species the dominant) determinant in the evolution of core loci within populations. Thus we might apply to bacteria something like Ernst Mayr's *Biological Species Concept* (BSC), in which HR maintains genetic cohesion within species, while barriers to HR between species ensure genetic divergence (Dykhuizen and Green 1991; Fraser et al. 2007).

HR-based species concepts focus our attention on the shared genomic core of the species (see Figure 4.4); the more fugacious set of accessory genes introduced by LGT have interestingly different but overlapping and interacting evolutionary dynamics, which call for a much more complex view of bacterial species and speciation than yet formally articulated by any theorist. Elements of such a model—in which HR and LGT are sometimes antagonistic and sometimes synergistic—have been sketched by Jeffrey Lawrence, Peter Gogarten, Jim Tiedje, and us (Gogarten et al. 2002; Lawrence 2002; Gogarten and Townsend 2005; Konstantinidis and Tiedje 2005; Doolittle and Papke 2006).

Paralogs and xenologs

Another consequence of fully embracing LGT as a dominant microevolutionary mechanism is that it becomes very hard to draw the line between paralogs (extra gene copies produced by within-genome duplication events) and xenologs (extra copies introduced by LGT). Lawrence has suggested that many extra gene copies that we have long seen as paralogs are in fact xenologs—that what we have thought of as duplication/divergence is often really LGT (see Gogarten et al. 2002). Orthologous genes in diverging lineages (strains or sister species) are expected to diverge functionally in their regulatory responses to changes in substrate availability or in the kinetic parameters of their products. Should the possession of genes of both types become advantageous to either lineage, then acquisition of a second, already "preadapted" gene through LGT from the sister strain or species is a simpler and more certain solution than duplication and accumulation of mutations, few of which will meet the new needs.

A sophisticated multigenome comparison recently undertaken by Lerat and colleagues (2005) indicates that, indeed, many cases of second and third copies of genes that we would until recently have taken as paralogs are *detectably* xenologous. In a similarly motivated study not specifically directed at distinguishing paralogs and xenologs, Pál and colleagues (2005) conclude that "most changes [15–32 versus 1] to the metabolic network of *Escherichia coli* in the past 100 million years are due to horizontal gene transfer, with little contribution from gene duplicates." [p. 372]

Adaptive LGT

There are countless anecdotal cases in which a major role for LGT in founding a novel ecotype, species, or indeed a whole adaptive radiation, can be plausibly inferred—that is, where both compelling evidence for LGT *and* a highly persuasive evolutionary scenario are available. Most examples come from clinical microbiology, but evolution by LGT is by no means the exclusive preserve of pathogens. Springael and Top (2004) have recently summarized evidence that LGT and "patchwork assembly" of genomic islands can occur in real time (during the course of an experiment) among "environmental microbes" exposed to never-before-encountered *xenobiotics*. Included are cases in which xenobiotics induce transfer and in which novel degradation pathways are assembled from components originally present in different genomes—the first steps in evolution of the self-transmissible xenobiotic-degrading genomic elements now well documented in Nature, and fully comparable to pathogenicity islands.

More extensive adaptive genomic remodeling by LGT can also be inferred from many whole-genome analyses (as in Figure 4.8). The completed genomes of haloarchaea reveal about the same number of genes with best BLAST hits to bacterial as archaeal genes, as if this clade was founded by an enormous LGT invasion of bacterial genes (many involved in respiratory metabolism) into the genome of a presumed anaerobic methanogenic ancestor. Recently, Omelchenko and colleagues (2005) have presented a detailed comparative genomic study of *Thermus thermophilus* and *Deinococcus radiodurans*. These bacteria are members of the same phylum, according to phylogenies based on rRNA and transcriptional and translational proteins. They also have generally similar physiologies, but the former is a thermophile relatively sensitive to radiation and other forms of oxidative stress, while the latter is a mesophile extraordinarily resistant to those challenges. Although the two bacteria still share many genes, including many determinants on what Omelchenko and colleagues take to be an ancestral megaplasmid, these authors conclude that "various aspects of the adaptation to high temperature in *Thermus* can be attributed to horizontal gene transfer from archaea and thermophilic bacteria … [while] by contrast, *Deinococcus* seems to have acquired numerous genes related to stress response systems from various bacteria." [p. 57]

Neutral LGT

Investigators continue to be astonished by the extent of between-strain gene content diversity among bacterial species for which several genomes are available, and by the apparent limitlessness of some species' pangenomes. The tendency has been to assume that strain-specific accessory genes are doing their possessors some good in whatever subspecific ecological niche they occupy. Otherwise, these genes would have accumulated inactivating nonsense mutations and deletions. But there are so many strain-specific loci that the situation seems reminiscent of that faced by population genetics in the 1960s, when protein electrophoretic studies indicated that there is far more gene sequence variation in natural populations than can be maintained by natural selection, and it was first mooted that such variation might be mostly neutral.

A recent study by Martin Polz and coworkers (Thompson et al. 2005) highlights the gene content variation conundrum. Among isolates of *Vibrio splendidus*—all nearly identical (<1% divergent) in 16S rRNA sequence and taken from the same site on the Massachusetts coast—they examined the diversity of *hsp60* gene sequences and genome sizes. They estimated a minimum of 436 *hsp60* alleles in the total sample (31.5 ml), all of these clustering within a single (>95% identity) clade. This sort of gene sequence *microdiversity* may represent neutral variation accumulated between selective sweeps in a single ecotype. But Thompson and coworkers found still greater diversity in genome size within this "ecotype," with pairs of isolates picked as identical in *hsp60* sequence often having significant size differences (170–800 kb). Thus, they estimated that their samples contained "at least a thousand different genotypes, each occurring at extremely low environmental concentrations (on average less than one cell per milliliter)." [p. 1311]

What can this mean? Are there really a thousand different ways for *V. splendidus* cells to make a living in these small and well-mixed volumes of relatively clean seawater? Are most cells just passing through, adapted in their gene content to some other microniche from which they have been torn by restless ocean currents? Or is gene content really of little selective consequence and the rate of acquisition of accessory genes by LGT (which must, of course, be balanced overall by loss) so high that it significantly exceeds (on a per genome basis) the rate of neutral substitution in genes? To say that most LGTs are neutral and that only a miniscule fraction will be fixed is not to deny the importance of those very few that are. Nor is it to say that this miniscule fraction cannot, in the long run, one gene at a time, replace an entire genome many times over. The neutral theory of molecular evolution, by analogy, does not contradict the theory of natural selection nor reduce its applicability to mutations of significant negative or positive effect.

LGT and Macroevolution

The most vigorously debated issue in lateral genomics is what LGT might mean for the universal Tree of Life, which is commonly understood to resemble Figure 4.6. This is arguably less a scientific issue than a philosophical one (Doolittle 1999; Doolittle and Bapteste 2007), but philosophy does inform science, so a brief sketch of the two main sides in the debate is warranted. The conservative, "committed verticalists" accept that LGT has been prevalent in Life's history, but see it as a kind of noise, beneath which we should still be able to detect the true tree-like phylogenetic signal, tracking evolutionary branchings since the time of LUCA. "Enthusiastic lateralists" hold that not only does LGT make this task impossible, it renders the effort meaningless: The Tree of Life is an artificial conceptual structure which verticalists inappropriately seek to impose on patterns and processes which are in essence reticulate or web-like (Doolittle and Bapteste 2007).

Verticalists have adopted two principle strategies to detect the signal that they believe lies under the noise. The simplest is to assume that there is a core of genes, however tiny (in one study, investigators were reduced to 14 genes), that have eluded LGT and whose weak signals, when concatenated, retrace the history of cell and organismal evolution. Whether or not this might be true was discussed earlier in this chapter. The second strategy, on the other hand, seeks to make a Tree of Life from as many genes in as many genomes as can be brought to bear, with the assumption that the noise from LGT is basically random—that is, that transfers come from many disparate sources and neutralize each other's topology-distorting effects (Wolf et al. 2001; Snel et al. 2005). Such "genome trees" take account of all the genes shared by any pair of genomes, with or without using measures of sequence similarity between the genes. They are generally said to agree with each other and with trees based on rRNA, which means that (1) taxa considered to be Archaea or Bacteria are indeed separated, and (2) taxa considered to be members of the same bacterial phylum are generally grouped together. There is seldom resolution among the bacterial phyla (which could mean either that they enjoyed an initial rapid "star" radiation or that there is little signal), and the various archaeal divisions are often quite variously displayed.

Snel and colleagues (2005), satisfied with this kind of agreement, take it to mean that genome trees put the lie to claims that LGT makes phylogeny impossible, that because "all are similar to one another and reflect the known species phylogeny … genome trees have yielded the fundamental insight that genome evolution is largely a matter of vertical transmission." [p. 205] Similarly, Kurland and colleagues (2003) write, of a genome tree constructed with 50 genomes, that "the congruence of this tree with the rRNA phylogeny from the same genomes is striking. It strongly suggests that Darwinian descent is the dominant mode of genome evolution for these 50 genomes." [p. 9658]

Lateralists would argue against this, asserting that there is another process, namely nonrandom LGT, that can generate such robust tree-like patterns. What creates the tree are differential patterns of gene exchange. For example, more frequent exchange between species A and B than between A and C or B and C would result in greater similarity in gene content and gene sequence between the first two, and C would look more distant, by any "phylogenetic" analysis. Ongoing differential gene exchange could generate tree-like patterns even if there were no deep historical tree-like process responsible for the apparent relationships between bacterial species. Thus a bacterium falling amongst archaea might, as *Thermotoga* has done, adopt many of their genes and ultimately group with them in gene content or many single-gene analyses (Gogarten et al. 2002).

Such a view, consistent with the available data, renders meaningless attempts to reconstruct very ancient scenes in the evolutionary play as if they involved its current actors. As well, although it remains meaningful to attempt to reconstruct and even date the last (most recent) common ancestral versions of contemporary gene families, it is not reasonable to suppose that these ancestors all existed at the same time in the same genome. Thus LUCA cannot be reconstructed, and, if by LUCA we mean a cell whose genome harbored the last common ancestor of all genes found in organisms today, it never existed (Doolittle 2000).

Eukaryotes, the Other Iceberg

Whenever it is argued that LGT could render meaningless the concept of a Tree of Life, many scientists object that surely our understanding of the relationships among primates, or between animals and plants, or eukaryotes more generally, should not be called into question. Almost certainly it should not. The frequency of LGT is probably extraordinarily low in the evolution of the nuclear genomes of most multicellular eukaryotes with sequestered germlines. The edifice of neo-Darwinian theory, built on research into the genetics, paleontology, and population genetics of such organisms, is not about to crumble. But the first eukaryotes were unicells, immersed in a world of prokaryotes and their DNA, as is much of the contemporary eukaryotic world. Although we have as yet no real idea of how big an iceberg lies below, there is much anecdotal and some systematic evidence for a considerable tip, in terms of prokaryote-to-eukaryote and eukaryote-to-eukaryote LGT. The evidence was recently summarized masterfully by Jan Andersson, from whose article (Andersson 2005) our Table 4.1 is updated.

LGT from inside: endosymbiotic gene transfer

Some of this iceberg of LGT is intracellular—gene transfer from organelle to nucleus (Timmis et al. 2004 provide an authoritative review). Such *endosymbiotic gene transfer* might only be considered LGT by courtesy, since it is lateral *within* a strictly vertical organismal context. But endosymbionts (and ingested bacteria) might be the principle vehicle by which eukaryotic nuclear genomes have acquired foreign genes; the original acquisitions of the symbionts that were to become mitochondria and plastids were arguably the two "big bangs" of LGT in the history of life. Thus endosymbiotic gene transfer is highly relevant to any general appreciation of LGT in eukaryotes. Much to the point here is the recent finding, by Hotopp and colleagues (2007), that large fractions (sometimes perhaps all) of the genomes of *Wolbachia* species (maternally inherited and often detrimental intracellular endosymbionts of insects and nematodes) have been transferred in large and small fragments to the nuclear genomes of many of their hosts.

An analysis of the yeast nuclear genome (Esser et al. 2004) showed that as many as 75 percent of its genes that have homologs among prokaryotes

Table 4.1 Some examples of lateral gene transfer involving eukaryotes[a]

Recipient lineage	Transferred gene or protein encoded by transferred gene	Reference
PROKARYOTE TO EUKARYOTE		
Protists		
Amoebozoa		
Dictyostelium discoideum	18 genes	Eichinger et al. 2005
Entamoeba histolytica	8 genes	Andersson et al. 2003
E. histolytica	IscS and IscU	van der Giezen et al. 2004
E. histolytica	Malic enzyme, acetyl-CoA synthetase, and alcohol dehydrogenase	Field et al. 2000
E. histolytica	96 genes	Loftus et al. 2005
E. histolytica	[Fe] hydrogenase	Nixon et al. 2003
E. histolytica	NADH oxidase, alcohol dehydrogenase 3	Nixon et al. 2002
Apicomplexa		
Cryptosporidium parvum	Inosine-5-monophosphate dehydrogenase	Striepen et al. 2002
C. parvum	Thymidine kinase	Striepen et al. 2004
C. parvum	24 genes	Huang et al. 2004b
Several species	Several genes	Huang et al. 2004a
Chlorarachniophytes (*Bigelowiella natans*)	Several plastid-targeted proteins	Archibald et al. 2003
Ciliates		
Entodinium caudatum	Glutamate dehydrogenase	Andersson and Roger 2003
Nyctotherus	[Fe] hydrogenase	Horner et al. 2000
Diplomonads		
Giardia intestinalis	Phosphoenolpyruvate carboxykinase	Suguri et al. 2001
G. lamblia	HMG-CoA reductase	Boucher and Doolittle 2000
G. lamblia	NADH oxidase, alcohol dehydrogenase	Nixon et al. 2002
G. lamblia and *Spironucleus salmonicida*	Sulfide dehydrogenase	Andersson and Roger 2002
G. lamblia, S. salmonicida	More than 80 genes	Andersson et al. 2003; Andersson et al. 2007
Dinoflagellates (several spp.)	Shikimate biosynthetic enzyme AroB and an O-methyltransferase (OMT)	Waller et al. 2006
Euglenozoa		
Leishmania major	2 MIF domain proteins, several peptidases, ecotin homologs	Ivens et al. 2005
Trypanosoma cruzi	Glutamate dehydrogenase	Andersson and Roger 2003
T. cruzi, T. brucei, L. major	47 genes	Berriman et al. 2005
Bodonids, trypanosomatids[b]	Dihydroorotate dehydrogenase	Annoura et al. 2005
Several species	Glyceraldehyde-3-phosphate dehydrogenase	Figge et al. 1999; Figge and Cerff 2001; Qian and Keeling 2001

Table 4.1   *Continued*

Recipient lineage	Transferred gene or protein encoded by transferred gene	Reference
Parabasalids		
Trichomonas vaginalis	*N*-Acetylneuraminate lyase	de Koning et al. 2000
T. vaginalis	Putative surface protein	Hirt et al. 2002
T. vaginalis	152 likely transferred genes	Carlton *et al.* 2007
Trichomonas	Alcohol dehydrogenase	Andersson et al. 2003
Trichomonas and others	Glyceraldehyde-3-phosphate dehydrogenase	Figge and Cerff 2001; Qian and Keeling 2001
Trichomonads	Class II fumarase	Gerbod et al. 2001
Parabasalids, diplomonads	Alanyl-tRNA and prolyl-tRNA synthetase	Andersson et al. 2005
Parabasalids, diplomonads	Glucokinase and glucosephosphate isomerase	Henze et al. 2001
Rhodophyta (red algae; *Cyanidioschyzon merolae*)	One of two DAHP II genes of the shikimate pathway	Richards et al. 2006
Trimastix pyriformis	At least four glycolytic enzymes	Stechmann et al. 2006
Fungi		
Microsporidia (*Nosema locustae*)	Catalase	Fast et al. 2003
Ascomycetes		
Nectria haematococca	Pea pathogenicity genes	Temporini and VanEtten 2004
Saccharomyces cerevisiae	Flavohemoglobin	Andersson et al. 2003
S. cerevisiae	10 likely recently transferred genes	Hall et al. 2005
Saccharomycetaceae[b]	Dihydroorotate dehydrogenase	Annoura et al. 2005
Aspergillus	Catalase	Klotz et al. 1997
Neurospora and others	DHQase II of the shikimate pathway	Richards et al. 2006
Penicilium and others	β-Lactam biosynthesis genes	Brakhage et al. 2005
2 Euascomycetes classes	β-Glucuronidase	Wenzl et al. 2005
Rumen fungi	Glycosyl hydrolase	Garcia-Vallve et al. 2000
Opisthokonts (fungi and animals, many species)	Tyrosyl-tRNA synthetase	Huang et al. 2005
Animals		
Bombyx mori	Glycerophosphodiesterase, and possibly sucrase, glucose-1-phosphatase	Mita et al. 2003
B. mori and other lepidopteran insects	Chitinase	Daimon et al. 2005
Ciona	Cellulose synthase	Matthysse et al. 2004; Nakashima et al. 2004
Meloidogyne	12 genes	Scholl et al. 2003
Vertebrates[c] (humans, pigs, mice)	*N*-Acetylneuraminate lyase clusters with *Vibrio* and *Yersinia*	Andersson et al. 2001
Nematodes, flies	Deoxyribose-phosphate aldolase, threonine dehydratase	Andersson et al. 2003

Table 4.1   *Continued*

Recipient lineage	Transferred gene or protein encoded by transferred gene	Reference
Plants		
Nicotiana	*Agrobacterium* proteins	Intrieri and Buiatti 2001
Several groups	Several genes of the shikimate pathway	Richards et al. 2006
Animals and plants (many lineages)	Signal transduction ATPases with numerous domains (STAND)	Leipe et al. 2004
EUKARYOTE TO PROKARYOTE		
Prosthecobacter	Tubulins	Jenkins et al. 2002; Schlieper et al. 2005
Proteobacteria (several spp.)	Deoxyribose-phosphate aldolase	Andersson et al. 2003
Geobacteraceae (several spp.)	Citrate synthase	Bond et al. 2005
Legionella pneumophila	A large number of likely Type IV secretion substrates	Nagai et al. 2002; Cazalet et al. 2004; de Felipe et al. 2005
Solibacter usitatus	DAHP II of the shikimate pathway	Richards et al. 2006
Several bacteria	Acyl-CoA binding protein	Burton et al. 2005
EUKARYOTE TO EUKARYOTE		
Protists		
Chlorarachniophytes (*Bigelowiella natans*)	Several genes encoding plastid-targeted proteins	Archibald et al. 2003
Ciliates		
Paramecium, Tetrahymena	Alanyl-tRNA synthetase	Andersson et al. 2005
Paramecium, Tetrahymena	Glucosamine-6-phosphate isomerase	Andersson et al. 2006
Dinoflagellates	Enolase	Keeling and Palmer 2001; Harper and Keeling 2004
Diplomonads or mycetozoa	Threonine dehydratase	Andersson et al. 2003
Entamoeba (several spp.)	Alanyl-tRNA synthetase	Andersson et al. 2005
Entamoeba histolytica	[Fe] hydrogenase	Nixon et al. 2003
Parabasalids (*Trichomonas*)[c]	*flp* gene	Steele et al. 2004
Animals		
Bruchid beetles	Mitochondrial *cytb*	Alvarez et al. 2006
Hydra (see parabasalids, above)[c]	*flp* gene	Steele et al. 2004
Plants		
Amborella trichopoda	26 mitochondrial genes	Bergthorsson et al. 2004
Botrychium virginianum	2 mitochondrial regions	Davis et al. 2005
Gnetum	Part of mitochondrial *nad1* gene	Won and Renner 2003
Rafflesiaceae	Mitochondrial *nad1 B-C*	Davis and Wurdack 2004
Many flowering plant species	Mitochondrial genes	Bergthorsson et al. 2003
Many lineages (protists, fungi, green algae)	EF-like protein	Keeling and Inagaki 2004

[a] LGT events of mobile genetic elements are not included, and the recent literature has not been exhaustively surveyed.

[b] Could also be intradomain transfer.

[c] Direction of the transfer unknown.

Source: Updated from Andersson 2005.

are more similar to bacterial than archaeal homologs. Such genes are found in all functional categories, but especially those of metabolic and biosynthetic processes. They are not restricted to serving mitochondrial functions, although they may be mitochondrial in origin. Acquisition of chloroplasts—by primary, secondary, and tertiary endosymbiosis—has provided further opportunities for organelle-to-nucleus gene transfer. An analysis of the *A. thaliana* genome (Martin et al. 2002) suggested that 18 percent of its genes have been transferred from the primary cyanobacterial ancestor of the plastid to the nuclear genome. The majority of these genes are targeted to cell compartments other than the chloroplast, implying that the endosymbiont copy in many cases probably replaced a preexisting host copy.

LGT from outside: prokaryotes to eukaryotes

Vertebrates with sequestered germ lines do not appear to acquire bacterial genes frequently; claims to this effect, which accompanied initial publication of the human genome sequence, have not been supported. However, some eukaryotes may, because of their biology and access to foreign DNA, obtain foreign genes as frequently as many prokaryotes. The phagotropic lifestyle of many unicellular eukaryotes, and the presence of genes of bacterial origin in these lineages, led Doolittle (1998) to suggest a ratchetlike mechanism in which prokaryotic genes (or eukaryotic genes) from food (or endosymbionts) inevitably replace ancient eukaryotic genes over time. This "you-are-what-you-eat" ratchet operates as does that described for orthologous replacement in Figure 4.7.

Gene transfers are also observed in nonphagotropic lineages (see Table 4.1), so other mechanisms must operate. For instance, the bacterial chitinase gene in the silkworm (*Bombyx mori*) was also found in baculovirus, suggesting that the virus could be a possible vector for the transfer. Another interesting possibility for transfer of prokaryotic (and eukaryotic) genes to eukaryotes is through *Agrobacterium tumefaciens*-like infections. Amazingly, the host range of the *A. tumefaciens* plasmid-encoded conjugation machinery can be extended to nonplant eukaryotic organisms: several fungi, as well as human culture cells, have been transformed by *Agrobacterium* (Lacroix et al. 2006). Spontaneous uptake and expression of extracellular DNA has been demonstrated for the nonphagotropic Apicomplexa, including the malaria parasite *Plasmodium falciparum*, in which several LGT events have been detected (Deitsch et al. 2001).

Many of the genes acquired by eukaryotes through LGT introduce new functions. For several eukaryotic lineages, genes acquired by LGT appear to support an anaerobic lifestyle (see for instance Andersson et al. 2003; Hall et al. 2005). Multiple transfers involving many different eukaryotic lineages are often seen in phylogenetic trees for these genes. Moreover, phylogenetic analyses showed that the anaerobic bacterial genes of *Giardia* have been acquired from many different prokaryotic sources, and these differ from the donors of the same genes in *Entamoeba* (Andersson et al. 2007). Rather than being an ancestral feature of eukaryotes, anaerobiosis is probably a derived trait that evolved several times independently, facilitated by LGT.

Several protist genome sequences have recently become available. These genome sequences have revealed additional instances of LGT from bacteria to protists (see Table 4.1), although not to the extent seen in prokaryotic genomes. The genome sequence of *Entamoeba histolytica*, for instance, revealed 96 genes for which LGT from bacteria is the simplest explanation (Loftus et al. 2005). Most of these (58%) encode metabolic enzymes, and the majority of the remaining genes (41%) encode proteins of unknown function. In the *Trichomonas vaginalis* draft genome, 152 cases of likely transfers from prokaryotes were detected (Carlton et al. 2007). These LGT genes affect several metabolic pathways, and many of the LGT genes appear to have been acquired from Bacteroidetes-related bacteria, which are abundant among vertebrate intestinal flora. Fewer transferred genes were observed in the other sequenced protist genomes: 47 genes in one or more of the three sequenced Trypanosomatid genomes (*Leishmania major, Trypanosoma brucei*, and *Trypanosoma cruzi* [Berriman et al. 2005]), and 18 genes in the genome of *Dictyostelium discoideum* (Eichinger et al. 2005).

LGT from prokaryotes into yeasts, since their divergence from a common fungal ancestor, has been suggested to be less frequent than that observed from prokaryotes into protists, with most of the bacterial genes in yeast being of mitochondrial origin, and less than one percent having been acquired by LGT. A comparative genomic study of two hemiascomycetes, *Saccharomyces cerevisiae* and *Ashbya gossypii*, agrees with this—it revealed only 11 potential cases of recent (after the divergence of these lineages) LGT from bacteria, 10 of these being in *S. cerevisiae* and amounting to only 0.2 percent of its genome (Hall et al. 2005; see Table 4.1). Nine of the ten genes identified in *S. cerevisiae* were located near the telomeres, which have been shown to harbor variable regions in *Saccharomyces* species, thus supporting the notion of their foreign origin.

Eukaryotes to eukaryotes

Phylogenetic relationships among many eukaryotic groups are still unresolved, making transfers between them difficult to pinpoint. Moreover, because relatively few eukaryotic genomes have been sequenced thus far, it is unlikely that sequences from both the donor and the recipient lineage would be available. Thus a lower observed rate of eukaryote intradomain transfer might not reflect a real biological pattern.

Nevertheless, there are some well-documented cases of transfers within the eukaryotes (see Table 4.1). For instance, Keeling and Inagaki (2004) found that eukaryotes harbor two distinct types of translation elongation factor 1: a canonical EF1α and an EF-like protein (EFL). Most organisms that have EFL lack EF1α. This suggests that EFL has replaced EF1α several times independently, most likely through eukaryote-to-eukaryote lateral gene transfer.

Very high rates of eukaryote-to-eukaryote transfer, between the mitochondria of donor and recipient, have recently been reported in higher plants by Jeff Palmer's group. In an analysis of one of the recipient mitochondrial genomes—that of *Amborella trichopoda*—Bergthorsson and colleagues (2004) found that it had acquired one or more copies of 20 of its 31 known mitochondrial genes, for a total of 26 foreign genes! Most of the *Amborella* transfers were from other angiosperm mitochondria, but there were also transfers from more distantly related plants, such as mosses. It is likely that frequent direct plant-to-plant contact is important for high rates of transfer between plants, as many transfers involve parasitic plants and their hosts. *Amborella* is found in tropical rainforests, an environment where such contacts are common. Taken together, the high rate of transfer observed between plant mitochondria and the high rate of gene transfer from mitochondria to the nucleus suggest that plant nuclear genomes probably are also significantly affected by LGT.

Eukaryotes to prokaryotes

Table 4.1 provides far fewer examples of transfer from eukaryotes to bacteria or archaea. There are many reasons this might be the case. There are fewer complete genome sequences of eukaryotes, and these are more uniform in gene content; thus there are fewer possibilities for a significant match in the databases. Introns in many eukaryotic genes will prevent their expression in prokaryotic genomes. Moreover, most of the metabolic diversity in life is observed in prokaryotes: eukaryotes have very little to offer them.

One example of acquisition of large numbers of eukaryotic genes is found in the intracellular pathogenic *Legionella* lineage. De Felipe and colleagues (2005) identified 46 genes with distinct eukaryotic motifs in *L. pneumophila* Philadelphia-1, and Cazalet and colleagues (2004) identified 62 with either eukaryotic domains or high similarity to eukaryotic genes in two other *L. pneumophila* strains.

Reuniting Process and Pattern

Darwin's goal, like that of many contemporary biologists, was to explain nature's patterns in terms of its processes. Indeed, Darwin's theory invoked an underlying natural process—lineage splitting and divergence—to explain the pattern of similarity and difference between organisms that systematists before him had already enshrined in tree-like hierarchies of "groups subordinate to groups." The Tree of Life is nothing more or less than a pre-Darwinian metaphor that Darwin adopted to encapsulate the theory (indeed calling it a "simile"). Nearly one hundred and fifty years later, molecular genetics (which focuses on process and microevolution) and molecular phylogenetics (which emphasizes pattern and macroevolution) are almost separate disciplines, at least as practiced by microbiologists. The debate over the meaning of LGT reflects this disciplinary split, but is healthy because it will in the end reunite them. Processes produce patterns in the natural world, not the reverse, but defenders of the Tree of Life have reified Darwin's simile. By rededicating ourselves to the study of what actually happens between genes and genomes and organisms in the real world in real time, we will better advance our science.

CHAPTER **5**

Evolution of Genomic Expression

Bernardo Lemos, Christian R. Landry, Pierre Fontanillas,
Susan C. P. Renn, Rob Kulathinal, Kyle M. Brown,
and Daniel L. Hartl

Introduction

GENOMIC REGULATION IS KEY TO CELLULAR DIFFERENTIATION, tissue morphogenesis, and development. Increasing evidence indicates that evolutionary diversity of phenotypes—from cellular to organismic—may also be, in large part, the result of variation in the regulation of genomic expression.

In this chapter we explore the complexity of gene regulation from the perspective of single genes and whole genomes. The first part describes the major factors affecting gene expression levels, from rates of gene transcription—as mediated by promoter–enhancer interactions and chromatin modifications—to rates of mRNA degradation. This description underscores the multiple levels at which genomic expression can be regulated as well as the complexity and variety of mechanisms used. We then briefly describe the major experimental and computational biology techniques for analyzing gene expression variation and its underlying causes. The final section reviews our understanding of the role of regulatory variation in evolution, including the molecular evolution and population genetics of noncoding DNA, as well as the inheritance and phenotypic evolution of levels of mRNA abundance.

The Complex Regulation of Genomic Expression

The regulation of gene expression is a complex and dynamic process. It is not a simple matter to turn a gene on and off, let alone precisely regulate its level of expression. Regulation can be accomplished through various mechanisms at nearly every step of the process of gene expression. Furthermore, each mechanism may require a variety of elements, including DNA sequences, RNA molecules, and proteins, acting in combination to determine the final amount, timing, and location of functional gene product. The complexity of regulation is even more evident when it is considered in the context of evolution and from the standpoint of integrated gene expression across the genome. The dynamic process of genomic expression is not strictly fixed but can be context dependent, responding to cellular or environmental influences. Through successive generations, the interplay of these mechanisms evolves, thus generating a selectively advantageous amount and/or location of functional gene product.

Most of the cellular elements regulating gene expression can be divided into two basic categories: *cis*- and *trans*-acting factors, from the Latin meaning "on the same side" and "on the opposite side," respectively. Both *cis*- and *trans*-acting regulatory elements may contribute to the various mechanisms of gene expression regulation. Strictly, *cis* and *trans* effects do not refer to the physical location of the regulatory element, but rather are operationally defined in terms of the way these regulatory elements segregate genetically with respect to the gene that is the target of the regulatory activity. Promoters, enhancers, regulatory introns, and 3' regulatory sequences are examples of *cis*-regulatory elements. This is because these elements are located within the gene locus itself or in close proximity to it, such that they are generally inherited together as a unit (i.e., the probability of recombination between the regulatory elements and the gene's structural sequences is virtually zero).

Trans-acting factors include proteins and RNAs derived from distant sites in the genome that act as regulatory elements. These can be either on different chromosomes or on the same chromosome far away from the gene locus, such that they can be independently inherited (i.e., the recombination frequency is virtually 50%). Specific factors necessary for initiating or blocking transcription, or proteins that allow for appropriate gene-specific mRNA trafficking and stability, are just a few of the myriad *trans*-acting factors that contribute to gene expression.

The definition of *cis* and *trans* with respect to segregation is important because these two types of elements work in concert and often there is no clear functional (or even positional) distinction between *cis* and *trans*. *Trans*-acting factors bind to or interact with *cis*-regulatory sequences of the DNA and RNA. It is therefore the interplay of *cis*- and *trans*-acting loci that determines the amount of functional gene product. Indeed, this level of complexity and interaction provides the very substrate for the evolution of regulation, as various *cis* and *trans* factors are brought together through segregation and recombination. When combined, elements that have evolved under differing selective pressures may produce novel phenotypes, thus offering novel substrates for natural selection.

Early work attempted to categorize genetic mutations affecting a particular phenotype as either structural or regulatory (e.g., Wilson et al. 1977). This separation was motivated by the intuition that some proteins have clearly defined structural roles (e.g., collagen), while the function of other proteins lies primarily in the regulation of other genes (e.g., muscle-specific transcription factors). However, it should be stressed that this classification of gene loci and their mutations breaks down for most genes. This is because many proteins (structural) act in *trans* to alter the expression level of other genes (regulatory), such that a single mutation can have both structural and regulatory consequences. Work by Yvert and colleagues (2003) clearly exemplifies this point. These authors mapped a large number of *trans*-acting loci affecting gene expression differences between two strains of yeast, and found that genes with a large variety of molecular functions, such as enzymes, signal transducers, and cytoskeleton-binding proteins, could influence gene expression levels. Accordingly, amongst *trans*-acting loci there is no specific enrichment for transcription factors.

In the following section, we outline the overall process of gene expression by indicating the mechanisms known to regulate gene product level, location, or timing at various stages of transcription, translation, and post-translation. Transcriptional regulation of gene expression can occur at the level of genomic DNA prior to transcription, or at the step of transcription, when the RNA is being produced. Post-transcription regulation occurs through several mechanisms affecting the processing, stability, and/or localization of the mRNA. The amount of functional gene product can also be regulated post-translationally. For each stage in the process of gene expression, we provide examples of a few well-studied mechanisms of regulation, and attempt to identify examples of both *cis*-acting and *trans*-acting elements involved in each of the regulatory mechanisms. Figure 5.1 gives examples of the different stages along the path to functional protein during which genomic expression may be regulated. In most instances, the evolutionary acquisition and consequences of these mechanisms have yet to be addressed. Most research regarding the *molecular evolution* of gene or genomic expression has focused on transcription factors (TF), promoter sequences, and TF-binding sites within promoters. Similarly, most research regarding the *phenotypic evolution* of genomic expression has focused on the evolution of the mRNA abundance phenotype, most often without an explicit connection to evolutionary variation in specific underlying mechanisms associated with variation in this phenotype.

Classical Transcriptional Regulation in *Cis* and *Trans*

Promoters, enhancers, repressors, transcription factors, and regulatory proteins

Promoters are among the most thoroughly studied and best understood regulators of gene expression (Ptashne and Gann 2002; Thomas and Chiang 2006). In eukaryotes, the promoter is defined as the DNA region within a few hundred base pairs upstream of the transcription start site, the section of DNA where the basic machinery of gene expression is assembled. The promoter region encodes various sequence motifs where transcription factors bind along with RNA polymerase II to initiate transcription (Smale and Kadonaga 2003). The initiation and rate of transcription depend on the interplay of many DNA and protein elements. Ultimately, a gene's transcription

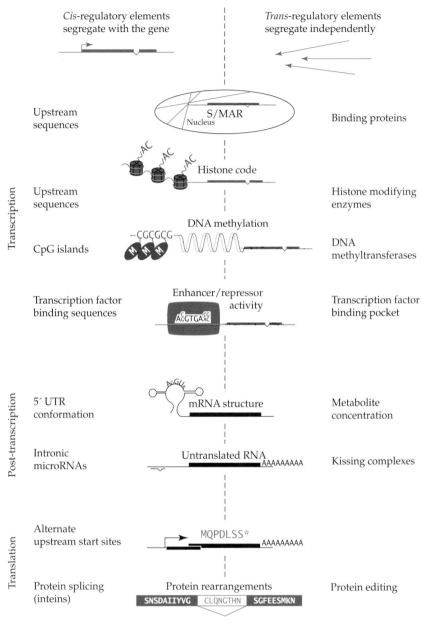

Figure 5.1 Some of the *cis* and *trans* elements associated with a variety of regulatory mechanisms acting at three levels of the gene expression cascade (see text for further details).

level reflects the interactions between various activating and inhibitory complexes assembled not only at the promoter region, but also at various enhancer and silencer sites along the chromatin (Wray et al. 2003).

The core promoter includes DNA elements that can extend about 35 nucleotides upstream and/or downstream of the transcription initiation site (Smale and Kadonaga 2003). The core is obviously the archetype of *cis*-regulatory factors. At least six core promoter elements have been discerned so far (Gershenzon et al. 2006): TATA box, Initiator (Inr), Downstream Promoter Element (DPE), TFIIB Recognition Element (BRE), Downstream Core Element (DCE), and Motif Ten Element (MTE). Although earlier studies suggested that the structure of the core promoter might be highly conserved throughout the eukaryotes, tremendous diversity is now evident. First, the sequence and the position of DNA motifs are variable within and between species (Bazykin and Kondrashov 2006; FitzGerald et al. 2006; Sandelin et al. 2007). Second, not all core promoter elements are systematically associated with all gene promoters; instead, combinations of a subset of promoter elements are more frequently observed. In *Drosophila melanogaster*, TATA box, Inr, DPE, and MTE are found in, respectively, 16, 66, 22, and 10 percent of the genes (Gershenzon et al. 2006). In mammalian promoters, the TATA box element is also present in a minority of genes, but shows substantial sequence conservation and is commonly associated with tissue-specific expression (Carninci et al. 2006). Surprisingly, the presence of the TATA box is also associated with elevated rates of gene expression divergence among yeast species (Tirosh et al. 2006). On the other hand, TATA-less promoters, which are often enriched in CpG islands, seem to be particularly rapidly evolving in mammals (Carninci et al. 2006).

The specific proteins that bind to the core promoters are perhaps the most fundamental *trans*-acting factors regulating gene expression. They include the general transcription factor TFIID, which recognizes the promoter and coordinates the assembly of the remaining general transcription factors (TFIIA, TFIIB, TFIIE, and TFIIH). TFIID is itself a large protein complex formed by the TATA-box binding protein (TBP), as well as several transcription-associated factors (TAFs; Muller et al. 2007). The latter include coactivators capable of propagating signals from distant enhancer or repressor elements to the promoter site. In a stereotyped sequence of protein binding, TBP, TFIID, TFIIA, and TFIIB must bind to the promoter region first in order to recruit RNA polymerase II, TFIIE, TFIIH, TFIIF, and other factors necessary to initiate transcription (Tjian 1996; Smale and Kadonaga 2003). These are the *trans*-acting partners to *cis*-occurring promoters and/or enhancer sequences. These proteins are remarkably conserved across the eukaryotes.

Eukaryotic enhancer and repressor elements are additional sites that influence gene expression and may occur several hundred to several thousand base pairs from the promoter site. In many cases, the various enhancer or repressor elements act independently of each other. Each such enhancer or repressor receives and integrates various signals from regulatory proteins that recognize specific binding sites; these signals are subsequently transmitted to the transcriptional machinery located at the promoter. The interaction of all enhancer elements, repressor elements, and other factors at the promoter results in the precise regulation of the timing, location, and level of gene expression.

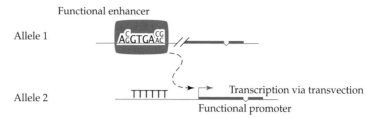

Figure 5.2 Transvection as demonstrated in Drosophila. The enhancer region from one allele can act in *trans* to affect the expression of the allele on the other chromosome. This has been demonstrated using one allele lacking a functional RNA polymerase binding site (so that transcription cannot be initiated) and another allele in which tissue-specific enhancer elements have been mutated. In combination, the functional portions of these genes are able to complement each other, and transcription on one chromosome is directed by enhancer sequences on the paired chromosome.

Although more has been learned about the role of promoters and enhancers in the regulation of transcription than about the role of any other regulatory element, the elusive phenomenon of transvection reminds us how little we know about even this most basic mechanism of gene regulation. Transvection was first described in 1954 by E. B. Lewis in the context of two mutations that complemented each other in spite of both being within the Drosophila *Ubx* locus (see Duncan 2002 and references therein). This was later recognized to arise from somatic pairing between one allele with a loss of function mutation in its regulatory sequence and another allele with a loss of function mutation in its coding sequence (Figure 5.2). It remains to be understood how regulatory elements in one allele regulate the expression of its homolog on the other chromosome.

Epigenetic Regulation and Chromatin Modifications

Appropriate chromatin conformation is required for access and binding of regulatory proteins to the DNA. The nucleosome is the fundamental repeating unit of chromatin. Made up of 146 base pairs of DNA wrapped around an octamer of conserved core histone proteins, nucleosomes are linked together to form a helical fiber. Each histone contains numerous sites for potential modifications, which have been hypothesized to act in a combinatorial code to mark a region for potential activation or silencing. These marks extend the information potential of the genetic code to provide a so-called epigenetic memory, which plays a major role in regulating cell fate decisions. These stable, epigenetic changes persist through mitosis and in some cases through meiosis. Consideration of epigenetic gene regulation has led to models for the inheritance of acquired epigenetic variations—models in which environmental stimuli induce heritable modifications that might result in adaptive responses to the stimuli (Jablonka and Lamb 1989; Jablonka and Lamb 2002; Gorelick 2005).

DNA methylation

DNA methylation is the most well-understood form of epigenetic gene regulation. First proposed in 1975 (Holliday and Pugh 1975), it has since been intensively characterized in mammals and plants (Jaenisch and Bird 2003; Scott and Spielman 2004). The establishment and maintenance of methylation is required for the normal development and cell differentiation of many organisms. In mammals, DNA methylation occurs predominantly at cytosines in CpG dinucleotides, and several enzymes (DNA methyltransferases) are responsible for de novo methylation and maintenance of methylation marks during mitosis and meiosis. Little is known, however, about the sequences and conditions that direct methylation activity. With respect to a gene under control of DNA methylation, the CpG islands and surrounding sequences that direct the specificity of methylation are *cis*-acting factors, while the methylation enzymes are *trans*-acting factors. In general, increased methylation is associated with down-regulation of gene expression (Wolffe and Matzke 1999), and unmethylated CpG islands lead to increased transcription, although exceptions to this have been described (e.g., Herman et al. 2003). In addition, DNA methylation states have been shown to be sensitive to environmental factors (Jaenisch and Bird 2003) and also to have longlasting effects on behavior (Weaver et al. 2004; Weaver et al. 2005; Feil 2006). Finally, DNA methylation is found in most transposable elements in Arabidopsis and primates (Lippman et al. 2004; Meunier et al. 2005), suggesting its role as a defense mechanism preventing the expression of these elements.

Yeast, worms, and flies have generally been thought to lack DNA methylation systems. In the case of fruit flies, this notion has been challenged by the availability of whole genome sequences. A single DNA methyltransferase has been identified in the genome of *Drosophila melanogaster* (Hung et al. 1999), and experimental work has detected low levels (<0.5%) of cytosine methylation in the fly (Lyko et al. 2000). Moreover, methylated sequences in the fly were found to be associated with CpT or CpA dinucleotides, in sharp contrast to the canonical CpG motif often found methylated in mammals (Kunert et al. 2003).

DNA methylation is particularly interesting in the context of parent-of-origin-dependent inheritance and genomic imprinting (Hajkova et al. 2002; Delaval and Feil 2004). In genomic imprinting, the expression of a subset of genes depends on "marked" maternal and paternal alleles that are recognized and differentially regulated by the transcriptional machinery. Methylation of CpG dinucleotides is the basis for the parent-specific mark. The evolutionary relevance of genomic imprinting has been examined in detail from a theoretical standpoint (see Wilkins and Haig 2003 and McDonald et al. 2005 for recent reviews). Taken together, these observations underscore the relevance of DNA methylation to the evolution of genomic expression.

Histone modifications

Five classes of eukaryotic histones are known (H1, H2A, H2B, H3, and H4), all of which are lysine/arginine rich and have a globular domain that facilitates their assembly with DNA to form chromatin. Histones also have a charged amino terminus (the so-called histone tail) that is subject to various post-translational modifications (e.g., acetylation, methylation, and phosphorylation), which influence chromatin structure and gene expression (Li et al. 2007). It has been conjectured (Jenuwein and Allis 2001; Felsenfeld and Groudine 2003; Fischle et al. 2003; Turner 2007) that the kind of histones, and their packing, location, and post-translational modification make up an "epigenetic code," which is used by the cell to specify a precise and stable gene expression profile.

Histone deacetylases, histone acetyltransferases, and histone methyltransferases make up some of the *trans*-acting elements involved in regulating histone modifications that ultimately influence gene expression. For instance, chromatin enriched for acetylated histones is generally thought to be "open" and accessible to transcription factors, thereby rendering its constituent genes transcriptionally active, or potentially so (Grewal and Moazed 2003). Conversely, chromatin enriched for nonacetylated histones is generally thought to be more "condensed," thereby making its constituent genes inaccessible to transcription factors and therefore silenced. Similarly, chromatin enriched for methylated histones is generally thought to be less transcriptionally active than other regions. Furthermore, DNA and histone methylation appear to maintain a repressed chromatin state in plants and vertebrates, although such links appear to be weaker in insects. Finally, we note that whereas histone modifications are known to persist through cell division, much of the histone code may be erased in meiosis; therefore, epigenetic memory is usually thought to result from other DNA and chromatin modifications.

As a cautionary note we emphasize that recent functional genomic work has questioned the specificity of the effects of histone modifications on gene expression. For instance, Dion and colleagues (2005) examined the effect of acetylation of four lysine residues in the tail of histone H4 in yeast. They constructed yeast strains containing up to three lysine-to-arginine mutations in the histone H4 tail, thus preventing acetylation while retaining the positive charge. All the single lysine-to-arginine substitutions showed quite similar gene expression changes, irrespective of the particular lysine altered. Similarly, all combinations of double mutants showed similar changes. This lead Dion and coworkers (2005) to propose that histone H4 acetylation has a simple cumulative effect on yeast gene expression, arguing against a more complex model in which combinations of acetylated lysines specify unique expression profiles. Furthermore, in *D. melanogaster*, a systematic mapping of methylation and acetylation of H3 and H4 histones suggested an "all-or-none" pattern (Shübeler et al. 2004; Liu et al. 2005): While the active genes tended to be marked by all the assayed modifications, the nontranscribed genes tended to have no histone marks. A similar, rather simple histone code has also been suggested for budding yeast (Kurdistani et al. 2004; Dion et al. 2005) and mammals (Bernstein et al. 2005). Nonetheless, yeast also shows a more subtle pattern of modification in which clustering of genes with similar patterns of acetylation distinguishes groups of coexpressed genes that are functionally related (Kurdistani et al. 2004). These studies are also modifying our views of the correlation between histone marks and gene expression by demonstrating that both hyper- and hypoacetylation of histones may be associated with gene activity (see also Barski et al. 2007; Heintzman et al. 2007; Zhang et al. 2007).

Chromosome territories and nuclear architecture

Gene expression is most often studied from the standpoint of promoters, enhancers, suppressors, and local epigenetic modifications. Less frequently studied, however, is the impact of higher-order chromosome structure and nuclear organization on gene expression. Nevertheless, a growing awareness of the relevance of spatial chromosome dynamics in genomic expression, together with technological developments, has stimulated a surge in interest and research in this area (e.g., Bolzer et al. 2005; Harmon and Sedat 2005; Pickersgill et al. 2006; Goetze et al. 2007).

The nucleolus is perhaps the best-known and most prominent structural feature in the nucleus of most plant and animal cells. This cytological structure is the site where the ribosomal DNA (rDNA) regions of several chromosomes come together and rRNA transcription takes place (see Santoro 2005 for a review). Interestingly, there is substantial natural genetic variation in the amount of methylation observed in rDNA regions of different Arabidopsis accessions (Riddle and Richards 2002). This natural variation in rDNA methylation may serve as yet another source of genetic variation in gene expression upon which natural selection can act.

In a less well-described regulatory mechanism, eukaryotic genomes are functionally compartmentalized by attachment to the supporting nuclear matrix (Bode et al. 2003). The overall dynamics of this structure are mediated in part by elements of 300 base pairs to several kilobases named scaffold/matrix-attachment regions (S/MARs)(e.g., Heng et al. 2004). In conjunction, S/MAR-binding proteins act in *trans* to regulate gene expression. Genome-wide predictions identify a large number of S/MARs (Frisch et al. 2002; Glazko et al. 2003; Evans et al. 2007) associated with enhancement as well as repression of gene expression.

On a smaller scale, locus control regions (LCR) can be identified that coordinately regulate promoters of several related genes spread over hundreds of kilobases. This is achieved by close juxtaposition of different chromosomal regions in the nucleus. This mechanism was recently shown to be relevant in the regulation of genes involved in T-helper cell differentiation (Spilianakis et al. 2005; Spilianakis and Flavell 2006). To address the evolutionary implications of such regulation, more information is needed about the genome-wide prevalence and impact of regulatory processes involving nuclear architecture, as well as a detailed mechanistic understanding of individual elements and their interactions. These processes undoubtedly depend on modification of chromatin structure, and they suggest that dispersed multigene complexes are coregulated in part by structural colocalization within the nucleus. How such structures vary across populations and species remains virtually unknown.

The mechanistic complexity of epigenetic gene regulation is perhaps best illustrated by paramutation—a phenomenon first described in peas (Bateson and Pellew 1915), most thoroughly studied in maize (Brink 1959; Brink 1973; Chandler et al. 2000; Hollick and Chandler 2001), and more recently discovered in mammals (Herman et al. 2003; Rassoulzadegan et al. 2006). Paramutation involves heritable changes in gene activity without changes in DNA sequence. The change in gene activity is mediated by heritable epigenetic modification induced by cross-talk between allelic loci (Figure 5.3). Paramutation and paramutation-like phenomena do not adhere to rules of

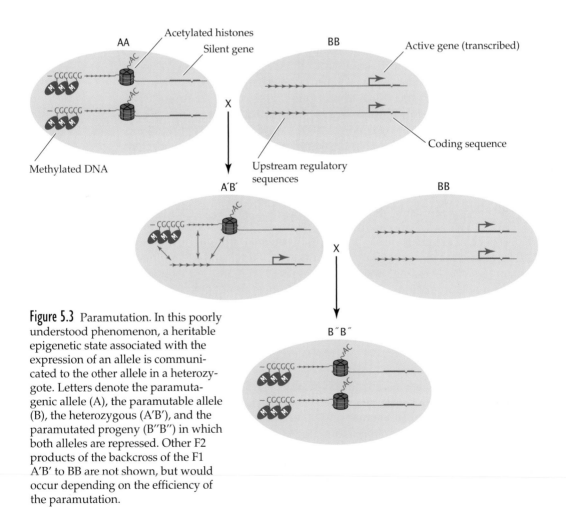

Figure 5.3 Paramutation. In this poorly understood phenomenon, a heritable epigenetic state associated with the expression of an allele is communicated to the other allele in a heterozygote. Letters denote the paramutagenic allele (A), the paramutable allele (B), the heterozygous (A′B′), and the paramutated progeny (B″B″) in which both alleles are repressed. Other F2 products of the backcross of the F1 A′B′ to BB are not shown, but would occur depending on the efficiency of the paramutation.

Mendelian inheritance. Our knowledge of the underlying mechanisms is limited, but evidence suggests a complex interplay of many epigenetic processes, such as RNA silencing, physical pairing of homologous chromosomal regions, and chromatin modifications. The term paramutation has come to describe many phenomena in which communication between two alleles or homologous sequences establishes distinct, heritable epigenetic states (Chandler and Stam 2004; Stam and Mittelsten Scheid 2005; Chandler 2007).

Post-Transcriptional Regulation

Once transcription has begun, the cell has a variety of post-transcriptional mechanisms to regulate the final amount of functional gene product. In this section we briefly outline some of the best-known mechanisms of post-transcriptional regulation of genomic expression, all of which involve the complex interaction of *cis-* and *trans*-acting factors.

The initial and most basic form of post-transcriptional regulation involves the premature termination of transcription. This gene regulatory mechanism is common in bacteria (Merino and Yanofsky 2005) and has been best studied in the transcription of the HIV-1 genome in host cells (Kessler and Mathews 1992). It may also play a role in gene regulation in eukaryotes, where premature termination of transcription typically results from the nascent mRNA folding into secondary structures that are recognized by the cellular apparatus (Muhlrad and Parker 1994; Arigo et al. 2006).

When a full-length RNA transcript is made, it is modified in several ways that determine its cellular fate. Immediately after transcription, a methylated guanine nucleotide is added to the 5′ end of all mRNA transcripts (5′ methylguanosine cap). This feature of mature mRNA is important for the initiation of translation (Alberts et al. 2002). On the 3′ end of the immature transcript, the RNA is cleaved at a specific site and a poly(A) tail is added. This poly(A) tail, usually about 100–200 nucleotides long, is important for regulating the export of the mRNA out of the nucleus, regulating the stability and half-life of the transcript, and ensuring efficient translation at the ribosome (Ross 1995; Alberts et al. 2001).

After transcription is complete, RNA splicing removes noncoding introns and joins together neighboring exons. Alternative splicing, by joining together different exons to create the mature transcript, can produce many different proteins from a single gene. An extreme example of this is the Drosophila *Dscam* axon guidance receptor gene, which can potentially generate 38,016 different protein isoforms (Graveley 2005; Crayton et al. 2006). Appropriately spliced and edited transcripts are then regulated further by transport to appropriate parts of the cell; this intracellular transport is often mediated by the untranslated 5′ and 3′ ends of mRNAs.

In a regulatory mechanism known as mRNA editing, the nucleotide sequence of the transcript can be changed at specific places. Ordinarily, this process leads to the production of protein variants differing in one or a few amino acids, but, in its most dramatic manifestations, can determine the splicing and ultimate cellular location of the final gene product (Maas and Rich 2000). An example of mRNA editing is the modification of adenosine to inosine, which is recognized as guanosine by the cellular machinery (Maas and Rich 2000; Barbon et al. 2003). This modification is mediated by the *trans*-regulatory activity of adenosine deaminases (Aphasizhev 2007) on an mRNA editing-site sequence, which represents the *cis*-regulatory component of this mechanism of regulation.

In addition, mRNA degradation plays an important role in post-translational gene regulation. For example, the rate of mRNA degradation has been shown to vary widely among genes (Wang et al. 2002; Foat et al. 2005). Degradation of functional, mature mRNA is regulated by mRNA binding proteins and is specified by various features, including sequence motifs and mRNA secondary structures, usually in the 3′ untranslated region of the mRNA (Ross 1995). Another recent study found that genes with tightly folded 5′ untranslated regions may have lower rates of translation, lower protein and mRNA abundances, and shorter half-lives (Ringner and Krogh 2005).

Typically found in the 5′ untranslated regions of bacterial mRNAs, riboswitches are structural elements that regulate gene expression post-transcriptionally (Winkler et al. 2002; Tucker and Breaker 2005; Winkler and Breaker 2005). Riboswitches regulate gene expression by binding to small metabolites, without the involvement of other cofactors. Riboswitches have also been identified in eukaryotes (Sudarsan et al. 2003) where they have been shown to be involved in the regulation of alternative splicing (Cheah et al. 2007). Thus, acting as sensors, these RNA sequences are capable of modifiying gene expression in a manner previously thought only to be carried out by proteins (Coppins et al. 2007).

Finally, small regulatory RNAs, known as microRNAs, interact with a complex set of cellular machinery to regulate targeted mRNAs in eukaryotes. Jacob and Monod (1961) proposed that untranslated RNAs might regulate gene expression in the *lac* operon. This idea was discredited with the discovery of protein transcription factors, and was largely forgotten until the discovery of microRNAs (Lau et al. 2001; Lee and Ambros 2001). These represent an entire class of genes producing small (21 bases) untranslated RNAs (Fire et al. 1998) in worms, flies, vertebrates, and plants (Bartel 2004). The microRNAs are transcribed from larger RNA genes and cleaved to their active form, in which they bind to target mRNAs either precisely (primary mode of action in plants) or with a few base pairs of mismatch, targeting the mRNA either for silencing or degradation. The target sequences in the mRNA can be considered as *cis*-acting elements, whereas the microRNAs and the enzymes that process them, as well as the proteins that perform the degradation, constitute the *trans*-acting factors in this mechanism of regulation.

In summary, a lot has been learned about the various mechanisms that regulate the differential expression of genomes and their genes. In some cases, where the mechanisms are well-described and the relevant genes identified, the availability of whole genome sequences allows for a rapid assessment of the conservation and phylogenetic distribution of these genes. Nevertheless, how these mechanisms impact the evolution of RNA abundance and genomic expression remains largely uncharacterized.

Measuring Attributes of Genomic Expression with Experimental and Computational Tools

The conservation of the genetic code across most of the tree of life facilitates the identification of protein-coding genes from raw DNA sequences. In contrast, the precise identification of regulatory regions and mechanisms has remained a difficult and elusive task. Unlike protein-coding sequences, regulatory domains are not readily identified by standard landmarks such as start and stop sites, open reading frames, and the splice sites that delineate introns from exons. Regulatory regions also do not possess characteristic genome-wide particularities such as similar codon biases or parallel rates of divergence among codon sites. In addition, regulatory modules are irregularly localized across the genome and the regulatory "code" appears to be considerably more degenerate than the genetic code.

Despite these inherent difficulties, researchers have made substantial inroads into identifying the sequences that control gene expression in a temporal, spatial, and quantitative manner. The recent integration of empirical analyses of gene expression and the high-throughput computational analysis of transcriptional regulation on a genome-wide scale is beginning to reveal how genomic regulation affects the transition from genotype to phenotype.

Typically, regulatory elements and motifs are found using two approaches, both of which are ideally combined with experimental validation of the sequences identified. First, homologous noncoding regions from different species are compared, and regions with unexpected conservation are targeted as being candidates for regulatory activity. Second, sequences from genes sharing a particular attribute (e.g., coregulation, similar rate of mRNA decay, and so forth) are compared in order to identify sequence features associated with that particular attribute. The most often used attribute is coregulation across a set of environmental treatments. In the next sections we outline major experimental and computational techniques used in studies of genomic regulation.

Experimental approaches

One of the most common experimental approaches used to identify and verify candidate regulatory regions is to assay for altered regulatory activity after mutation, a procedure generally known as "promoter bashing." Typically, putative regulatory regions are cloned, mutated, and then checked for differences in gene expression by reporter assays either in vivo or in vitro (Stanojevic et al. 1991).

A technique that has become central to delineating the precise region of transcription factor binding is DNAse footprinting (Brenowitz et al. 1986).

In DNAse footprinting, a binding site sequence is bound to its transcription factor in vitro in order to protect it from partial DNAse cleavage. Partial sequences of the protected fragments are then determined on a sequencing gel and the precise location of binding identified. Electrophoretic mobility shift assays (EMSA) can also be employed to identify transcription factor binding sites. EMSA works on the principle that bound DNA migrates at a different rate than unbound DNA. After two decades of footprinting experiments, there are now species-specific databases, such as the Drosophila DNase I Footprint Database (Bergmann et al. 2004), and more general databases, such as ORegAnno (Montgomery et al. 2006), that serve as important curated repositories for information on transcription factor binding sites.

The principles of transcriptional control have been elucidated by detailed studies on individual genes. However, the global architecture of the regulatory network only began to be understood with the advent of microarray-based methods (van Steensel and Henikoff 2003). These techniques permit a genome-wide mapping of protein–DNA interactions, chromatin packaging, and epigenetic modifications, such as DNA methylation and histone modifications. We outline these technologies and their main contributions to our understanding of regulatory networks.

GENE EXPRESSION LEVELS   Measuring gene expression level is the most fundamental requirement for studying the evolution of genomic expression. Several techniques are available for this purpose, widely varying in terms of cost, throughput, time investment, and practicality. Most recent techniques make use of array technologies in which the abundance of a given message is assessed by hybridizing a labeled cDNA sample to spotted microarrays.

DNA microarrays can vary extensively in the length of the DNA sequence in each spot (from cDNA clones, to single-exon PCR products, to medium-sized oligonucleotides of about 60–70 nucleotides, to very short oligonucleotides of less than 35 nucleotides), as well as in terms of genomic coverage. Tiling arrays, for instance, can cover both protein-coding as well as non-protein-coding sequences and provide a high-resolution sliding window view of expressed sequences in a particular genomic region or—at lower resolution—over the entire genome (Mockler et al. 2005). Use of tiling arrays has uncovered a large number of noncoding expressed sequences, many of which are transcribed from intergenic sequences. Figure 5.4 summarizes the important steps in the microarray analysis of gene expression levels. A limitation of oligonucleotide arrays is that the intensity of the signal may be overly sensitive to sequence mismatches. This will cause difficulties if there is genetic variation between the samples being contrasted, an obvious issue in evolutionary comparisons.

Two sequencing-based methods are still used to study genomic expression, but are being superseded by array technologies, even for nonmodel organisms. Expressed sequence tags (ESTs) were often used in early studies of genome-wide gene expression. ESTs are sequences obtained by random sequencing of clones from cDNA pools and were mostly produced in paral-

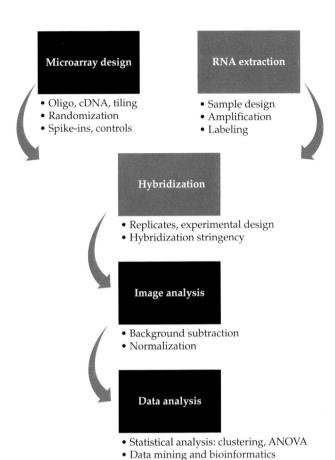

Figure 5.4 Microarray experimental design. The main steps of a typical experiment are highlighted. Specific procedural operations are also listed. Steps performed away from the lab bench are labeled in red. Specific control DNA might be spotted in the array if spike-ins of foreign RNA of known concentration is to be used for data normalization.

lel with genomic projects in model organisms (Hatey et al. 1998). Because cDNA pools are often normalized by subtractive hybridization, such that low-copy transcripts are enriched and highly abundant transcripts are removed, they should generally be avoided as a guide to mRNA transcript abundance. Serial analysis of gene expression (SAGE) is yet another technique for estimating levels of gene expression, which makes use of the fact that a sequence of about 15 nucleotides already contains enough information to identify (tag) the genomic sequence in which it is embedded. In this technique, a large number of DNA molecules, each consisting of a series of short expressed tags linked together in a single chain, are sequenced (Yamamoto et al. 2001). Each tag is then assigned to its gene of origin, and the number of times a tag for a given gene is found is used as a measure of the gene's expression level.

PROTEIN–DNA INTERACTIONS   The use of DNA microarrays is not limited to the measurement of RNA abundance. They can also be used to explore other aspects of gene expression that have to do with the state of the gene being expressed (Figure 5.5). These techniques have been instrumental in

Protein-DNA interactions

(A) ChIP-on-chip

Cross-link proteins/DNA + shearing

Immunoprecipitation

Reverse cross-link

DNA labeling

Hybridization

(B) DamID

DNA extraction

*Dpn*1 digestion

Purification/amplification of adenine-methylated DNA

DNA labeling

Hybridization

(C) PBM

Transcription factor | Tag

Transformation

E. coli

Expression purification

Tag

Hybridization

Tag labeling

DNA methylation

(D) MSO

DNA extraction

—TA*m*CTTCAG—
—TA*m*CTT*m*CAG—
—TACTTCAG—

Sodium bisulfite treatment

—TA*m*CTTUAG—
—TA*m*CTT*m*CAG—
—TAUTTUAG—

PCR

—TACTTTAG—
—TACTTCAG—
—TATTTTAG—

Hybridization

—ATGAAATC—
—TACTTTAG—

—ATGAAGTC—
—TACTTCAG—

—ATAAAATC—
—TATTTTAG—

(E) MSRE

DNA extraction, random shearing

Undigested   Digested with *Mcr*BC

Large fragment purification

DNA labeling

Hybridization

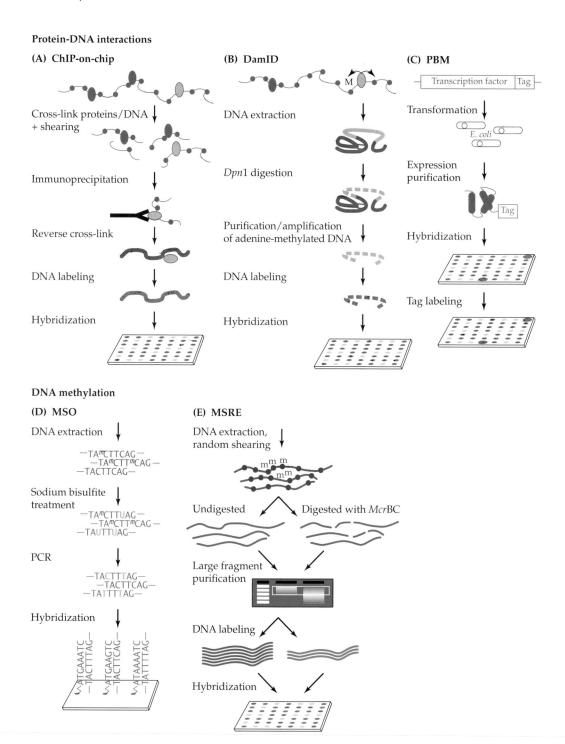

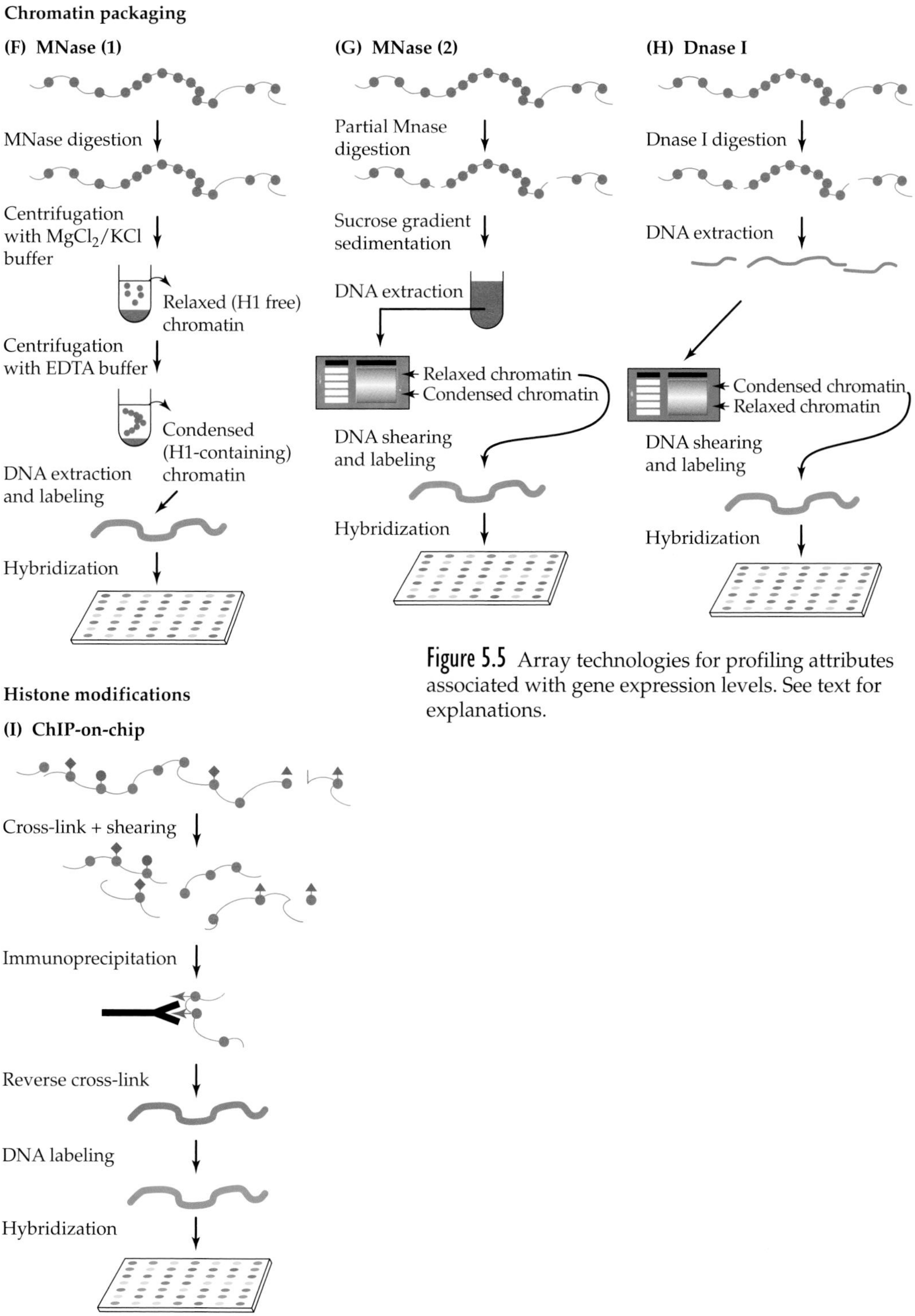

Chromatin packaging

(F) MNase (1)

MNase digestion

Centrifugation
with MgCl$_2$/KCl
buffer

Relaxed (H1 free)
chromatin

Centrifugation
with EDTA buffer

Condensed
(H1-containing)
chromatin

DNA extraction
and labeling

Hybridization

(G) MNase (2)

Partial Mnase
digestion

Sucrose gradient
sedimentation

DNA extraction

Relaxed chromatin
Condensed chromatin

DNA shearing
and labeling

Hybridization

(H) Dnase I

Dnase I digestion

DNA extraction

Condensed chromatin
Relaxed chromatin

DNA shearing
and labeling

Hybridization

Histone modifications

(I) ChIP-on-chip

Cross-link + shearing

Immunoprecipitation

Reverse cross-link

DNA labeling

Hybridization

Figure 5.5 Array technologies for profiling attributes associated with gene expression levels. See text for explanations.

deciphering the genomic architecture of gene expression, and promise much for future use in evolutionary biology. ChIP-on-chip is the more commonly used method to analyze protein–DNA interactions. The technique combines chromatin immunoprecipitation (ChIP) and microarray analysis (Weinmann and Farnham 2002). Cells are treated with a chemical agent (typically formaldehyde) that cross-links the protein complexes in situ to DNA. The chromatin is then fragmented and immunoprecipitated using specific antibodies that recognize the protein of interest. To identify the DNA sequence of the binding site, the cross-link is reversed, and the DNA fragments are labeled with fluorescent dye and hybridized to microrrays. Two other, complementary approaches, DamID (van Steensel and Henikoff 2000) and PBM (protein-binding microarray; Mukherjee et al. 2004), employ direct in situ labeling (methylation) of the bound DNA and in vitro identification of transcription factor binding sites, respectively.

With these approaches, basic questions about the regulatory network can be answered. For instance, how many genes are under control of one particular transcriptional factor? In *S. cerevisiae*, systematic studies of 106 transcription factors showed that, on average, about 40 genes are targeted by any given factor, the upper limit being 180 genes (Lee et al. 2002). In higher eukaryotes, it seems that some transcription factors also interact with even larger sets of promoters (Orian et al. 2003). A complementary question is: how many transcriptional factors link to a particular gene? More than one third of the genes in *S. cerevisiae* are bound by two and more factors (Lee et al. 2002), and some genes can have more than 12 transcription factors bound. These figures are expected to be underestimates because they depend on an arbitrarily chosen, conservative statistical threshold. In addition, almost all studies have focused on the upstream regions of genes, but downstream regions and introns also play an important role in the regulation of gene expression (e.g., Martone et al. 2003). These techniques have not yet been used to ask how transcription factor interaction patterns change between related species.

CHROMATIN PACKAGING   Microarray technologies have also been used to investigate chromatin states along chromosomal regions. One set of methods takes advantage of the resistance of condensed chromatin to nuclease digestion, followed by separation of different-sized chromatin fragments by sedimentation in a sucrose gradient. The relaxed chromatin in each fraction is extracted after fractionating in an agarose gel (Gilbert et al. 2004). The DNA fragments can then be labeled and hybridized to microarrays. A second approach uses DNase I and separation in agarose gels to isolate condensed or relaxed chromatin (Sabo et al. 2006). Such studies have revealed that relaxed chromatin is tightly correlated with high gene density in the chromosomes (e.g., Gilbert et al. 2004). Perhaps more surprisingly, no correlation was found between gene expression level and the distribution of relaxed chromatin. Genes may be active even in condensed chromatin regions, and inactive in relaxed regions (Gilbert and Bickmore 2006).

DNA METHYLATION   Another DNA modification that affects a gene's transcriptional state is DNA methylation. Several techniques have been developed to map the distribution of 5-methylcytosine (^{m5}C) in eukaryotic genomes. In the MSO (methylation-specific oligonucleotide microarray) technique, genomic DNA is treated with sodium bisulfite, which converts unmethylated cytosine, but not methylated cytosine into uracil (Adorján et al. 2002). During subsequent PCR, the DNA polymerase reads uracil as thymine and cytosine as guanine. To discriminate methylated and unmethylated cytosine at specific nucleotide positions in the original DNA, a specially designed oligonucleotide microarray is used: it contains a set of oligonucleotides with different combinations of guanine (to detect unmethylated cytosine) or adenine (to detect methylated cytosine) substituted at the cytosine positions (Adorján et al. 2002; Gitan et al. 2002). Another method uses methylation-sensitive restriction enzymes (MSRE). For instance, the enzyme *Mcr*BC cuts methylated, but not unmethylated DNA. After shearing, *Mcr*BC-treated genomic DNA samples are depleted in the high-molecular-weight fraction if the original DNA was methylated. Microarray hybridizations can then identify the methylated fragments. Many variations of this approach have been reported (e.g., Lippman et al. 2005).

Computational approaches

With the recent availability of genome sequences and transcription profiles, along with advances in the field of computational biology, researchers have begun to successfully investigate regulatory regions at structural and functional levels. The power to detect, describe, and model conserved motifs both within and between species has increased substantially. Researchers are no longer restricted to analyzing a small subset of sequences from a particular gene (usually promoter regions), but can now apply motif-finding principles to large regions around many genes (Nardone et al. 2004). The combination of computational and empirical tools has resulted in powerful approaches to the discovery of regulatory regions across genomes. There are now a large number of online resources for the analysis of regulatory sequences, including searchable databases (Table 5.1) and computational biology tools (Table 5.2).

While finding promoters in a genome is assisted somewhat by knowing the positions of genes, promoter discovery from raw nucleotide sequences is, in many ways, a much more difficult task than finding protein-coding sequences. Promoters comprise a large and diverse set of sequences and show no clear defining signature. Also, because many mammalian genes have large noncoding 5′ exons, promoters are located at variable distances from the gene that they regulate. As a result, promoter prediction algorithms must strike a balance between finding real regions of interest and falling victim to false positives. Fortunately, empirical studies of gene regulatory elements have generated a large and diverse template for computational biologists to construct genome-scanning algorithms (e.g., Fickett and Hatzigeorgiou 1997).

Table 5.1   Online regulatory resources: searchable databases

Database	Description	Web address
Promoters[a]		
DoOP	Database of orthologous clusters of promoters	http://doop.abc.hu/
EPD	Eukaryote POLII promoter database	http://www.epd.isb-sib.ch/
Worm DB	*C. elegans* promoter database	http://rulai.cshl.edu/cgi-bin/CEPDB/home.cgi
Mammal DB	Mammalian promoter databases	http://rulai.cshl.edu/CSHLmpd2/
SCPD	Promoter database for *S. cerevesiae*	http://rulai.cshl.org/SCPD/index.html
PlantProm DB	Plant promoter database	http://mendel.cs.rhul.ac.uk/mendel.php?topic=plantprom
Motifs[b]		
Transterm	Translational signal database (mRNA motifs)	http://guinevere.otago.ac.nz/transterm.html
PLACE	Plant *cis*-acting regulatory DNA elements database	http://www.dna.affrc.go.jp/htdocs/PLACE/
Transcription Factors[c]		
TRANSFAC	TF database	http://www.gene-regulation.com/pub/databases.html#transfac
ooTFD	Object-oriented TF database	http://www.ifti.org/ootfd/
ProteinLounge TFdb	TF database	http://www.proteinlounge.com/trans_home.asp
MIRAGE	Resource for the analysis of gene expression	http://www.ifti.org/
PRODORIC	Prokaryote database of gene regulation	http://prodoric.tu-bs.de/
TFdb	RIKEN mouse TF database	http://genome.gsc.riken.jp/TFdb/
AGRIS	Arabidopsis gene regulatory information server	http://arabidopsis.med.ohio-state.edu/

TRANSCRIPTION FACTORS BINDING SITES AND SEQUENCE MOTIF DISCOVERY
The experimental elucidation of binding sites for individual transcription factors has provided computational biologists with an indispensable tool. From a set of known aligned binding sites, positional matrices (sometimes called profiles or position-specific scoring matrices) of base pair frequencies can be generated. Different algorithms use various implementations of this signal-based data, including position weight matrix (PROMOTER2.0) and neural net (ProScan). Motifs can be short and contiguous or bipartite and long. The latter includes, for instance, palindromic sequences separated by a spacer element that is usually variable in length. Regulatory sequences can be readily visualized using a sequence logo format, which transforms matrix data into visual information.

Table 5.1 *Continued*

Database	Description	Web address
Transcription Factors[c] *(continued)*		
RARTF	RIKEN Arabidopsis TF database	http://rarge.gsc.riken.go.jp/rartf/
RiceTFDB	Rice TF database	http://ricetfdb.bio.uni-potsdam.de/
DBTBS	Transcriptional regulation database for *B. subtilis*	http://dbtbs.hgc.jp/
RegulonDB	*E. coli* K-12 database for transcriptional regulation	http://www.cifn.unam.mx/ Computational_Genomics/regulondb/
Ecoli TFDB	*E. coli* TF database	http://bayesweb.wadsworth.org/binding_sites/
Transcription Factor Binding Sites[d]		
FlyReg	Drosophila DNase I footprint database	http://www.flyreg.org/
ORegAnno	The open regulatory annotation	http://www.bcgsc.ca:8080/oregano/Index.jsp
JASPAR	TF binding profile database	http://jaspar.cgb.ki.se/cgi-bin/jaspar_db.pl
MAPPER	Multigenome analysis of positions and patterns	http://bio.chip.org/mapper
DNASTAR	TF binding site database	http://www.dnastar.com/web/r50.php
TFSEARCH	Search TF binding sites	http://www.cbrc.jp/research/db/TFSEARCH.html
TESS	Predicting transcription binding sites	http://www.cbil.upenn.edu/tess/

[a]These publicly available databases contain a curated list of promoter regions from various species. Data can be downloaded in bulk or individually visualized by searchable IDs such as GenBank accession numbers or gene names.

[b]Motifs of various kinds are available from these interactive databases. Motifs can be searched against specific sequences and genomes.

[c]A useful set of curated databases of known transcription factors (TFs) and their protein domains. TFs may be downloaded in bulk. Some sites offer in-house BLAST portals.

[d]These databases, some species-specific, are generated from literature reports of footprinting experiments. Transcription factors, their binding sites in the genome, and sometimes also their binding affinities, are reported. Depending on the website, phylogenetic conservation of the binding sites may also be reported.

In contrast to these signal-based methods, another way to find sequence motifs is based purely on the sequence content. Characteristic patterns of conserved sequences may be found among coregulated genes or among orthologous sequences of different species, for example, overrepresentation (relative to random noncoding sequence) of putative sequence motifs. The human genome contains approximately 1850 distinct transcription factors and the number of potential combinations of any number of these acting upon a particular gene is enormous.

COMPARATIVE GENOMICS AND PHYLOGENETIC FOOTPRINTING   A powerful and popular approach to decoding regulatory sequences is comparative genomics. By finding sequences that are conserved across species, one can

Table 5.2 Online regulatory resources: search tools

Database	Description	Web address
Promoter Prediction[a]		
McPromoter	The Markov Chain Promoter Prediction Server	http://genes.mit.edu/McPromoter.html
NNPP	Promoter prediction by neural network	http://www.fruitfly.org/seq_tools/promoter.html
TRES	Comparative promoter analysis	http://bioportal.bic.nus.edu.sg/tres/
PromoterWise	Compares 2 DNA sequences, ideal for promoters	http://www.ebi.ac.uk/Wise2/promoterwise.html
PromoSer	Batch retrieval of proximal promoters	http://biowulf.bu.edu/zlab/PromoSer/
Motif Searching[b]		
MOTIF Search	Search motifs	http://motif.genome.jp/
EZRetrieve	Sequence retrieval tool	http://siriusb.umdnj.edu:18080/EZRetrieve/
Possum	Detect *cis*-elements in DNA sequences	http://zlab.bu.edu/~mfrith/possum/
CorePromoter	Core-Promoter Prediction Program	http://sciclio.cshl.org/genefinder/CPROMOTER/
Gene Express	Analysis of genomic regulatory sequences	http://wwwmgs.bionet.nsc.ru/systems/GeneExpress/
Phylogenetic Footprinting[c]		
Phylofoot	Portal to phylogenetic footprinting	http://www.phylofoot.org/
UCSC	UCSC Genome Browser	http://genome.ucsc.edu/cgi-bin/hgGateway
TraFaC	Finds conserved *cis*-elements across species	http://trafac.cchmc.org/trafac/index.jsp
PipMaker	Aligns similar regions of sequence	http://pipmaker.bx.psu.edu/pipmaker/
VISTA	Suite of programs that aligns genomic sequences	http://genome.lbl.gov/vista/index.shtml
LAGAN	Comparative genomic alignment programs	http://lagan.stanford.edu/lagan_web/index.shtml
FootPrinter	Phylogenetic footprinting of orthologous sequences	http://bio.cs.washington.edu/software.html
Bayesaligner	Phylogenetic footprint using a Bayesian approach	http://bayesweb.wadsworth.org/cgi-bin/bayes_align12.pl

quickly infer whether they are functionally important without using costly molecular or biochemical procedures. Phylogenetic footprinting aims to find functionally important regulatory regions by identifying conserved orthologous sequences (Gumucio et al. 1993; Hardison et al. 1997). This approach has been very successful with the advent of complete genome sequences from model genetic organisms. Deep-rooted phylogenetic taxa can be used to find invariant regions indicative of constrained function, and closely related sister taxa can be used to find regions of sequence conservation, as well as genus-specific regulatory units, among species with a more shallow ancestry (i.e., phylogenetic shadowing; Boffelli et al. 2003).

Table 5.2   *Continued*

Database	Description	Web address
TF and TF Binding Sites[d]		
ConSite	TF binding sites via aligned genomic sequence	http://mordor.cgb.ki.se/cgi-bin/CONSITE/consite
TFSEARCH	Transcription factor search	http://www.cbrc.jp/research/db/TFSEARCH.html
MSCAN	Find functional clusters of TF binding sites	http://mscan.cgb.ki.se/cgi-bin/MSCAN
Weeder Web	TF binding sites in sequences via coregulated genes	http://159.149.109.16
SITECON	Conserved physicochemistry in TFBS alignments	http://wwwmgs.bionet.nsc.ru/mgs/programs/sitecon/
POBO	TF binding site verification with bootstrapping	http://ekhidna.biocenter.helsinki.fi:9801/pobo/
DTFAM	Explores TF associations through text-mining	http://research.i2r.a-star.edu.sg/DRAGON/TFAM/
Fly Enhancer	Finds clusters of binding sites in Drosophila	http://flyenhancer.org/Main
AliBaba	Prediction of transcription factor binding sites	http://www.alibaba2.com/

[a]A set of promoter prediction software publicly available online. Your sequence of interest can be uploaded onto each webserver in order to identify promoter regions using a wide variety of approaches.

[b]Motifs in uploaded sequences can be detected using a multitude of methods from these websites. These tools represent just a handful of available online resources to identify motifs.

[c]These online alignment sites allow one to find conserved sequences in regulatory regions. Some of these websites already contain precomputed alignments of regulatory regions from sequenced genomes.

[d]These sites contain tools that allow you to search your sequence of interest for transcription factors and their binding sites.

Since the highlighting of conserved regions has become such an important tool, the major genomic databanks have started to provide genome browsers that allow one to easily visualize conserved regions. NCBI's Map Viewer (National Center for Biotechnology Information), EMBL-EBI (European Molecular Bioinformatics Laboratory), and the UCSC Genome Browser (University of California at Santa Cruz) possess excellent graphical interfaces for users to search for conserved orthologous regions. In addition, VISTA and PipMaker are popular phylogenetic footprinting tools.

The Evolution of Genomic Expression: What Do We Know?

Evolutionary genomics is beginning to address the molecular evolution of regulatory sequences and the phenotypic evolution of mRNA abundance. Ultimately, this approach may provide a comprehensive understanding of the magnitudes and patterns of evolutionary variation in regulatory sequences and relevant phenotypes. Transcription is the first step in the mapping of genetic variation to higher-level phenotypes, and, therefore, diversity in gene expression levels is one of the most direct phenotypic outcomes of regulatory variation. This diversity is now being widely documented and interesting patterns are being discovered. Understanding how genomic regulation translates into variation in gene expression levels is thus the first step towards understanding variation in more complex organismic features. Because of this, gene expression levels are likely to become a model phenotype for testing current methods and assumptions in our understanding of the dynamics of polymorphism and divergence in natural populations, including ecologically relevant and disease-related variation. However, virtually nothing is known about evolutionary variation in attributes that affect variation in gene expression (e.g., rates of mRNA degradation, levels of methylation, nucleosome positioning, and so forth). In this section we review our current understanding of the patterns and processes affecting the evolution of genomic expression.

Genomic expression and morphological evolution

We know a lot more about variation in the coding sequence of genes than about variation in regulatory regions or in gene expression patterns. Yet variation in gene expression is likely to account for a large fraction of the phenotypic diversity observed within and between species. Abundant examples of regulatory variation contributing to phenotypic diversity at the morphological, behavioral, and physiological levels illustrate the role of regulatory evolution in phenotypic and adaptive diversification.

An early and now classical demonstration of the relevance of regulatory variation in evolution was provided by Cherry, Case, and Wilson in 1978. These authors took metrics of shape typically used by systematists interested in distinguishing between species of frogs to measure morphological differences between humans and chimpanzees. The striking result was that, morphologically, human and chimps are much more different from each other than are species of frogs belonging to different suborders. This result stands out because, while suborders of frogs show considerable differences in protein-coding DNA sequences, humans and chimps are remarkably similar to each other at the DNA level. This suggests a substantial role for regulatory evolution in the human–chimp divergence.

In another example, the pattern of hairs on the first instar larva varies among closely related species of the *Drosophila melanogaster* group, and the evolution of *cis*-regulatory elements of the *ovo/shaven-baby* gene is responsible for hair patterning in *Drosophila sechellia*, distinguishing it from its closest relative (Sucena et al. 2003). Similarly, the cuticular hydrocarbon pheromones involved in mating preference in *D. melanogaster* display geographic variation in the 5,9-heptacosadiene/7,11-heptacosadiene ratio. This polymorphism is caused by a deletion in the promoter region of a desaturase gene, changing its expression pattern and causing knock-on effects on pheromone production (Takahashi et al. 2001). Closely related species of cichlid fishes have different visual spectral sensitivities, which probably has important consequences for the foraging behavior of these fishes and their mate choice (based on male coloration). Differences in visual sensitivity are often achieved by shifts in chromophore usage or in opsin coding sequences, but in this case they are caused by changes in opsin gene expression (Carleton and Kocher 2001). Finally, the colors and patterns of eyespots on butterfly wings are yet another system in which a connection between variation in the expression of specific genes and higher-level evolutionary changes has been established (Brunetti et al. 2001; Beldade and Brakefield 2002).

Molecular evolution of regulatory sequences

The study of protein-coding sequences has been boosted by the explosion of comparative data coupled to new statistical methods for analyzing and interpreting coding sequences. Current methods have incorporated a number of factors regarding rates of coding sequence variation (e.g., transition/transversion ratios, position heterogeneity), and have allowed several genome-wide analyses of the selective forces acting on protein sequences (e.g., Nielsen et al. 2005a). Most of these analyses were based on models and statistical tests on the ratio of nonsynonymous (dN) to synonymous (dS) nucleotide substitutions. It is noteworthy that genome-wide analyses have also challenged the very assumptions that underlie using dS as a reliable proxy for the neutral mutation rate (Wyckoff et al. 2005), which suggests that classical interpretations may need to be reevaluated.

Historically, the molecular evolutionary analysis of regulatory sequences has largely remained outside the mainstream of such analyses of coding sequences. This is not because the relevance of regulatory evolution has been underappreciated, but is rather due to the major challenges inherent to the evolutionary analysis of noncoding DNA. First, despite the large number of mechanisms of gene regulation already described, qualitatively novel mechanisms are still being discovered. These elusive mechanisms relate to mRNA genes, microRNAs, S/MARs, and nuclear organization, to cite a few. Similarly, several ultra-conserved noncoding DNA sequences have been identified (e.g., Bejerano et al. 2004) whose functional role is unknown. Hence, developments in the last ten years have uncovered a variety of regulatory mechanisms whose evolution has yet to be investigated, let alone modeled and fully incorporated into mainstream studies of molecular evolution. It is clear, nevertheless, that a huge variety of elements and regulatory phenomena influence genomic expression, and a truly comprehensive theory for the evolution of regulatory sequences should include all these mechanisms.

Second, little is known about the evolution even of noncoding DNA with apparently obvious relevance to genome expression. For example, there is now a vast amount of data regarding promoter function and its regulation by other elements, such as enhancers. This is particularly well-illustrated in the case of gene regulation in the human and mouse immune systems. In spite of this wealth of information, a comprehensive understanding of gene regulation from the perspective of promoter activity is still missing. As a consequence, although promoters have long been recognized as the site for transcriptional regulation, few evolutionary analyses have been carried out.

In spite of these challenges, a few recent studies have attempted to define structural features of regulatory sequences (Dermitzakis et al. 2003; Chin et al. 2005), as well as to develop metrics for measuring regulatory divergence of these sequences (Castillo-Davis et al. 2004; Chin et al. 2005). These studies are also complemented by analysis of patterns of substitution in noncoding DNA within and between species (Andolfatto 2005; Borneman et al. 2007). All in all, these studies provide new venues to explore the evolution of regulatory sequences, and suggest promising directions for future research.

Dermitzakis and coworkers (2003), for instance, found that conserved noncoding sequences are often under stronger selective constraint than proteins and noncoding RNAs. Furthermore, the patterns of evolutionary variation in conserved noncoding sequences are distinguishable from those observed for protein-coding sequences. Substitutions in noncoding sequences were more clustered along the sequence compared to those in protein-coding sequences (Dermitzakis et al. 2003), presumably because purifying selection pressure is unevenly distributed along regulatory sequences. Moreover, noncoding sequences showed more symmetric rates of divergence (i.e., A → T and T → A, or C → G and G → C) than coding sequences. Recent work by Moses and colleagues (2003) attempted to characterize the pattern of evolution within transcription factor binding sites in yeast, which were shown to evolve more slowly than background sequences. In addition, they found substantial position-specific variation in rates of sequence evolution within transcription factor binding sites. While some positions within transcription binding motifs were highly conserved, rates of evolution in less important sites could not be distinguished from background rates. Furthermore, Moses and colleagues (2003) found a strong correlation between positional rate variation in a single genome and that observed between genomes. Work by Chin and coauthors (2005) is also illustrative. These authors used a hidden Markov model (HMM) to break down promoters into selectively neutral regions and evolutionarily constrained regions under purifying selection. The latter contained an overabundance of regulatory motifs. Chin and coauthors (2005) estimated that about 30 percent of the promoter sites in yeast are evolving under purifying selection, whereas the remaining 70 percent are accumulating mutations at the neutral rate. Another interesting study by Keightley and coworkers (2005b) compared the extent to which sequence conservation differs between two pairs of species. They found that the conservation of noncoding sequences upstream (5′) of the coding region was substantially greater in mouse–rat comparisons than in human–chimpanzee comparisons. Based on this observation, the authors argued that purifying selection on regulatory variation has been less efficient in primates than in rodents.

Stabilizing selection, positive selection, and neutrality of gene expression levels

It has long been realized that stabilizing selection (i.e., purifying selection) is a pervasive force in the evolution of higher-order morphological phenotypes, as well as protein-coding sequences. Gene expression levels are no exception to this pattern, and increasing evidence suggests a fundamental role for stabilizing selection in restricting evolutionary variation in transcript levels (Denver et al. 2005; Jordan et al. 2005; Lemos et al. 2005b; Rifkin et al. 2005; Landry et al. 2007).

The high conservation of gene expression levels across species is particularly striking in view of the ample supply of mutations expected to influence gene expression across evolutionary timescales, as measured by the experimental accumulation of mutations and their effects on gene expression levels (Figure 5.6). In particular, Denver and colleagues (2005) and Rifkin and colleagues (2005) used mutation accumulation lines of worms and flies, respectively, to experimentally estimate the neutral mutation rate for gene expression levels at about 10^{-5}. Although this rate is about two orders of magnitude below the typical value found for a number of morphological and enzyme activity traits (Lynch 1988), mutation accumulation lines still have more dispersion in mRNA abundances than is observed across genotypes segregating in natural populations (Denver et al. 2005). Accordingly, it has been suggested that gene expression divergence between yeast gene dupli-

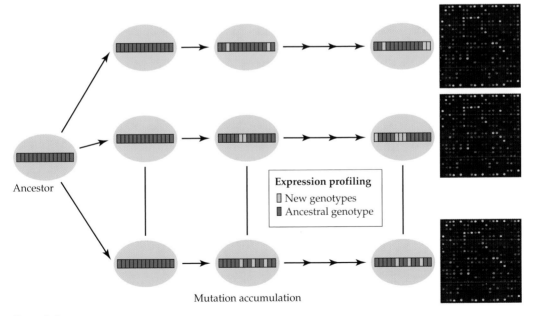

Figure 5.6 Mutation accumulation design for studying the neutral rate of regulatory divergence.

cates (Oakley et al. 2005) or between orthologous genes in different species (Lemos et al. 2005b) does not follow a phylogenetic model, in that closely related genes are not more likely than distantly related genes to show similar expression levels. Instead, the comparisons suggest that gene expression evolution proceeds so rapidly that the magnitude of divergence quickly saturates—due solely to the high mutation rates associated with mRNA abundances—even in the absence of diversifying or positive natural selection. Diversifying and positive selection would only further increase the rate of gene expression evolution, thereby leading to an even more rapid saturation of the evolutionary signal. Indeed, Ferea and colleagues (1999) and Toma and colleagues (2002) showed that large differences in gene expression can be accomplished in only a few generations of artificial selection.

These conclusions are in sharp contrast with a suggestion that gene expression variation may be unconstrained (Khaitovich et al. 2004). This suggestion was motivated by the finding that the divergence of translated mRNAs did not seem to be lower than that of transcribed pseudogenes, whose mRNA is not capable of producing a functional protein. However, results from comparisons between translated mRNAs and untranslated mRNAs derived from pseudogenes should generally be interpreted with caution. This is because untranslated mRNAs that happen to be expressed across timescales as long as that observed between species are unlikely to be nonfunctional.

An important question whose answer is not yet completely apparent regards the major forces that *produce* evolutionary differences in gene expression levels. Although the prevalence of stabilizing selection appears to be beyond doubt, it remains to be established whether gene expression differences between species or populations arise mainly through fixation by positive selection of distinct expression alleles in different environmental contexts or, alternatively, through fixation by random genetic drift of selectively equivalent expression states. The neutral theory of molecular evolution developed by Kimura (Kimura 1983) and others promoted the view that most segregating variation within species, as well as most fixed differences in protein sequences between species, arise from selectively equivalent alleles whose small differences have negligible effects on organismal fitness. Consequently, most differences observed between species and populations would result from random fixation of equivalent or nearly equivalent alleles. It should be stressed that the prevalence of stabilizing selection on transcription levels does not imply that a gene's mRNA abundance is at its optimum or that it has a negligibly small environmental or mutational variance.

Furthermore, we must note a few shortcomings of many analyses of gene expression divergence and polymorphism. First, most statistical methods for identifying gene expression differences test each gene independently of all others, one gene at a time. However, genes are often coordinately regulated as gene expression modules, and these modules, rather than the genes themselves, may often be the relevant targets for evolutionary analysis.

In fact, evolutionary biologists have long debated the proper multivariate description of biological complexity, a problem by no means restricted to gene expression level (Lande and Arnold 1983). It is therefore important to consider whether any gene-by-gene metric of gene expression variability and divergence is an adequate descriptor of the biological complexity. This is because gene expression profiles might be more meaningfully examined as a single complex character (e.g., a network) instead of a simple collection of single-gene similarities and differences. The quantification of biological divergence and polymorphism on a truly systemic scale remains elusive, and the development of biologically meaningful multivariate descriptors of gene expression states remains a challenge. Such descriptors would integrate across entire pathways, functional groups, and interrelated modules.

The normalization of mRNA abundance data across samples is another difficult area in the analysis of gene expression variation across evolutionary timescales. This is because normalization methods commonly assume that total mRNA abundance is constant across samples or, in other words, that only a small number of genes differ in expression levels between samples (Quackenbush 2002). This is a fundamental problem because evolutionary comparisons often involve a large number of differences in gene expression levels; therefore, the assumption of similar mean expression level across samples may not always hold. The potentially confounding effect of sequence divergence on estimates of mRNA abundance is another fundamental issue when analyzing and interpreting polymorphism and divergence data. Along these lines, Gilad and colleagues (2005) examined how normalization procedures interact with sequence divergence to produce biased estimates of gene expression levels. They found significant effects even between samples that are as similar as humans and chimpanzees.

Genomic attributes and rates of gene expression evolution

Understanding the constraints imposed upon evolutionary variation in gene expression levels is a major research goal of evolutionary genomics. This goal includes not only identification of the sources of constraints on expression levels, but also, equally challenging, reconstruction of the selective landscape underlying evolutionary variation in gene expression levels. Although direct causation is elusive and hard to establish, several factors have so far been shown to be associated with evolutionary variation in gene expression levels (Koonin and Wolf 2006). Furthermore, we note that, whatever the determinants of gene expression polymorphism and divergence might be, there seem to be remarkable commonalities in the patterns and relative magnitudes of evolutionary variation in gene expression and protein-coding sequences— so much so that a positive correlation between these two modes of evolution can be detected (Nuzhdin et al. 2004; Khaitovich et al. 2005; Lemos et al. 2005b; Liao and Zhang 2006).

Parisi and coworkers (2003) and Ranz and coworkers (2003) found extensive differences in whole-organism transcriptional profiles of male and female adult fruit flies. These authors showed that about half the genome is differentially expressed between the sexes and argued for the relevance of sex-dependent evolution of gene regulation. Moreover, by classifying genes as male- or female-biased (i.e., gene expression higher in males or females, respectively), Ranz and colleagues (2003) and Meiklejohn and colleagues (2003) showed that male-biased genes have higher levels of gene expression polymorphism and divergence than female-biased and unbiased genes.

The functional class or biological process of a gene product is another attribute relevant to evolutionary variation in gene expression levels. Expression variation in genes from functional classes closely related to transcriptional regulation (e.g., transcription factors) might be expected to influence gene expression more strongly than variation in genes from functional classes more distantly related to transcriptional regulation (e.g., metabolic enzymes). This prediction was verified in a number of studies (e.g., Rifkin et al. 2003; Lemos et al. 2005b), suggesting that the expression levels of transcription factors are indeed more tightly controlled than the expression levels of metabolic enzymes. Classically, two sets of genes have been shown to evolve rapidly at the level of the protein-coding sequence: immune system–related genes and male reproduction–related genes. As discussed above, genes that tend to be more expressed in males also tend to show higher levels of polymorphism and divergence in expression. Genes of the immune system also appear to show the same trend towards faster regulatory evolution. Genes of the major histocompatibility complex (MHC) have been known for a long time to be under balancing selection (reviewed in Bernatchez and Landry 2003), which acts to maintain high levels of polymorphism in the coding sequences of these genes. A recent study (Loisel et al. 2006) on the evolutionary history of *cis*-regulatory regions of a MHC gene (*DQA1*) in primates shows that balancing selection also acts on transcription factor binding sites to maintain functional nucleotide variation with consequences for gene regulation.

Physical attributes such as the number of protein–protein interactions and mRNA abundance may also be relevant for determining the magnitude of evolutionary variation in gene expression levels. For instance, proteins that interact might impose mutual stoichiometric constraints on the amount of variation permitted in their concentrations. This is because a change in the concentration of one protein might result in a stoichiometric cost in its interacting partners. Following this, it is expected that the concentration of proteins whose function depends on direct interaction with a large number of partners should be more evolutionarily constrained than proteins with fewer interacting partners. This prediction has been confirmed using evolutionary variation in gene expression levels as a proxy for evolutionary variation in protein concentration (Lemos et al. 2004). Absolute mRNA abundances may be another physical factor highly relevant to evolutionary variation in gene expression levels. It has been suggested that highly expressed genes might show higher levels of expression polymorphism and divergence (Lemos et al. 2005a). However, it remains unclear to what extent this observation depends on the particular metric used and the accuracy of gene expression assays across a wide range of absolute values (Lemos et al.

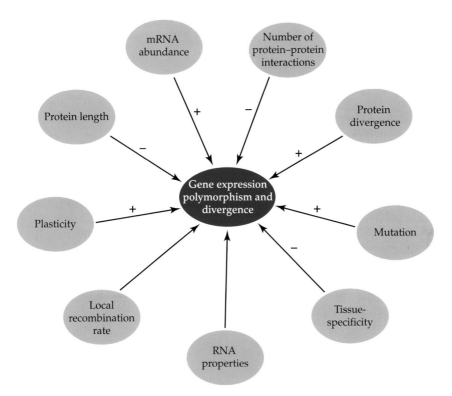

Figure 5.7 Genomic attributes associated with evolutionary variation in gene expression levels. Hypotheses for negative and positive associations are noted.

2005b). Finally, it can be predicted that genes not essential for organismal survival will be less constrained to vary in expression level. This prediction is supported by the observation that nonessential genes show greater genetic variation for gene expression than essential genes among natural isolates of wine yeasts (Landry et al. 2006). Finally, we note that the presence of a TATA-box motif in the promoter of genes has been positively associated with a gene's level of expression polymorphism and divergence, whereas TATA-less genes show decreased levels of evolutionary variation (Tirosh et al. 2006). Figure 5.7 shows some attributes associated with evolutionary variation in gene expression levels, as well as the sign of the effect.

Inheritance of gene expression levels: Regulatory variation in cis and trans

An understanding of the evolution of gene regulation and particularly of the evolutionary forces that control and direct it—mutation, genetic drift, and natural selection—requires the study of genetic variation for gene

Table 5.3 Some genomic studies reporting levels of polymorphism in gene expression[a]

Organism	Number of genotypes and / or individuals	Number of genes assayed
Insects		
Drosophila melanogaster	8 inbred strains	~5000
D. simulans	10 heterozygous strains	~8000
Fungi		
Saccharomyces cerevisiae	9 strains	~6000, whole genome
S. cerevisiae	4 strains	~6000, whole genome
Fish		
Fundulus heteroclitus/F. grandis	10/5 individuals within populations	~1000
Atlantic salmon	12 individuals	
Mammals		
Human	35 CEPH cell lines	~5000
Plants		
Arabidopsis thaliana	5 accessions	~8000
Worm		
C. elegans	5 natural isolates	~5500

[a]mRNA abundances are measured across genotypes raised in a controlled environment.

expression in natural populations. Large-scale gene expression profiling of organisms as diverse as yeast, fruit flies, and humans provides an unequivocal result: Heritable genetic variation for gene expression level is abundant in nature (Table 5.3).

Since gene expression levels represent multifactorial quantitative traits, they can be studied using the tools and theoretical models developed by quantitative genetics. A formal quantitative genetic study of mRNA abundance thus requires a good understanding of its genetic architecture. The genetic architecture of a trait refers to its characterization in terms of the direct effects of genes and environment, as well as the genetic and environmental interactions affecting the trait expression. The first aspect of the genetic architecture of transcriptional variation is the contribution of *cis*- and *trans*-acting genetic variation (Wray 2007). Several factors have contributed to the recent surge of interest in the distinction between *cis* and *trans* effects (e.g., Pastinen and Hudson 2004; Wittkopp et al. 2004; Landry et al. 2005). For instance, genetic variation in *cis*-acting elements is thought to be less likely to have pleiotropic effects than genetic variation in *trans*-acting molecules. In contrast, a single transcription factor may regulate the expression level of dozens of genes such that a mutation in it may affect many of its targets. Furthermore, *cis*-regulatory variation is also of particular interest because its effect

Number of genes showing significant variation	Reference	Tissue	Platform	Statistics
218–928 between pairs of strains	Meiklejohn et al. 2003	Whole flies, males	cDNA	Bagel
1136 ($P < 0.05$); 218 ($P < 0.001$)		Whole flies, males	Affymetrix	ANOVA
241 ($P < 0.01$)	Fay et al. 2004	Cells grown in rich medium	Oligo array	ANOVA
	Townsend et al. 2003a	Cells grown in rich medium	cDNA	Bagel
~161 ($P < 0.01$)	Oleksiak et al. 2002 Aubin-Horth et al. 2005	Heart	cDNA	ANOVA
	Cheung et al. 2003	Lymphoblastoid cells	cDNA	
1525 ($P < 0.01$)	Chen et al. 2005	Leaf tissue	Affymetrix	ANOVA
118 ($P < 0.01$)	Denver et al. 2005	Whole worms	cDNA	ANOVA

on gene expression is potentially easier to identify as individual regulatory mutations than *trans*-regulatory variation. This is because individual mutations in *cis* are clear QTL candidates, whereas *trans* effects often represent an aggregate of effects across multiple sites dispersed throughout the genome, and thus are much harder to identify (Yvert et al. 2003).

The first question one might ask about the architecture of gene expression variation is how much variation is found in *cis* and in *trans*? This has been assessed in vivo using traditional genetics and a variety of novel molecular biology tools. The first approach combines gene expression profiling with genetic markers in F2 progenies to map linkages (expression QTL or eQTL) of the transcription phenotypes. *Cis*-acting eQTLs are identified by binning the genome in small regions (physical or genetic distance) and locating each eQTL relative to the gene being regulated. If they both fall in the same bin, the eQTL is said to be *cis*-acting. Depending on the density of markers used and the number of meiosis events analyzed, the size of the bin may vary and thus limit the resolution of this approach. In the case of the study conducted by Morley and colleagues (2004), these bins were five megabases long. The effects of identified *cis*-regulatory variants can be confirmed by in vitro approaches such as transient transfection assays (e.g., Rockman and Wray 2002). This method has attracted much attention

1. Produce F2 progeny between two lines of interest

2. Establish the genotypes of the lines SNPs, AFLPs, SSR

3. Perform transcription profiling

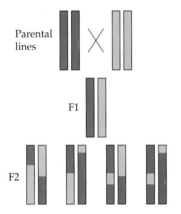

Parental lines

F1

F2

4. Map the expression profiles onto the linkage map (QTL mapping)

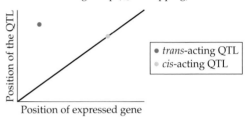

Figure 5.8 Mapping genetic determinants of gene expression variation can be achieved through genetical genomics, the steps of which are illustrated here.

recently and has given rise to a field named "genetical genomics" (de Koning and Haley 2005) (Figure 5.8).

Another approach relies on the fact that *cis*-acting variation is a property of an allele of a gene. In an individual heterozygous for an exonic SNP (single nucleotide polymorphism), *cis*-regulatory divergence can be estimated by measuring the relative concentration of mRNAs containing the two alternative nucleotides, usually using genomic DNA as a control. Since the two alleles and their *cis*-regulatory elements share the same pool of *trans*-acting factors, unequal abundance of transcripts of the two alleles would suggest the presence of genetic variation acting in *cis*. In cases where crosses are performed between two inbred lines or closely related species, the divergence in gene expression level between parental lines can be compared to the difference between alleles in the F1 generation. Any difference between the parental lines that is not assigned to *cis* divergence is then assigned to divergence in *trans*. This approach was used in a study of hybrids between *Drosophila melanogaster* and *D. simulans*, which showed that 28 of 29 genes studied had divergent *cis*-regulatory elements (Wittkopp et al. 2004).

The two approaches just described have been used to assess the relative contribution of *cis* and *trans* factors in a variety of species. Because these studies sampled different numbers of genotypes, used different numbers of markers, and used different molecular techniques, it is difficult to compare their results directly (e.g., de Koning and Haley 2005). However, the results lead to some general conclusions. First, *cis*-acting eQTLs typically represent a smaller proportion of the total (~30%) than *trans*-acting eQTLs. Second, eQTLs that have the strongest effects tend to be *cis*-acting. For instance, increasing the stringency for statistical significance of eQTL effects increases the proportion of *cis*-acting eQTLs observed (e.g., Schadt et al. 2003). Finally, there are clusters of eQTLs in *trans* that affect the expression of large numbers of genes—in other words, there are portions of chromosomes that affect a larger number of genes than expected by chance alone. For instance, eight such clusters were identified in a cross between two strains of yeast (Brem et al. 2002). In the human lymphoblastoid cell lines studied by Morley and colleagues (2004), two regions of five megabases each contained six or more eQTLs out of the 142 most significant ones.

Genotype-by-environment interactions, sex-biased genes, and epistasis

Another important aspect of the genetic architecture of quantitative traits is how alleles at different loci interact with each other. Given that gene expression regulation involves numerous molecular interactions, one might expect epistasis for fitness (i.e., the contribution of nonadditive gene interactions to fitness) to be an important factor in how selection acts on gene expression. Consider, for instance, the effect of a mutation in a *cis*-regulatory sequence. If the effect of this mutation depends on the genetic background (variation in *trans*, for example), we predict an epistatic interaction.

Mathematical models of gene expression regulation have predicted that epistasis will be an important factor contributing to gene expression variation (Gibson 1996, Landry et al. 2005). In their review of *cis*-regulatory variation in humans, Rockman and Wray (2002) identified many cases of *cis*-by-*trans* interactions, where the effect of a *cis* variant depends on the genetic background. They also identified *cis*-by-*cis* interactions, where the effect of a *cis* variant depends on other *cis*-acting variants. Many interactions fell into these two categories in recent large-scale studies of the genetic architecture of the yeast, eucalyptus, and Drosophila transcriptomes (Brem et al. 2005; Kirst et al. 2005; Landry et al. 2005). Finally, a survey of allelic expression in interspecific hybrids of *Drosophila* reveals that many *cis*-by-*trans* interactions have accumulated since the divergence between *Drosophila melanogaster* and *Drosophila simulans* (Landry et al. 2005). Several of the approximately 30 genes studied displayed a pattern consistent with *cis*-by-*trans* interaction, which means that the divergence in gene expression between the two species was smaller, or in the opposite direction, than the divergence between alleles measured in the hybrid background. All these studies have used crosses between closely related species to identify *cis*-by-*trans* interactions, and it remains to be shown that the same interactions are common contributors to epistatic genetic variation within species.

As mentioned above, genetic variation for gene expression is extensive in nature, providing abundant raw material for evolution. Since this genetic variation will be parsed by natural selection, any factor that influences this variation will affect the course of evolution. Nongenetic sources of variation in gene expression, such as the environment or development, may also contribute to modification of gene expression. Most importantly, these effects may interact with the genotypes to shape the amount of genetic variation observed. Surveys have revealed that genetic variation in gene expression can be dependent on the sex (genotype-by-sex interaction), the environment (genotype-by-environment interaction, or genetic variation for phenotypic plasticity), and developmental stage (age-by-genotype interaction). For instance, genetic variation among strains of *D. melanogaster* is dependent on the sex in which it is measured; some genes that display genetic variation in gene expression in males may not be variable in females (Jin et al. 2001). If an eQTL experiment were performed for those genes that show genotype-by-sex interaction, different numbers and locations of eQTLs would be identified in each sex. Similarly, Landry and colleagues (2006) showed that genetic variation in gene expression among strains of *Saccharomyces cerevisiae* depends on the environment in which it is measured and that substantial genotype-by-environment interactions are evident (Figure 5.9). Understanding how these interactions are maintained, mechanistically and

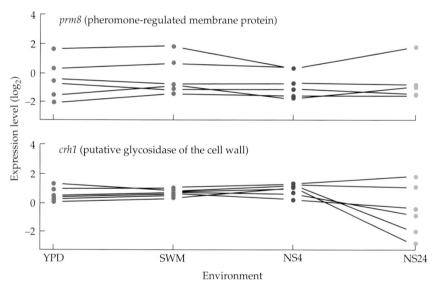

Figure 5.9 Plasticity and genotype-by-environment interaction in gene expression. The expression levels of two genes, *prm8* and *crh1*, were measured in six different yeast strains grown in four different environments. *prm8* (top graph) shows genetic variation for gene expression, but the six strains show the same gene expression differences across growth conditions and, therefore, no genotype-by-environment interaction. *crh1* (bottom graph) shows genotype-by-environment interaction, in that gene expression in the six strains responds differently to different environments (e.g., compare NS4 to NS24). YPD, SWM, NS4, and NS24 are different growth media. (Data from Landry et al. 2006.)

evolutionarily, will therefore be important in deciphering the forces acting on the regulation of gene expression.

Gene Regulatory Networks, Subnetworks, and Modules

Gene regulatory networks are yet another level of organization of gene expression and evolution. A network can be represented as a graph—a collection of nodes connected by edges, interacting as a system (Barabasi and Oltvai 2004). A subnetwork is a subset of the whole network. A module is a coherent subnetwork whose structure and function is largely independent of interactions with members of other subnetworks. These general concepts have been fruitfully applied at many levels of the biological hierarchy, from molecules to species. Usually, the nodes represent biological units (proteins, cells, organs, individuals, and so forth), and the edges represent interactions between them. For example, a trophic network depicts prey/predator relationships and the energy flow through a food web. The species are the nodes and the edges are "who eats whom" interactions. At the other end of the scale, a gene regulatory network describes the transcriptional and translational web and its regulation (by proteins, RNAs, or environmental signals) within a cell. In this case, the nodes are genes and the edges represent regulatory interactions between pairs of genes or a shared regulatory element.

Biological systems are complex and the information must often be simplified to gain useful insights. In the simplest case, networks are considered as Boolean objects, represented by uniform nodes and connected by undirected edges (for more complex models, see Proulx et al. 2005). Boolean nodes have only two discrete states, *on* or *off*, and the nodes interact through logical or Boolean functions. In this modeling framework, only the topology of the network is retained. Even with these simplifications, network behaviors are still extremely rich. A straightforward and well-established indicator of network topology is the distribution of the connectivity (the number of edges of a particular node). A second indicator is the clustering coefficient, which measures the degree of connection between nodes connected to the same specific node. Nodes densely connected to each other define clusters, or modules. In a gene regulatory network, the connectivity and the modularity provide direct insights about important concepts in evolutionary biology, like epistasis, canalization, and plasticity. Wagner (1996) modeled the evolution of transcriptional regulatory networks and concluded that more densely connected networks are more insensitive to disruption by mutations.

Much excitement has been generated by high-throughput experimental techniques such as genome-wide expression profiling and location analysis (ChIP-on-chip), because these techniques promise to allow for rapid reconstruction of regulatory networks. Large-scale gene perturbation experiments generate valuable information about the number of genes whose expression is affected by an environmental or a gene perturbation (mutation, overexpression, or knockout). Regrettably, such perturbation experiments cannot distinguish direct from indirect interactions. On the other hand, protein–DNA interaction data are produced at a slower rate, but directly identify the binding sites of transcriptional factors. Accordingly, Lee and coworkers (2002) performed a genome-wide analysis to determine the binding distribution of 106 known regulatory proteins along the yeast genome. The results suggest that about ten percent of these regulatory genes are autoregulated. This proportion seems substantially smaller than in *E. coli*, where most genes (52% to 74%) are autoregulated (Shen-Orr et al. 2002). Also, about 37 percent of the yeast regulators are involved in feed-forward loop motifs; these contain a regulator controlling a second regulator that acts together with the first one to bind a common target gene. In *E. coli*, this type of motif was found to control about 240 genes.

In another approach, Stuart and colleagues (2003) and Bergmann and colleagues (2004) used published perturbation experiments to compare the network topologies of evolutionarily distant organisms, such as *A. thaliana*, *C. elegans*, *D. melanogaster*, *E. coli*, *H. sapiens*, and *S. cerevisiae*. They found that, for all these organisms, the connectivity is distributed as a power law, with negative exponents of similar magnitudes. These power law connectivity distributions indicate that most genes have few connections, while a few genes have many connections. Moreover, there is a significant enrichment of highly connected genes as compared to random networks. Power law distributions have been attributed to dynamically evolving networks and to systems that are optimized to provide robust performance in uncertain environments. For gene networks, gene duplication could be the proximal mechanism explaining this enrichment of connected genes (Teichmann and Babu 2004). This study also found that expression networks are highly clustered. In yeast, the network comprises from 5 to 100 independent gene modules, depending on the analysis methods and an arbitrarily defined threshold. However, modules and interactions may vary significantly between organisms.

Conclusion

Clearly, there are myriad regulatory interactions and mechanisms that play a role specifying the location, timing, and level of gene expression. This provides a rich stage where hypotheses can be unambiguously stated and the variables of evolutionary genetics (epistasis, pleiotropy, plasticity, and so forth) can be more concretely defined. Also, the impact of a number of attributes on variation in genomic expression can be assessed. Accordingly, investigators have described many determinants of variation in genomic expression within and between species, including membership in functional classes, pattern of sex and tissue biases, properties of the protein–protein interaction network, protein attributes, and mRNA abundance itself. Also being disentangled are the effects of mutations and selection on levels of gene expression polymorphism and divergence. Interpreting data integrated from disparate sources is likely to remain a key challenge to understanding the overall picture of the regulation and evolution of genomic expression.

CHAPTER **6**

The Evolution of Proteome Complexity and Diversity

László Patthy

Introduction

IN THE LAST DECADE THE COMPLETE SEQUENCES OF THE GENOMES of hundreds of bacteria, more than two dozen archaea, several unicellular eukaryotes, fungi, plants, and invertebrate and vertebrate animals have been determined. This has permitted the identification of all proteins encoded by these genomes and analysis of the evolution of protein complexity and protein diversity in the three domains of life.

In parallel with the various genome sequencing efforts, structural genomics projects have been initiated in order to expand our structural knowledge of all proteins encoded in the genome of a given organism. A key aspect of structural genomics projects is their emphasis on a high-throughput approach, which lowers the average costs of structure determination by X-ray crystallography and NMR spectroscopy. Large-scale cloning, expression, and purification of proteins are carried out in dedicated centers of structural genomics and the great impact of these projects is illustrated by the fact that they now account for half the novel structures contributed to the Protein Data Bank (Levitt 2007).

Significant progress has also been made in the characterization of interactomes, whole sets of molecular interactions of cells. Thanks to high-throughput methods, massive amounts of information have accumulated on protein–protein interaction networks of key organisms such as yeast, fly, worm, and human.

These developments provide an unprecedented opportunity for analyzing the evolution of the complexity of proteins, proteomes, and protein interactomes, and the impact of these evolutionary changes on the evolution of organismic complexity.

Evolution of Protein Complexity

The complexity of a protein is a function of the number (and number of types) of amino acids, secondary structural elements, and domains they are constructed from, as well as the number, type, and dynamics of interactions among these constituents. At one extreme of the protein complexity scale we find small, low-complexity proteins with biased amino acid composition and few secondary structural elements. At the other extreme are large proteins constructed from multiple types of structural domains (Figure 6.1).

Since the probability of de novo formation of a protein is inversely proportional to its complexity, low-complexity proteins could have arisen independently many times during evolution. The independent evolution of structurally similar antifreeze proteins in Arctic and Antarctic fishes illustrates this point (Chen et al. 1997a). The gene for the antifreeze protein of one Antarctic fish, made up of a simple sequence of repeating tripeptide units, evolved from the pancreatic enzyme trypsinogen by extensive duplication of an ancestral 9 bp unit encoding a Thr-Ala-Ala element. A protein with a similar antifreeze function in Arctic cod also has a Thr-Ala-Ala tripeptide repeat, but evolved independently and has no relationship with the trypsinogen gene.

De novo creation of short structural elements of proteins (e.g., α-helices, β-sheets, reverse turns, transmembrane helices) also has a relatively high probability. Although a short secondary structure is usually unstable in itself, it is relatively easy to increase stability by repeating these elements. Proteins with leucine-rich repeats (LRRs) provide examples of this phenomenon. The basic leucine-rich motifs of these proteins are short (usually containing 20–29 amino acid residues) and consist of a short β-strand and an α-helix approximately parallel to each other. Several (8–30) tandem repeats of this motif are used to build the unique LRR domain. A single leucine-rich motif does not appear to fold into a stable, defined structure; several LRRs are needed to form the LRR domain (see Figure 6.1). The individual repeats are arranged consecutively and parallel to a common axis, forming a horseshoe-like structure with a protein-binding site located in the cavity (Kobe and Deisenhofer 1995). In view of the relative simplicity of this architecture, it has been suggested that the same type of structure may have emerged independently several times during evolution.

More complex protein folds could also have arisen de novo. This was shown by the presence of such proteins in a library of synthetic proteins randomly generated from sequences of leucine, glutamine, and arginine residues (Davidson and Sauer 1994). However, evolution has more typically remodelled duplicates of existing complex protein folds than invented them de novo. The examples of de novo creation described above appear to be rare and limited to relatively simple structures. Most extant proteins appear to have arisen from a relatively small number of ancient protein folds by repeated duplication and divergence. According to some estimates, the majority of proteins belong to less than 1000 types of protein folds (Brenner et al. 1997).

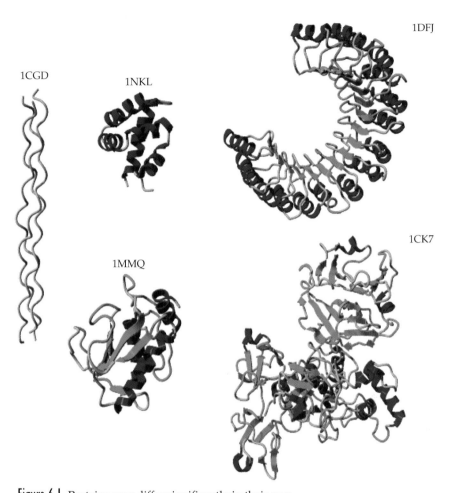

Figure 6.1   Proteins may differ significantly in their complexity. The figure shows the three-dimensional structure of proteins, ranging from simple to complex, found in the Protein Data Bank (PDB) database, http://www.rcsb.org/pdb. The simplest is a short, triple-helical, collagen-like segment (PDB code 1CGD) consisting of tandem repeats of the tripeptide Pro-Hyp-Gly. The small saposin fold (PDB code 1NKL) contains only α-helices, whereas the metalloprotease fold of matrilysin (PDB code 1MMQ) contains both α-helices and β-strands. In the large LRR-domain of the RNAse inhibitor (PDB code 1DFJ), each of the individual leucine-rich repeats consists of a short β-strand and an α-helix. The repeats are arranged consecutively and parallel to a common axis, forming a horseshoe-like structure. The matrix metalloproteinase MMP-2 (PDB code 1CK7) is a multidomain protein that consists of five distinct structural domains: a metalloprotease domain homologous to that of matrilysin, a hemopexin domain, and three tandem fibronectin type 2 domains.

The evolutionary significance of protein evolution by gene duplication is that it gives rise to a redundant duplicate (paralog) of a gene within a genome. This then accumulates divergent mutations and eventually emerges as a new gene with novel or significantly modified functions (Ohno 1972). This may happen if one of the duplicated genes retains its original function while the other accumulates molecular changes adapting it to perform a different task. In many cases, both duplicates acquire functions that are different from that of their common ancestor: they may specialize in different subfunctions of their ancestor.

Most frequently, new functions arise by combining the minor contributions of several point mutations in diverging paralogs; continuous modification of the original function may eventually lead to a novel function. This point may be illustrated by the pancreatic proteases (elastase, chymotrypsin, and trypsin) involved in the digestion of proteins present in foodstuffs. Despite their close structural similarity, these proteases differ markedly in their primary sequence specificity. Elastase cleaves in the vicinity of amino acids with small nonpolar side-chains (Ala, Val, Ser, etc.), chymotrypsin cleaves bonds next to bulky hydrophobic residues (Trp-X, Phe-X, Tyr-X, etc.), whereas trypsin cleaves only at Arg-X or Lys-X bonds. The biological advantage of having multiple digestive proteases with different sequence specificity is that their combined activities ensure more efficient degradation of proteins than any one of them would be capable of alone. It should be emphasized that paralogs that have diverged through accumulation of point mutations and short insertions/deletions do not differ significantly in complexity: the number (and number of types) of amino acids, secondary structural elements, and domains, and the number, type, and dynamics of interactions among these constituents are very similar.

Protein paralogs, however, may diverge not just through point mutations but also through insertion, deletion, and duplication of larger protein segments (including entire domains), as well as fusion to other proteins. Unlike point mutations, domain duplication, domain shuffling, and domain fusion mutations lead to major changes in protein complexity. For example, the regulatory proteases of the blood coagulation and fibrinolytic cascades, in which various domains are joined to a trypsin-like serine protease domain, are much larger and more complex proteins than their single-domain paralogs elastase, chymotrypsin, and trypsin (Figure 6.2).

In summary, the above examples indicate that the major mechanisms for increasing complexity of existing proteins include internal duplication of domains, shuffling of domains, and fusion of proteins, thereby creating large multidomain proteins. Surveys of protein structure databases have indeed revealed that a significant proportion of the protein world arose by these mechanisms: the average size of a protein domain of known crystal structure is about 175 residues, and more complex proteins that are larger than about 200–300 residues usually consist of multiple protein folds (Gerstein 1997). The individual structural domains of such multidomain proteins are compact folds that are relatively independent; the interactions within one domain are usually more significant than those between domains.

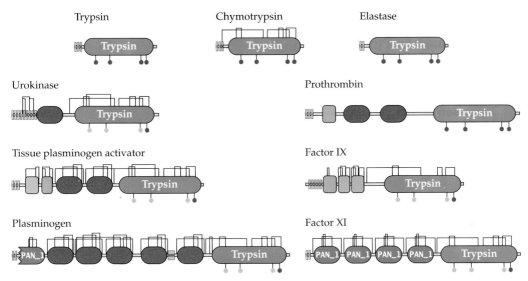

Figure 6.2 Single-domain versus multidomain proteins. Schematic representations of some single-domain members of the trypsin family (elastase, chymotrypsin, and trypsin), as well as representative members of the multidomain regulatory proteases of the blood coagulation and fibrinolytic cascades. (Adapted from the Pfam protein family database, http://www.sanger.ac.uk/Software/Pfam/.)

Evolution of Proteome Complexity

The complexity of the complete set of proteins—proteome—of a given entity (cell, organism) is a function of the *number* and *complexity* of proteins and the *number and dynamics of interactions* between these proteins and other cellular constituents. Having numerous completely sequenced genomes from all three domains of life, it is now possible to study how proteome complexity evolved.

Comparative genomic studies have revealed that the size of bacterial and archaeal genomes varies within a relatively narrow range, from about 0.6 megabase pairs (Mbp) in the obligatory intracellular parasite *Mycoplasma genitalium*, to more than 6 Mbp in some free-living bacterial species. A noteworthy feature of bacterial and archaeal genomes is that gene density is very high and relatively constant. Gene sequences usually cover more than 85 percent of the prokaryotic genome, with about 900 genes per Mbp of genomic DNA. As a corollary, genome size, gene number, and protein number change in parallel in the Archaea and Bacteria, and their proteomes are encoded by about 500–5000 protein-coding genes, depending on genome size (Table 6.1).

Although some unicellular eukaryotes (e.g., the yeast *Saccharomyces cerevisiae*) have a small genome quite similar in size to those of the Bacteria and Archaea, the genome size of eukaryotes is usually much larger than those of prokaryotes (see Table 6.1). For example, the genomes of *Caenorhabditis*

Table 6.1   Genome size, number of protein-coding genes, and gene density
in some representatives of different evolutionary groups

Evolutionary group	Genome size (bp)	Number of genes	Gene density (genes/Mb)
BACTERIA			
Mycoplasma genitalium	5.8×10^5	479	831
Borrelia burgdorferi	9.1×10^5	853	937
Chlamydia pneumoniae	1.2×10^6	1052	856
Treponema pallidum	1.1×10^6	1041	914
Rickettsia prowazekii	1.1×10^6	834	751
Aquifex aeolicus	1.5×10^6	1512	975
Campylobacter jejuni	1.6×10^6	1654	1008
Helicobacter pylori	1.7×10^6	1590	935
Haemophilus influenzae	1.8×10^6	1727	959
Thermotoga maritime	1.9×10^6	1877	1008
Streptococcus pyrogenes	1.9×10^6	1877	946
Neisseria meningitides	2.2×10^6	2121	971
Pasteurella multocida	2.3×10^6	2014	892
Xylella fastidiosa	2.7×10^6	2904	1083
Listeria monocytogenesis	3.0×10^6	2853	969
Deinococcus radiodurans	3.3×10^6	3187	970
Vibrio cholerae	4.0×10^6	3885	963
Caulobacter crescentus	4.0×10^6	3767	937
Bacillus subtilis	4.2×10^6	4100	976
Mycobacterium tuberculosis	4.4×10^6	4000	907
Escherichia coli	4.6×10^6	4288	932
Yersinia pestis	4.7×10^6	4012	863
Salmonella enterica	4.8×10^6	4599	956
Pseudomonas aeruginosa	6.3×10^6	5570	884
Sinorhizobium meliloti	6.7×10^6	6204	927

elegans, *Drosophila melanogaster*, and *Arabidopsis thaliana* are about a hundred-fold larger than a typical bacterial genome and there is a thousandfold difference in size between the human genome and most bacterial genomes. The increase in genome size in higher eukaryotes is not accompanied by a comparable increase in gene number: whereas the size of the human genome is a thousandfold larger than a typical bacterial genome, the number of human protein-coding genes (about 20,000–27,000) is only about fivefold higher than is found in bacteria. On the other hand, thanks to the presence of introns in eukaryotic genomes and to alternative splicing, intron-rich eukaryotes may have a significantly higher number of unique protein sequences than protein-coding genes (Claverie 2001).

Systematic comparisons of proteins encoded in complete genomes have led to the genome-scale delineation of protein superfamilies (Tatusov et al.

Table 6.1 *Continued*

Evolutionary group	Genome size (bp)	Number of genes	Gene density (genes/Mb)
ARCHAEA			
Thermoplasma acidophilum	1.6×10^6	1509	964
Methanococcus jannaschii	1.7×10^6	1738	1022
Archaeoglobus fulgidus	2.2×10^6	2436	1107
Halobacterium sp. NRC-1	2.6×10^6	2630	1023
Sulfolobus solfataricus	3.0×10^6	2977	995
EUKARYA			
Protists			
Entamoeba histolytica	2.4×10^7	9938	414
Leishmania major	3.3×10^7	8272	250
Plasmodium falciparum	3.0×10^7	6300	210
Fungi			
Saccharomyces cerevisiae	1.3×10^7	5800	446
Aspergillus oryzae	3.7×10^7	12074	326
Neurospora crassa	4.3×10^7	10082	234
Plants			
Arabidopsis thaliana	1.2×10^8	25498	221
Oryza sativa	4.2×10^8	50000	119
Metazoa			
Caenorhabditis elegans	1.0×10^8	19000	196
Drosophila melanogaster	1.8×10^8	13600	76
Anopheles gambiae	2.8×10^8	14000	50
Mus musculus	3.0×10^9	24174	8
Homo sapiens	3.0×10^9	27000	9

1997). In this classification, each member of a superfamily has evolved from a single ancestral gene through a series of duplication and divergence events. As expected, these studies show that the mean number of proteins per superfamily increases with gene number in a genome.

Structural classification of proteomes from the three domains of life show that the proportion of basic protein structural types (all-alpha, all-beta, alpha-beta, irregular structures, etc.) are quite similar in the Bacteria, Archaea, and Eukarya, consistent with the view that increased gene numbers were primarily achieved by duplications of an ancient set of genes, rather than by de novo creation of novel types of folds. Nevertheless, a structural census of protein sequences indicates that the known protein folds are not equally represented in different groups of organisms, suggesting lineage-specific expansion of certain fold types (Gerstein 1997). For example, the most common folds in the Metazoa (such as the immunoglobulin, globin, and serine protease folds) are rare in bacteria, whereas in bacteria and plants, protein folds found in metabolic enzymes are the most common. Within the eukaryotic lineage, the percentage of small protein folds (such as the fibronectin type III domains) is much larger in complex multicellular animals than in simple eukaryotes (protists, fungi). Moreover, the frequency of such small protein folds increases as we move from simple metazoa to more complex metazoan organisms.

Multidomain Proteome Complexity

There are significant differences in the complexity of proteomes of the Archaea, Bacteria, and Eukarya with respect to the proportion and size distribution of more complex, multidomain proteins. The proportion of multidomain proteins was found to increase in the order Archaea < Bacteria < Fungi < Plantae < Metazoa. At one extreme we find the Archaea, where only 23 percent of the entries contain more than one Pfam-A domain; the Metazoa represent the other extreme, where 39 percent of the entries correspond to multidomain proteins (Tordai et al. 2005). Furthermore, the multidomain proteins of the Metazoa tend to be larger than those in the Archaea, in harmony with earlier conclusions that the average protein length is considerably greater in eukaryotes than in prokaryotes (Tordai et al. 2005).

Mathematical analyses of the distribution of multidomain proteins according to the number of different constituent domains revealed that their distribution follows a power law—in other words, single-domain proteins are the most abundant, whereas proteins containing larger numbers of domains are increasingly less frequent in all types of organisms. Koonin and colleagues (2002) also showed that the likelihood of formation of multidomain proteins increases in the order Archaea < Bacteria < Eukarya, and there is a significant excess of larger multidomain proteins in eukaryotes. Wuchty (2001) has come to a similar conclusion: higher organisms tend to have more complex, multidomain proteins.

Our recent analyses of the size distribution of multidomain proteins show that, even within the eukaryotic group, there are noteworthy differences between animals, plants, and fungi: the likelihood of domain joining was found to be significantly greater in the Metazoa than in plants and fungi (Tordai et al. 2005). To test whether this reflects the fact that the population of extracellular multidomain proteins has significantly expanded in the Metazoa (Patthy 2003), we analyzed the size distribution, separately, of extracellular and intracellular multidomain proteins of the Metazoa. These analyses confirmed that in the Metazoa there was a significantly greater propensity to form large extracellular rather than intracellular multidomain proteins. It has also been shown that this results from the exceptional mobility of extracellular class 1-1 modules (modules flanked on both boundaries by phase 1 introns, such as the EGF, immunoglobulin, CUB, and sushi domains). These observations are consistent with earlier suggestions that exon shuffling of class 1-1 modules has favored the creation of larger (primarily extracellular) multidomain proteins in the Metazoa (Patthy 2003).

The complexity of the multidomain protein repertoire of different organisms may be characterized not only by the proportion and size distribution of multidomain proteins, but also by the number of domain combinations and the number of domain architecture types (i.e., distinct domain organizations). Using graph theory–based tools to compare protein domain organizations of different organisms, Ye and Godzik (2004) showed that the number of domains, the number of domain combinations, and the size of the largest component of the domain network (measured by the number of domains it consists of) of each organism all increase as we move from prokaryotes to higher eukaryotes. Our analyses also revealed that the number of architecture types increases in parallel with the evolution of higher organisms of greater organismic complexity: the Archaea have the lowest values for this parameter, whereas the Metazoa, particularly chordates, have the highest number of architecture types (Tordai et al. 2005).

The domain combination analysis of multidomain proteins also shows that during the evolution of eukaryotes significant changes occurred in the structural organization of the domain networks (Tordai et al. 2005). In the case of the Protozoa, fungi, and plants, the largest connected component of the domain networks consists of two major subclusters containing nuclear and cytoplasmic signaling domains. The domain network of the Metazoa is strikingly different from those of other eukaryotes as it has a third large distinct subcluster consisting of extracellular (primarily class 1-1) modules (Figure 6.3). The extracellular subcluster of the domain network is connected to the cytoplasmic signaling subcluster through multidomain transmembrane proteins, such as receptor tyrosine kinases and G protein–coupled receptors. Comparison of domain networks of different eukaryotes thus confirms that the evolution of increased organismic complexity in the Metazoa is intimately associated with the generation of novel extracellular and transmembrane multidomain proteins that mediate the interactions among their cells, tissues, and organs (Patthy 2003).

Proteome Interaction Networks

One of the most crucial aspects of the complexity of the proteome is the complexity of the network of interactions of proteins with other cellular constituents, including other proteins, DNA, RNA, carbohydrates, and so forth (Figure 6.4). The full range of functional complexity and diversity in biological systems arises from the interactions among these biological entities. The architecture and organization of such interactions are best represented as networks—for example, networks of interacting proteins that reflect biochemical pathways and genetic regulations.

The interactions among proteins—in multisubunit proteins and in transient complexes among proteins that also exist independently—are funda-

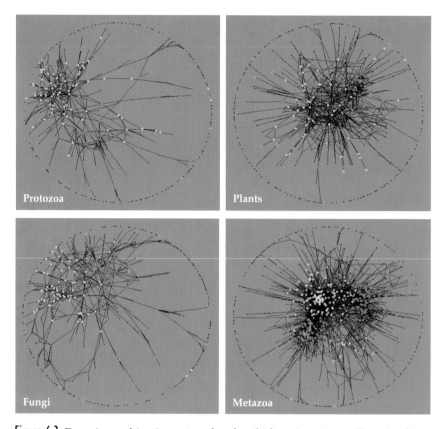

Figure 6.3 Domain combination networks of multidomain proteins. Domains (vertices) are colored according to their subcellular locales. Green dots correspond to nuclear domains, yellow dots indicate signaling/cytoplasmic domains, and magenta dots identify extracellular domains. Class 1-1 modules (mostly extracellular domains) are identified by red dots. Note that the domain network of the Metazoa is strikingly different from those of other eukaryotes, inasmuch as it has a third large distinct subcluster consisting of extracellular (primarily class 1-1) modules. (From Tordai et al. 2005.)

mental to cellular functions. Experimentally, protein–protein interactions are usually defined by the yeast two-hybrid method, by in vitro experiments (e.g., glutathione S-transferase pulldown) or by in vivo experiments (e.g., coimmunoprecipitation) (Box 6.1). Although these techniques may capture meaningful protein–protein interactions and complexes, they still suffer from some serious technical problems. As a consequence, some valid protein–protein interactions will be missing and some artefactual interactions will be included in the experimental interactome data sets. As pointed out by Wuchty and Almaas (2006), the heavy burden of both false positive and false negative protein–protein interaction data may cast doubt on the usefulness of these interaction data sets.

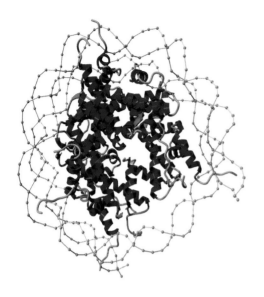

Figure 6.4 3-D structure of a nucleosome. Proteins fulfill their biological function through interactions with other proteins and other cellular constituents, such as DNA, RNA, and carbohydrates. The figure shows the structure of a nucleosome (Protein Data Bank code 1ID3; http://www.rcsb.org/pdb/). Nucleosomes contain two copies of each histone protein (H2A, H2B, H3, and H4) and 146 base pairs of DNA. Each histone has the characteristic histone fold; these interact to form the heterodimers H2A-H2B and H3-H4, which in turn interact to form the histone octamer, which then interacts intimately with DNA.

Although the interactomes defined by these approaches must be treated with caution, they may still provide valuable insight into the complexity of protein–protein interactions. For example, the strong enrichment of nuclear–nuclear, cytoplasm–cytoplasm, cytoskeleton–cytoskeleton, and endoplasmic reticulum–endoplasmic reticulum interactions observed in experimentally determined human, yeast, worm, and fly interaction data sets is in harmony with the biological expectation that protein–protein interactions should be most frequent between proteins within the same cellular compartment (Giot et al. 2003; Gandhi et al. 2006). It is gratifying that—despite the technical problems—the yeast two-hybrid network maintains a signature of cellular topology.

The uncertainties of protein interaction maps generated by the yeast two-hybrid approach can be illustrated by comparing the results of two studies on the interactome of *Drosophila melanogaster* that used different versions of this technology (Giot et al. 2003; Formstecher et al. 2005). The two protein–protein interaction maps are surprisingly different, with little overlap between the data sets. It is noteworthy that signal peptides are significantly depleted from the map of Formstecher and coworkers, probably reflecting the fact that the yeast two-hybrid technology is not ideal for detecting the interaction of extracellular domains (likely to be associated with signal peptides). The same bias probably contributes to the fact that only intracellular interacting domains are enriched in binding domains: the most frequently encountered domains are PDZ, PH, ANK, SH3, RING, and WD40, with the most enriched being PDZ, RA, ANK, 14-3-3, FERM, and LIM (Formstecher et al. 2005).

Protein interaction networks can be also described as graphs (Figure 6.5), where nodes and edges represent proteins and their interactions, respectively (Barabasi and Oltvai 2004). The distribution of protein connectivity in such networks also follows a power law—proteins with few interactions

Figure 6.5 An interaction network of yeast proteins. The figure illustrates that a few highly connected nodes (hubs) hold the network together. (From Jeong et al. 2001.)

are the most abundant, whereas proteins interacting with larger numbers of other proteins are increasingly less frequent (Giot et al. 2003; Li et al. 2004; Gandhi et al. 2006). For example, the connectivity distribution of the human protein–protein interaction network decreases slowly, closely following a power law (Stelzl et al. 2005). On average, the 1705 proteins in the human interaction network each have 1.87 interaction partners; however, 804 have only one partner, whereas there are 24 highly connected "hubs" with more than 30 partners.

Persico and coworkers (2005) recently assembled an inferred human protein interaction network where interactions discovered in model organisms (yeast, fly, worm) were mapped onto the corresponding human orthologs. The authors concluded that the human protein interaction network has a scale-free topology with its connectivity distribution not differing substantially from those of the interactomes of model organisms. It should be pointed out, however, that this orthology-based approach inherits the errors and limitations of the model organism data sets; therefore, the similarity may simply reflect this. In other words, this map overemphasizes similarities and underestimates unique features of the human interactome.

Since detection of protein–protein interactions by experimental methods has a relatively high error rate and is definitely not as fast as genome sequencing, computational prediction methods based on known protein structural interactions may be useful for the analysis of large-scale protein–protein interaction studies on complete genomes. Park and colleagues (2005) recently used this approach to carry out a comparative interactomics analysis of protein family interaction networks. They showed that the interactomes of 146 species are scale-free networks and share a small core network of 47 protein family interaction pairs related to indispensable cellular functions. The functions of protein families constituting this core network are mostly related to protein translation, ribosomal structure and

BOX 6.1
Interactomics

Interactomics is the discipline that aims to define the whole set of interactions among various molecules within a cell: the interactome. Depending on the types of molecules involved in the interactions we may distinguish protein–protein, protein–carbohydrate, protein–DNA, protein–RNA, DNA–RNA, DNA–DNA, and RNA–RNA interactomes, among others.

The terms interactomics and interactome are frequently used in a narrower sense, just to refer to the most widely studied interactome, the protein–protein interactome. In this chapter we discuss only protein–protein interactomes.

The interactions among proteins—in large macromolecular complexes, in multisubunit proteins, and in transient complexes among proteins that also exist independently—are fundamental to most cellular functions. Since information about protein–protein interactions may give important clues about the function of proteins, a major goal of proteomics is to identify such interactions. Experimentally, protein–protein interactions are usually defined by the yeast two-hybrid method, by affinity experiments, or by coimmunoprecipitation. These approaches have different virtues and weaknesses with respect to the sensitivity and specificity of the method.

1. *Yeast two-hybrid analysis.* The yeast two-hybrid system detects interaction between artificial fusion proteins inside the nucleus of yeast. This approach involves the creation of two hybrid molecules, one in which the "bait" protein is fused with a transcription factor, and one in which the "prey" protein is fused with a related transcription factor. If there is interaction between the bait and prey proteins, then the two factors fused to these two proteins are brought into proximity and a specific signal (transcription of a target yeast gene) is produced. The method has a very high false-positive rate, making it necessary to verify the interaction by other techniques. Another limitation of the technique is that it is not ideal for the detection of interactions among extracellular domains that do not fold properly in the milieu of the nucleus.

2. *Affinity chromatography analysis.* These techniques are more reliable than the two-hybrid approach (they generate fewer false positives) and they are suitable for both extracellular and intracellular proteins, but—unlike the yeast two-hybrid system—require purified proteins.

3. *Coimmunoprecipitation analysis.* In these techniques, the protein of interest is isolated with a specific antibody, and the interaction partners that bind to this protein are subsequently identified (e.g., by mass spectrometry). A limitation of this technique is that—unlike the yeast two-hybrid system—it requires pure protein and specific antibodies.

biogenesis, DNA binding, and ATP metabolism. The results confirmed previous studies that all species share the same basic protein families and family interactions critical to basic cellular functions. These data also support the notion that the core protein family network shared by all types of life forms was formed in the very early stage of evolution, and performs the core biochemical processes of life. A gradual attachment of the interactome seems to have occurred as novel functions, such as cell motility, extracellular structure organization, and so forth were needed.

Park and coworkers (2005) have also characterized the interaction networks of different species by calculating the total number of connections of each interacting protein family. These analyses have confirmed that the distribution of protein connectivity follows a power law in all species (i.e., proteins with few interactions are the most abundant, whereas proteins interacting with larger numbers of other proteins are increasingly less frequent). However, there was a major difference between eukaryotes and prokaryotes, as the Eukarya have a much larger number of highly connected hub families than the Archaea and Bacteria. It is also significant that the hub families have more interaction partners in the Eukarya (ranging from 24 to 38) than in the Bacteria (ranging from 11 to 17) or in the Archaea (ranging from 9 to 11).

In the interactome networks the number of interaction pairs increases linearly with the number of protein families from species to species, suggesting that the size of a protein interaction network per se does not determine the topology of the evolving network (Park et al. 2005). The topology is primarily influenced by the presence and number of extreme hub families that are capable of continuous growth, suggesting that the addition of new protein families to the evolving protein interaction network is driven by a selective rather than random process. For example, the number of interacting partners of G proteins is very high in the Eukarya and it seems very likely that the growth of this hub is closely associated with the evolution of multicellularity and the involvement of G proteins in signal transduction.

Barabasi and Oltvai (2004) have suggested that growth and preferential attachment have a common origin rooted in gene duplication. Since duplicated genes produce identical proteins that interact with the same protein partners, each protein that is in contact with a duplicated protein gains an extra link. Highly connected proteins thus have a natural tendency to gain more new links if a randomly selected protein is duplicated.

Another explanation of these patterns rests on the higher frequency of multidomain proteins in the Eukarya. Wuchty and Almaas (2006) have recently investigated the relationship between domain co-occurrence networks (of multidomain proteins) and domain interaction networks. These studies have confirmed earlier observations that the innermost cores of the domain co-occurrence networks gradually expand as one moves from single-celled to multicellular eukaryotes. Combining domain co-occurrence and interaction information, they found that the co-occurrence of domains in the innermost cores strongly coincides with physical interaction. From these analyses they concluded that domains with numerous interaction partners have an elevated chance of being fused in a higher eukaryote and suggested that the driving force behind the fusion of a domain pair is not their frequent interactions, but rather the engagement of the two domains in a multitude of interactions with other domains.

To put it in another way, the similarity of the two types of networks is due primarily to the fact that selection has favored the creation of multidomain proteins from domains (modules) that play crucial roles in protein–protein interactions: these domains are therefore likely to be hubs in both the domain co-ocurrence and protein–protein interaction networks. For example, in complex cellular signaling pathways there is a great demand for domains that mediate interaction with other constituents of the pathways (e.g., PDZ, PH, ANK, SH3, WD40 domains). Thus, selection may have favored the spread of these modules to other multidomain proteins. It is noteworthy that PDZ, PH, ANK, SH3, WD40, and other such domains are hubs in both domain combination networks (e.g., Tordai et al. 2005) and domain interaction networks (e.g., Formstecher et al. 2005).

Recently Rual and colleagues (2005) have used a high-throughput yeast two-hybrid system to test pairwise interactions among the products of 8100 human open reading frames and detected approximately 2800 interactions. Their observation that messenger RNAs corresponding to interacting protein pairs are likely to be coexpressed lends support to the validity of this interactome map. To gain an insight into the evolution of the interactome, they classified proteins in this interaction network as "eukaryotic," "metazoan," "mammalian," or "human" and asked whether proteins specific to different evolutionary classes tend to interact with one another. Importantly, the network appeared to be enriched for interactions between proteins of the same evolutionary class, but not for interactions between proteins from two different evolutionary classes. This observation indicates that the interactome has evolved through the preferential addition of interactions between lineage-specific proteins. As will be discussed below, the validity of this principle may be illustrated by the fact that the various multidomain proteins that evolved in and are unique to vertebrates interact preferentially with each other.

Multidomain Proteins and Organismic Complexity

The conclusion that the huge human genome might contain only about 20,000–27,000 protein-coding genes came as a surprise, since earlier estimates ranged up to 150,000 genes. This means that the organismic complexity of humans and other higher organisms is achieved by a "surprisingly" small number of genes. Claverie (2001) has referred to this lack of a simple linear correlation between gene number and organismic complexity as the N-value paradox. The most plausible explanation for the N-value paradox is that biological complexity is a function of the complexity of interactions among genes and their products—a feature that is not a simple linear function of actual gene numbers.

The complexity of the architecture of an organism's genetic networks depends on the number of protein–DNA, protein–RNA, and protein–protein interactions in the different cells, tissues, and organs. Thus, if we wish to assess the contribution of a protein to organismic complexity, a relevant aspect is its connectivity (the number of interactions with other molecules): highly connected proteins are more likely to increase complexity than proteins with only a small number of links.

Multidomain proteins form a special group among proteins since they are more likely to participate in multiple interactions with multiple partners and thus may be more highly connected than single-domain proteins. In complex multidomain proteins a large number of functions (catalytic activities, different binding activities) may coexist, making such proteins indispensable constituents of regulatory or structural networks where multiple interactions (protein–protein, protein–carbohydrate, protein–DNA, protein–RNA, etc.) are essential. For example, the domains that constitute multidomain proteins of intracellular and extracellular signal transduction pathways mediate multiple interactions with other components of the signaling pathways. Similarly, the coexistence of different domains with different binding specificities is also essential for the biological function of multidomain proteins of the extracellular matrix: multiple, specific interactions among matrix constituents are indispensable for the proper architecture of the extracellular matrix. As a corollary of their involvement in multiple interactions, formation of novel multidomain proteins is likely to contribute significantly to the evolution of increased organismic complexity.

According to this view, increased organismic complexity of higher eukaryotes is not only associated with a larger repertoire of complex multidomain proteins, but also a direct result of their complex interactions. The importance of the creation of novel multidomain proteins is best appreciated if we analyze the correlation between the appearance of novel multidomain proteins and major changes in organismic complexity.

There is now overwhelming evidence that the formation of novel extracellular and transmembrane multidomain proteins is intimately associated with the emergence of novel biological functions and innovations in the Metazoa (Patthy 2003). For example, one of the basic aspects of the biology of multicellular animals is that they require mechanisms for intercellular communication and cohesion. Organization of all multicellular animals depends on molecules that mediate adhesion of cells to other cells, to an extracellular matrix or to a basement membrane. Some key multidomain proteins of the basement membrane (e.g., laminins) are present in sponges, hydra, worm, fly, and vertebrates, indicating that these fundamental constituents of basement membranes were formed very early in the evolution of multicellular animals.

Similarly, linkage of the extracellular matrix to the cytoskeleton via multidomain transmembrane proteins of the integrin family is a very ancient mode of connection between intracellular and extracellular spaces. Consistent with this, integrins are also highly conserved in organisms ranging from sponges, corals, nematodes, fly, and echinoderms to mammals. The fact that many types of multidomain transmembrane proteins involved in cell adhesion in vertebrates are already present in the worm and fly (e.g., cadherins, members of the NCAM family, and netrin receptors) also indicates that they were formed in an early stage of metazoan evolution (Hutter et al. 2000; Hynes and Zhao 2000).

The multidomain receptor tyrosine kinases and receptor tyrosine phosphatases that are absolutely essential for intercellular communication in metazoa also appear to have emerged very early in metazoan evolution. Phylogenetic analysis of the evolutionary history of metazoan protein tyrosine phosphatases has led to the conclusion that there was a period of explosive gene duplication before the parazoan–eumetazoan split, leading to the formation of most major families of receptor tyrosine phosphatases (Ono et al. 1999). Similarly, evolutionary analyses of receptor protein tyrosine kinases have shown that most of the present-day subtypes had been established in a very early stage of the evolution of animals before the parazoan–eumetazoan split (Suga et al. 1999).

There is also strong evidence that the formation of novel extracellular multidomain proteins was responsible for the appearance of novel biological functions unique to vertebrates. In contrast with the wide evolutionary distribution and highly conserved structure of basic multidomain protein components of the extracellular matrix, the various types of multidomain collagens typical of vertebrates are missing from invertebrates. Vertebrates use a wide variety of multidomain collagens to construct their endoskeletons (bone, cartilage), the tendons that connect bones, and the interstitial connective tissue that provides structure for vertebrate tissues. It thus appears that the multidomain collagens have evolved in the vertebrate lineage and their formation must have been a sine qua non for chordate evolution. Several other multidomain protein constituents of the endoskeleton/extracellular matrix are also missing from invertebrates and have been formed during vertebrate evolution: the cartilage aggregating proteoglycan, aggrecan (and other members of this family); cartilage link protein; cartilage matrix protein (and other matrilins); and thrombospondins and the related cartilage oligomeric matrix protein (Patthy 2003).

There are many novel multidomain proteins associated with regulation of the sophisticated vertebrate hemostasis and defense mechanisms. These are regulated primarily by complex interactions among multidomain proteases and cofactors of the coagulation, fibrinolytic, and complement cascades. It is now clear that most multidomain proteases of mammalian plasma effector systems arose in vertebrates: orthologs for the majority of protease and nonprotease components of vertebrate-type blood coagulation, fibrinolyis, kinin, and complement pathways are missing from the *Drosophila melanogaster* and *Caenorhabditis elegans* genomes (Patthy 2003).

In summary, assembly of novel multidomain proteins played a very significant role in vertebrate evolution. Many unique aspects of vertebrate biology rely on these novel multidomain proteins.

Even though the Metazoa and vertebrates provide the most striking examples of how the formation of novel (primarily extracellular and transmembrane) multidomain proteins contributed to increased organismic complexity, it is now clear that the same process also played a critical role in the rise of eukaryotes. Multidomain proteins involved in cytoplasmic or nuclear signaling processes of single-celled eukaryotes, plants, and metazoa have some striking similarities, suggesting that they originated in early eukaryotes and are primarily responsible for the increased organismic complexity of eukaryotes (Aravind and Subramanian 1999; Copley et al. 1999).

Factors Favoring the Formation of Multidomain Proteins

The propensity of different domain types to form multidomain proteins shows great variation, ranging from "static" domains that rarely or never occur in multidomain proteins, to "mobile" domains that are frequently used to build multidomain proteins (Tordai et al. 2005). As for any other type of genetic change, the probability of joining any given domain type to others reflects the probability of such a genetic change and its fixation.

Accordingly, it is likely that the most mobile modules have acquired this status as a result of a combination of special structural, functional, and genomic features. First, certain structural features of domains may facilitate their preferential proliferation in multidomain proteins. For example, the stability and folding autonomy of domains in multidomain proteins may be of utmost importance for their mobility, since this minimizes the chance that folding is disrupted by neighboring domains. It thus seems very likely that the most widely used domains have been selected according to the rate, robustness, and autonomy of folding. Functional aspects may also contribute to the proliferation of certain domains. For example, in complex cellular signaling pathways there is a great demand for domains that mediate interaction with other constituents of the pathways (e.g., protein kinase, SH3, and PDZ domains). Thus, selection may favor the spread of these modules to other multidomain proteins.

Finally, special genomic features may significantly facilitate the shuffling or duplication of domains. Soon after the discovery of introns, it was suggested that they may have played a general role in promoting the assembly of novel genes/proteins from parts of old ones through exon shuffling (Gilbert 1978). During the past decade, complete genome sequences of numerous bacteria, archaea, unicellular eukaryotes, plants, and animals have been determined, permitting assessment of the possible impact of intronic recombination on protein evolution. These studies have shown that bacterial and archaeal genomes do not contain large quantities of nongenic DNA and introns are very rare or absent. In eukaryotes, in contrast, increased genome size is accompanied by decreased gene density and a concomitant increase in the proportion of nongenic and intronic DNA. The intimate association of introns with eukaryotes has led Martin and Koonin (2006) to suggest that one of the major functions of the nuclear envelope was to allow mRNA splicing, which is slow, to go to completion, so that translation, which is fast, would occur only on mRNA with intact reading frames. According to their hypothesis, the rapid spread of introns in early eukaryotes created strong selective pressure that eventually led to nucleus–cytosol compartmentalization.

Since intron-rich genomes suitable for exon shuffling only evolved in higher eukaryotes, this mechanism could play a role only in the formation of multidomain proteins of higher eukaryotes (Patthy 1999). Actually, the first proofs for exon shuffling came from studies on proteases of the vertebrate blood coagulation and fibrinolytic systems (Bányai et al. 1983; Ny et al. 1984; Patthy 1985). Subsequent studies have shown that this evolutionary mechanism has been widely used in the creation of a variety of multidomain proteins unique to animals, including the nonprotease constituents of the plasma effector cascades (see Figure 6.2), most constituents of the extracellular matrix, extracellular parts of the transmembrane proteins, and receptor proteins discussed above (Patthy 2003).

As a signature of exon shuffling, the exon–intron organization of genes produced by this mechanism shows a clear correlation with the domain organization of the multidomain proteins they encode. Furthermore, since only symmetrical modules (i.e., modules flanked by introns of the same phase) are suitable for exon duplication and exon insertion (Patthy 1987), intermodule introns show a characteristic intron phase distibution in the genes of proteins created by exon shuffling (Patthy 1987; 1991; 1994). Since the vast majority of the modules used in the assembly of extracellular multidomain proteins are class 1-1 modules (i.e., both introns flanking the module are phase 1), the intermodule introns split the reading frame in phase 1 (Patthy 1991).

Although exon shuffling by intronic recombination is by far the most powerful mechanism of modular protein evolution, this does not mean that it is the only way to exchange domains among protein-coding genes (Patthy 1996, 1999). Most intracellular modular proteins show no clear evidence of exon shuffling. This may simply be due to the fact that they were formed through recombination within exons. It must also be remembered that most multidomain proteins involved in intracellular signaling were formed in the early eukaryotes (Aravind and Subramanian 1999)—that is, in organisms with intron-poor genomes.

The prominant role of class 1-1 modules and exon shuffling in the evolution of multidomain proteins unique to the Metazoa reflects the fact that evolution is an opportunistic process: the preexisting constitution of the organisms determines in what direction it will go. The ease with which the versatile class 1-1 modules could be used to build novel multidomain proteins has favored formation of multidomain proteins from a rather limited number of class 1-1 module types.

Conclusion

The availability of numerous completely sequenced genomes from all major groups of organisms now permits a detailed analysis of the evolution of protein complexity and protein diversity, and their impact on the evolution of organismic complexity.

These studies have revealed that the proportion of basic protein structural types (all-alpha, all-beta, alpha-beta, irregular structures, etc.) are similar in the Bacteria, Archaea, and Eukarya, indicating that increased gene numbers in higher eukaryotes were primarily achieved by duplications of an ancient set of protein-coding genes, rather than by creation of novel types of folds. The general observation that the mean number of proteins per protein superfamily in a genome increases with gene number also implicates gene duplication as the main mechanism for increasing the number of protein-coding genes. Although the genomes of higher eukaryotes may encode many more proteins than those of prokaryotes, there is a general consensus that differences in gene and protein number can't account for the marked differences in the complexity of these organisms.

Comparison of the proteomes of the Archaea, Bacteria, and Eukarya, however, revealed striking differences with respect to multidomain proteins: the proportion, size, and complexity of multidomain protein architectures increased as we moved from prokaryotes to higher eukaryotes. This parallelism between the evolution of more complex multidomain protein repertoires and evolution of increased complexity of multicellular organisms is not just a coincidence. If we define organismic complexity as a function of the number and types of cells, tissues, and organs an organism is constructed from—as well as the number and dynamics of their interactions—then it is clear that multidomain proteins are uniquely suited to mediate the intracellular and intercellular interactions essential for complex multicellular organisms. The fact that the appearance of novel biological functions crucial for increased organismic complexity of metazoa, vertebrates, and so forth is intimately associated with the formation of the appropriate novel extracellular and transmembrane multidomain proteins underlines the validity of this conclusion.

It must be emphasized that we are still far from understanding the structure and dynamics of the whole (intercellular, extracellular, and intracellular) interactome of complex organisms. As discussed above, intracellular protein–protein interactions constitute just one section of the interactome, but relatively little is known about intercellular and extracellular protein–protein interactomes and interactomes of other macromolecules. In view of the rapid development of interactomics technologies, it is likely that the next few years will see increased exploration of the significance of these interactomes in the evolution of organismic complexity.

An important question remains to be answered: what drives evolution in the direction of increased organismic complexity (and formation of multidomain proteins that support complexity)?

According to the r/K selection theory, selective pressures drive evolution in one of two stereotypical directions: r- or K-selection. In the case of r-strategists, natural selection favors maximum reproductive and/or growth rate. Typically, r-selected species have a short lifespan and produce large numbers of offspring. In unstable environments, r-selection predominates, since the ability to reproduce quickly is crucial and there is little advantage in adapting to a changing environment. Traits that are characteristic of r-strategist organisms include small size and short generation time.

K-selected species, on the other hand, produce fewer offspring, each of which has a greater chance of survival. *K*-selection predominates in relatively stable environments, since adaptation to the environment and the ability to compete successfully for limited resources is crucial. Traits that are characteristic of *K*-strategist organisms include large size, long generation time and lifespan, and the production of fewer offspring that are well cared for.

Although *r/K* selection has a continuous spectrum (i.e., the majority of organisms fall between these two extremes), prokaryotes are prototypic examples of *r*-selected species, and eukaryotes display more and more traits characteristic of *K*-strategists as we move to higher eukaryotes (e.g., from fly through fish to human). It therefore seems safe to conclude that the increased organismic complexity of higher eukaryotes is the result of *K*-selection: increased organismic complexity ensures improved adaptation to the environment and improved ability to compete for limited resources.

Acknowledgments

The author wishes to acknowledge support from the BioSapiens project, funded by the European Commission within its FP6 Programme, under the thematic area "Life sciences, genomics and biotechnology for health," contract number LHSG-CT-2003-503265. The work was also supported by grant OTKA/TS049890 from the Hungarian Research Fund.

CHAPTER **7**

Genomic Redundancy and Dispensability

Laurence D. Hurst and Csaba Pál

Introduction

PERHAPS ONE OF THE MOST STRIKING AND SURPRISING DISCOVERIES of modern molecular genetics is the extent to which, at first sight, organisms appear not to "need" many of their genes. Early reviews of single-gene knockout mutants of mice, for example, found that only about 30 percent of the mutated genes were necessary for viability or fertility (Hurst and Smith 1999). This tallies with a more recent systematic analysis suggesting that fewer than 20 percent of mouse gene knockouts result in embryonic lethality (Wilson et al. 2005). Comparable experiments have been performed on a genomic scale in bacteria (Figure 7.1), yeast (Giaever et al. 2002; Steinmetz et al. 2002; Decottignies et al. 2003), worm (Kamath et al. 2003), and Arabidopsis (Alonso et al. 2003). In nearly all cases the conclusion is the same—namely, that the great majority of genes appear not to be essential for growth and/or viability (see Figure 7.1A for bacterial data). In yeast, for example, approximately 80 percent of the more than 5700 single-gene knockout strains are capable of efficient growth (Steinmetz et al. 2002). In worm, RNAi experiments on 16,000 genes suggest a phenotype for less than 10 percent of them (Kamath et al. 2003). The most dramatic exception is the highly reduced genome of the intracellular parasite *Mycoplasma genitalium*, in which more than half of the genes are required for growth under laboratory conditions (Hutchison et al. 1999; Glass et al. 2006).

These findings prompt a series of interrelated questions. First, if a gene is considered to be redundant in such experiments, does this really mean it is of no utility to the organism? Second, what are the mechanisms of redundancy? Third, given that dispensability is common, is it likely that it resulted from natural selection that specifically favored redundancy? Alternatively, might dispensability just be an incidental side product of other evolutionary processes?

(A)

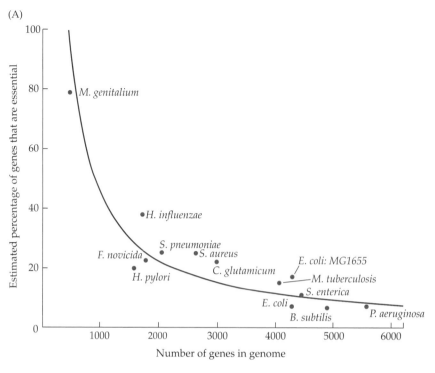

(B)

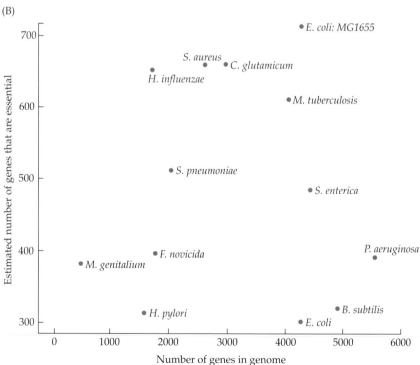

◀ **Figure 7.1** (A) The proportion of protein-coding genes that are essential in a bacterial genome as a function of the number of genes in the genome. The blue curve indicates the percentage expected if a constant 450 genes are essential in every genome. (B) The estimated number of genes that are essential as a function of number of genes in the genome. These estimates are derived from estimates of the proportion of genes that are essential and the number of protein-coding genes in the genome. (A, data for *B. subtilis* from Kobayashi et al. 2003; *P. aeruginosa* from Jacobs et al. 2003 and Liberati et al. 2006; *E. coli: MG1655* from Gerdes et al. 2003; *E. coli* from Baba et al. 2006; *S. enterica* from Knuth et al. 2004; *H. pylori* from Salama et al. 2004; *S. aureus* from Ji et al. 2001, Forsyth et al. 2002 and Ji et al. 2002; *H. influenzae* from Akerley et al. 2002; *M. genitalium* from Hutchison et al. 1999 and Glass et al. 2006; *S. pneumoniae* from Zalacain et al. 2003; *M. tuberculosis* from Sassetti et al. 2003; *F. novicida* from Gallagher et al. 2007; *C. glutamicum* from Suzuki et al. 2006.)

Are "Dispensable" Genes Really Dispensable?

The language involved in this debate, to say the least, is confusing. In the biological literature it is common to hear genes referred to as being nonessential, dispensable, or redundant, treating all three terms as synonyms. Using redundant and dispensable as synonyms is confusing. In engineering, a redundant system is specifically one that has backup provided by (nearly) identical copies of the same functional element. A lift (elevator) may, for example, be held on two wires—one taking the strain, the other as a backup should the first break. Would it also not appear to be a contradiction in terms to say dispensable genes are not needed? Surely this is what dispensability means (Brookfield 1997)? In this sense, for a gene to be dispensable to the organism would require that the sequence could be lost without any effect, and a deletion or novel dispensable gene should spread by drift (neutral evolution) alone (Sliwa and Korona 2005).

There are many reasons that laboratory assessments of dispensability do not imply dispensability in the eyes of natural selection. First and foremost, laboratory-based assays typically score mutants as either being viable or not (but see Thatcher et al. 1998; Deutschbauer et al. 2005). This is the crudest of distinctions and fails to detect even relatively large effects on fitness. Indeed, a substantial fraction of yeast mutant strains that lack obvious defects (i.e., are viable), are nevertheless at a significant selective disadvantage growing under normal conditions (Thatcher et al. 1998; Deutschbauer et al. 2005).

More generally, laboratory-based measures have little bearing on the probable fate of a deletion allele. A gene that is essential in the laboratory is unlikely to be lost in the near evolutionary future of a species; however, for a weakly deleterious allele in a diploid to reach fixation by drift alone would require it to have a fitness effect less than $\dfrac{1}{2N_e}$ (where N_e is the effective population size). Even for organisms with small effective populations (e.g., humans), this means a fitness effect no greater than about 10^{-5}. For bacteria and yeast a more realistic figure might be 10^{-9}. Laboratory-based fitness measures are simply not sensitive enough to detect such subtle effects (Brookfield 1997). It would be an error to suppose, therefore, that viability of a knockout strain in the laboratory equates to evolutionary dispensability.

Moreover, laboratory conditions will often fail to detect genes required under special environmental conditions and may, therefore, wrongly classify genes that are sometimes essential for viability as nonessential. For example, stress response genes may be superfluous in the laboratory but vital in the field (Fang et al. 2005). This problem has been examined both *in silico* and through empirical analysis. The *in silico* approach is best applied to metabolic systems that are both well described and amenable to modeling of flux. We, for example, addressed the issue by analyzing flux balance in yeast (Papp et al. 2004). In this analysis, we started with a network of 809 metabolites as nodes (including external metabolites), connected by 851 different biochemical reactions (including transport processes). We then defined a solution in which fluxes of all metabolic reactions in the network satisfied the governing constraints (i.e., steady state of metabolites, flux capacity, direction and possible reversibility of reactions, and nutrients available in the environment). Next, we used various optimization protocols to find the optimal use of the metabolic network, among all possible solutions, to produce major biosynthetic components for growth.

With such a network, one may then ask whether a given gene appears nonessential simply because the flux through its pathway is zero (the enzyme the gene encodes isn't doing anything) under one set of conditions, while the gene may be needed (flux is non-zero) under different conditions. Importantly, the model indicates that condition-specificity is the dominant explanation for apparent dispensability in the laboratory, and could account for 37–68 percent of dispensable genes (Papp et al. 2004) (Figure 7.2). For more than one-half of the latter genes nonessential under nutrient-rich conditions, it is possible to predict environmental conditions under which they will be essential. As might be expected, such condition-specific genes have a more restricted phylogenetic distribution than genes necessary under all conditions (but see later discussion for caveats). Similarly, analysis of the growth phenotypes of *Escherichia coli* mutants (Glasner et al. 2003) showed that most gene knockout strains show severe fitness defects only under a small fraction (10%) of the 282 different growth conditions investigated (Papp et al. 2004). These genes also have a limited phylogenetic distribution (Papp et al. 2004).

This notion that most "dispensable" genes are just not employed under laboratory conditions has recently gained experimental support from direct measurements of metabolic flux in yeast (Blank et al. 2005). This study also reports that about 50 percent of apparently dispensable genes are simply inactive under laboratory conditions. Indeed, a recent analysis (Harrison et al. 2007) suggests that at least 20 percent of the 5000 apparently nonessential genes in *Saccharomyces cerevisiae* make a large contribution to fitness under other environmental conditions (see also Dudley et al. 2005). Dele-

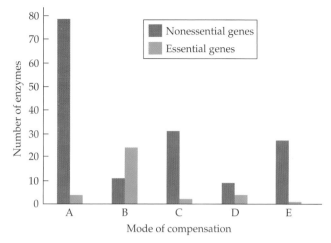

Figure 7.2 Mechanisms of gene dispensability in metabolic networks of yeast (*S. cerevisiae*). The metabolic capabilities of the *S. cerevisiae* network were calculated using computational flux balance analysis. Gene activities and the impact of gene deletions were calculated under nine nutrient conditions (Papp et al. 2004). When the model is initiated to mimic the growth conditions used in experimental studies, it predicts gene deletion phenotypes with over 80% accuracy (Papp et al. 2004). Based on the presence of isoenzymes (gene duplicates that encode enzymes that catalyze the same reactions) and predictions of the model, enyzmes were divided into five nonoverlapping classes: A, nonessential enzymes that appear to be essential under other nutrient conditions; B, single-copy enzymes with no alternative pathways; C, enzymes with isoenzymes; D, single-copy enzymes with alternative pathways; and E, enzymes with isoenzymes and alternative pathways. The figure shows the number of experimentally verified essential and nonessential genes in these categories. The model predicts that 33–68% of the experimentally verified nonessential genes should be important under different nutrient conditions (class A). Redundant gene duplicates (classes C and E) can explain 15–28% of gene dispensability in the network, while alternative pathways (classes D and E) have a relatively small role (4–17%). As expected, the fraction of experimentally verified essential genes is much higher among enzymes with no backups (class B) compared to all others.

terious phenotypes are generally restricted to a small fraction of the tested environments (Figure 7.3). Intracellular bacteria are unusual in having a majority of their genes needed for viability; this may be because they inhabit a stable environment and so do not need or have many condition-specific genes (Papp et al. 2004; Blank et al. 2005).

All of the above analyses, however, examine unicellular organisms in which, with the exception of intracellular bacteria, the need for metabolic flexibility might be high. It is an open question as to whether multicellular organisms have high rates of dispensability for similar reasons.

Figure 7.3 Distribution of condition-ally essential genes. Many seemingly nonessential genes in yeast (*S. cerevisiae*) make a contribution to growth under specific conditions. Of 4823 genes not essential for growth on nutrient-rich medium (YPD), 963 exhibited lethality or a strong growth defect under some other conditions. Moreover, most of these conditionally essential genes make a contribution to growth under only one or few of the 31 conditions examined. Gene deletions showing conditional growth phenotypes were compiled from published large-scale screens (Data from Harrison et al. 2007.)

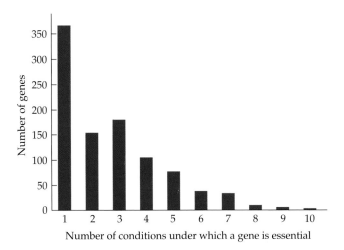

Nonessential genes are typically under strong purifying selection

Another way to determine whether "dispensable" genes really are evolutionarily dispensable is to ask about their rates of evolution. Under the premise that a gene that is irrelevant will not be under purifying selection, we might expect the gene to evolve at very high rates. We can be even more specific. If we examine the orthologous gene sequence in two species, we can partition nucleotide changes as being either amino acid changing (nonsynonymous) or not (synonymous). If the gene is neutrally evolving we can expect the number of nonsynonymous changes per nonsynonymous site (K_a) to be equal to the number of synonymous changes per synonymous site (K_s). All analyses of "nonessential" genes suggest that this is not so and that K_a is much lower than K_s (Hurst and Smith 1999; Hirsh and Fraser 2001; Jordan et al. 2002; Castillo-Davis and Hartl 2003; Pál et al. 2003; Yang et al. 2003; Rocha and Danchin 2004; Wall et al. 2005; Zhang and He 2005; Liao et al. 2006). This reinforces the suggestion that viability of a gene under the laboratory setting is not a good predictor of the importance of either a gene knockout or other mutations in the wild.

Moreover, and perhaps more surprisingly, it is not even clear that nonessential genes are under weaker purifying selection than essential ones. Some researchers have suggested that, just as active sites in proteins might be under stronger constraint, so too essential genes might have more sites at which nonsynonymous changes would be opposed by selection ($s \gg \frac{1}{2N_e}$) and thus evolve more slowly (Kimura and Ohta 1974; Wilson et al. 1977; Hirsh and Fraser 2001). The first broadscale genomic analysis in rodents found that K_a/K_s was significantly higher for nonessential genes (Hurst and Smith 1999), and the same has also been reported in a larger recent survey (Liao et al. 2006). A similar higher rate has been claimed for nonessential

(A)

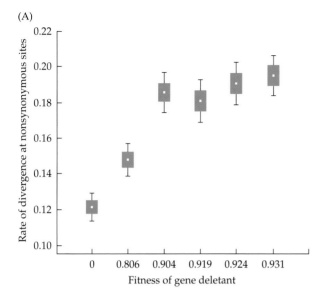

(B)

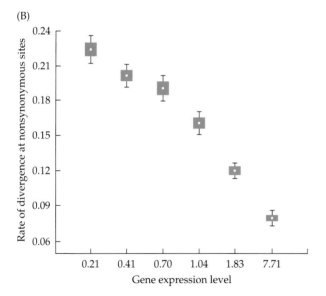

Figure 7.4 The impacts of gene dispensability and gene expression level on protein evolution. (A) Only a relatively weak relationship was found between the severity of the fitness effect of gene deletions and the rate of protein evolution in yeast (*S. cerevisiae*). (B) In contrast, gene expression level (measured by mRNA abundance) correlates strongly with the rate of protein evolution: highly expressed genes evolve slowly. As essential genes are generally highly expressed, the relative contributions of gene dispensability and expression level on protein evolution have remained a subject of intense debate (e.g., Pál et al. 2003; Wall et al. 2005). Protein evolution rates (nonsynonymous divergence) were calculated by Wall et al. (2005) using sequence information on closely related species of the *Saccharomyces* genus. Fitness estimates denote relative growth of gene deletants compared to wild type. For more details on data sources and statistical treatments, see Pál et al. 2003 and Pál et al. 2006a. Boxes show mean ± standard error, whiskers show ± 2*standard error.

genes in yeast (Hirsh and Fraser 2001). However, rate differences either disappear or are very small when other factors are taken into account (Figure 7.4) (Hurst and Smith 1999). Any difference between essential and nonessential genes may be owing to faster evolution of genes with duplicates (Yang et al. 2003), many of which are dispensable. The lower rate of evolution of essential genes in bacteria (Jordan et al. 2002) also looks as though it is best explained as a covariate to expression rate (Rocha and Danchin 2004) rather than a function of dispensability per se.

On the other hand, in some some taxa (Castillo-Davis and Hartl 2003), and when using particular comparator species (Zhang and He 2005), essential genes appear to evolve more slowly than nonessential ones, even allowing for possible covariates (Wall et al. 2005; Liao et al. 2006). However, the statistical methods employed by many of these researchers were criticized by Drummond and coworkers (2006), as the covariates are not themselves independent. Instead, using principal component analysis and numerous possible predictor variables, this team concluded that dispensability has little or no influence on the rate of protein evolution (Drummond et al. 2006). In summary, whether essential genes experience stronger purifying selection than nonessential ones—allowing for covariates—remains uncertain. Nonetheless, the above complex sets of results all point to the same conclusion—namely, that genes that appear to be dispensable in the laboratory are needed by the organism, and most sites within the encoded proteins are typically under strong purifying selection.

It may yet be the case, however, that genes more dispensable in the laboratory are also more dispensable over evolutionary time. Unfortunately, the current attempts to answer this question are problematic. This is for the simple reason that attempts to find orthologs typically rely on sequence matches (e.g., via BLAST), a method known to miss quickly evolving genes (Wolfe 2004). Sequence similarity methods for detecting the presence of orthologs are inherently biased toward finding long, slowly evolving genes and missing short, quickly evolving genes. Put differently, suppose that all genes persist over the same length of evolutionary time, but that they differ in their rates of evolution. For any arbitrarily chosen similarity cutoff value applied to define an ortholog, one will find that slowly evolving sequences have a broader taxonomic distribution. Hence, the claim that essential genes have a broader taxonomic distribution (see Gerdes et al. 2003; Krylov et al. 2003), while an intuitively reasonable result, needs to be treated with caution. A simulation study (Elhaik et al. 2006) suggests that the same problems undermine the claim (Alba and Castresana 2005) that older genes tend to evolve more slowly. That "young" genes tend also to be short (Alba and Castresana 2005) may well derive from the same bias. Likewise, attempts to relate expression patterns of genes (known to covary with evolutionary rate; Duret and Mouchiroud 2000; Lercher et al. 2004; Zhang and Li 2004) with phylogenetic age (Subramanian and Kumar 2004; Freilich et al. 2005) need to be considered with caution.

By What Mechanisms Are Nontrivially Dispensable Genes Dispensable?

While some genes are dispensable only in the sense that laboratory studies fail to show the conditions under which they are needed for growth, there remains a set of genes that are not actually vital for growth in the laboratory, yet are still being expressed and doing something (Papp et al. 2004;

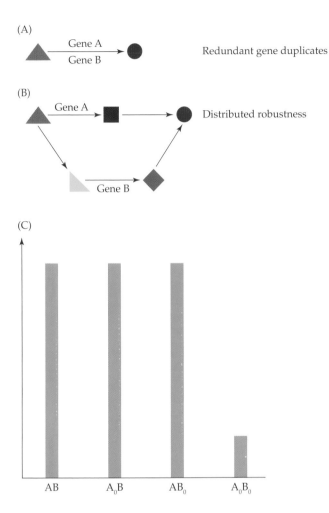

Figure 7.5 A model for the two causes of robustness in metabolic networks. (A) The first of the two causes is redundancy of a system's parts: a gene may be dispensable if the genome contains other copies of the same gene with overlapping functions. In our example, both gene A and gene B encode enzymes that can catalyze the same reaction independently of each other, producing a key metabolite (red circle). (B) Alternatively, robustness may be distributed across pathways: the same key metabolite (red circle) can be produced through parallel metabolic routes. (C) In both cases, because production of the key metabolite is little affected, single-gene deletant genotypes (A_0B or AB_0) have relatively high fitness, comparable to that of wild type (AB). In contrast, deleting both copies of a duplicated gene or genes sitting on alternative pathways (A_0B_0) is expected to have a drastic effect on fitness.

Blank et al. 2005). These genes may be considered as being nontrivially dispensable genes.

Two main mechanisms to account for nontrivially dispensable genes are regularly discussed (Figure 7.5). The first involves what an engineer would classify as redundancy: there is at least one copy of the knocked out gene somewhere else in the genome that, it is assumed, can take the strain, so to speak, when the paralog is lost (Gu et al. 2003). Many possible examples could be given. In yeast, for example, there are three TPK genes, which encode catalytic subunits of cyclic AMP–dependent kinase, a key molecule in cell signaling. Any two of the three genes are dispensable for cell growth (Toda et al. 1987).

This mechanism of dispensability through gene duplication can be contrasted with an alternative, namely, distributed robustness (Wagner 2005a). In this instance, to carry on with the lift (elevator) analogy, a faulty lift need not stop us getting to where we want to go, as we could take the stairs. Distributed robustness might then be considered as an alternative routing or possibly a rearrangement of what is left after a knockout. In the cell this might mean, for example, routing a signal through a parallel pathway.

Such processes are harder to demonstrate than the dispensability associated with gene duplication. Nonetheless, a few good examples can be given. In mammals, bile acids are the degradation products of cholesterol. The same bile acids act to repress the genes necessary for cholesterol degradation by activating two or more alternative and mutually redundant pathways that involve different transcription factors (based on activation of the xenobiotic receptor PXR or the c-Jun N-terminal kinase JNK; Wang et al. 2002). A similar routing through multiple pathways to achieve one end has been demonstrated for the gene products necessary for genome stability in yeast (Kolodner et al. 2002). In particular, there are two partially redundant branches of the intra-S checkpoint, one of which requires RAD17, RAD24, and other genes, and one that involves SGS1. Inactivation of each individual intra-S checkpoint branch has little effect on genome instability while inactivation of both leads to very high levels of genome rearrangements. In many cases of distributed robustness, the mechanism can be understood in terms of alternative means to the same end. In *E. coli*, glucose-6-phosphate dehydrogenase is nonessential, with loss-of-function mutants growing at near wild-type rates. How can this be when this enzyme is part of the pentose phosphate shunt, which produces two-thirds of a cell's NADPH? By contrast, most of the cell's NADH is produced by the tricarboxylic acid cycle, and this holds the key. When the pentose phosphate shunt is blocked by a loss-of-function mutation, there is increased flux through the tricarboxylic acid cycle, generating more NADH. A massively increased flux through the transhydrogenase reaction permits this NADH to be converted to NADPH. The route may be substantially adjusted, but the end product (high NADPH) is the same (Wagner 2005a).

While at first sight these two mechanisms of dispensability (duplication and distributed robustness) look distinct, it may be better to consider gene duplication as a special case of distributed robustness. Consider, for example, a pair of isozymes (two sequence-related proteins doing the same metabolic job). Delete one and the job gets done by the other. This would be a fine example of dispensability by gene duplication. However, look at it another way and it is an example of distributed robustness; it just so happens that the alternative pathway is short, being just one protein (the remaining isozyme), and that the protein is sequence-related to the one deleted. Put differently, if the same mechanism of dispensability involved, instead, two functionally related genes that were *unrelated* in sequence, then we would be obliged to consider this an example of distributed robustness. Nonetheless, the distinction remains important in so much as it allows further clarification of physiological and evolutionary mechanisms. As regards distributed robustness, we are led to ask about its relationship to network architecture. As regards duplicates, we can ask how duplicates provide backup.

Network architecture and distributed robustness

Understanding the mechanism of distributed robustness, as the above example illustrates, is nontrivial, as it cannot be pinned down to just one gene. Rather, we need to consider a protein's role within a network.

Initial claims suggested that a gene's dispensability might be satisfactorily explained by its topological position within a network. Indeed, theoretical studies claim that the robustness of most cellular networks to perturbation follows from their scale-free nature (Albert et al. 2000; Jeong et al. 2000). Unlike random networks, in which the number of connections between entities follows a Poisson distribution, in a scale-free network the distribution of the number of connections follows a power law, such that a few members, so-called hubs, have a very large number of connections. While this framework is elegant, it is unlikely to satisfactorily explain robustness in biological networks, for two reasons. First, the data are too crude, and, therefore, it is difficult to see if the assumptions of these models hold. Second, these models have very limited predictive power when it comes to gene dispensability. We discuss each of these problems in turn.

Consider, for example, the argument of Jeong and colleagues (Jeong et al. 2000):

- Metabolic networks generally have the same topological scaling properties and show striking similarities to the inherent organization of complex nonbiological systems.

- Theory suggests that scale-free networks are robust against perturbations.

- Selection has favored this organization to provide error-tolerant architecture.

Such a broad-brush approach is problematic, as it obscures numerous details. For example, in metabolic pathways, in contrast to expectations of graph-theoretical models (Albert et al. 2000), essentiality of reactions at a node is not correlated with node connectivity (Mahadevan and Palsson 2005). This lack of correspondence makes some sense in that some essential reactions/genes have very low connectivities, as they are on linear pathways with no backups (Mahadevan and Palsson 2005). This lack of correspondence between node connectivity and dispensability is also reported in analyses of other biological networks (Hahn et al. 2004a).

In the case of metabolic networks, the topological approach has several weaknesses. First, unlike more dynamical models, most structural analyses of metabolic networks ignore the possibility that many reactions are condition-specific and hence only a fraction of the network is actually realized in a given environment. Second, what is true for some networks need not be true of metabolic networks, as their representations may be different. In metabolic networks, metabolites are represented as nodes and reactions as links (Jeong et al. 2000). This representation is different from that of protein–protein interaction networks, where the nodes are gene products and the links correspond to interactions. Finally, simple graph-based representations tend to be misleading. The reason for this is that pathways computed by graph theoretical analysis (Jeong et al. 2000) typically do not correspond to traditional biochemical pathways, owing to the way such analyses treat common metabolites (Arita 2004).

The situation is not much better in the case of protein–protein interaction and transcriptional regulatory networks. Some studies suggest that these networks are scale-free and that hubs tend to be the essential parts of the network, the rest being dispensable (Jeong et al. 2001; see also Chen and Xu 2005; Hahn and Kern 2005; Pereira-Leal et al. 2005). However, many of these analyses rely on highly biased data sets with a large fraction of false-positive interactions (von Mering et al. 2002). Both the scale-free nature of the yeast protein–protein interaction network (Han et al. 2005), and the relationship between dispensability and position in the network (Coulomb et al. 2005) may be artifacts of biased data. However, analysis of literature-curated protein–protein interactions and multivalidated data sets suggests that hubs do tend to be essential in yeast (Batada et al. 2006a) and in mammals (Liang and Li 2007). Moreover, those hubs that partner many other hubs are even more likely to be essential (Batada et al. 2006b).

Assuming hubs in protein–interaction networks do tend to be essential, the question then is: Why? He and Zhang (2006c) propose a simple null explanation, namely that hubs have large numbers of protein–protein interactions and therefore high probabilities of engaging in essential interactions. However, controlling for the absolute number of interactions, hubs with a higher proportion of connections to other hubs are in turn more likely to be essential (Batada et al. 2006b), suggesting network architecture is important and that the simplest null explanation (He and Zhang 2006c) is not entirely adequate.

Similar problems apply to transcriptional regulatory networks (Evangelisti and Wagner 2004). While transcription factors with many targets tend to be essential, target genes regulated by many transcription factors are usually not essential (Yu et al. 2004). This latter pattern is likely to reflect the fact that essential genes (measured under nutrient-rich conditions) encode housekeeping functions, and therefore need few regulators (Yu et al. 2004).

For all the reasons given above, it is dubious that essential genes or reactions could be defined purely by consideration of their topological position in the network, without referring to the environmental conditions and specific biochemical roles played by the genes. Recent dynamical metabolic networks based on strong empirical footing attempt to handle these problems. The most successful, in terms of predicting dispensability, employ dynamic models rather than relying on simple topological characterization (e.g., see Edwards and Palsson 1999; Edwards and Palsson 2000b; Edwards et al. 2001; Ibarra et al. 2002). For example, these models accurately predict close to 90 percent of the knockout phenotypes of metabolic genes in yeast (Forster et al. 2003a) and *E. coli* (Edwards and Palsson 2000a). Extension of these dynamic models to protein–protein and transcriptional networks is an obvious next step (Covert et al. 2004; Herrgard et al. 2006).

How do duplicates provide backup?

The underlying mechanism of backup provided by duplicates may not be as obvious as it first appears. Perhaps the simplest way of thinking about duplicate gene backup is by analogy with single-gene knockouts in a diploid heterozygote, which typically show dominance of the wild type (Papp et al. 2003a). The classical explanation for dominance, as applied, for example, to metabolic systems, supposes that the relationship between enzyme concentration and flux through a metabolic pathway follows the law of diminishing returns (Kacser and Burns 1981). That is, for each extra unit of enzyme the increase in flux gets ever smaller (for evidence see Charlesworth 1979; Orr 1991) (Figure 7.6). Hence, assuming the enzyme concentration is halved in the heterozygote knockout, this would reduce flux by less than half—often very much less than half. The phenotypic effects of this as regards fitness are then dependent on the relationship between flux and fitness. Empirical analyses suggest that changes in flux have a linear relationship with fitness (Dykhuizen et al. 1987; Dykhuizen and Dean 1990), which permits dominance to follow as a logical corollary of enzyme kinetics (see also Keightley 1996).

Using this model, one might also imagine that if we had two proteins doing similar things and being expressed at the same time in a haploid, then removing one will leave the other doing the same job at the same time. We might also expect there to be a correlation between the sequence similarity of two duplicated genes and the effect of mutations in one of these genes (Wagner 2000). Such a model would provide a simple explanation for why mutational robustness is greatest for closely related gene duplicates, large gene families, and similarly expressed genes in *C. elegans* (Conant and Wagner 2004). In this context, recent evidence suggesting that duplicate genes are as essential as singletons in mammals is unexpected (Liao and Zhang 2007).

While this model may well account for some instances, it cannot be the whole truth. If one analyzes paralogous pairs (i.e., pairs of duplicates) of yeast genes that also have some degree of coexpression, there is, paradoxically, no relationship between the degree of coexpression and the likelihood a gene will be dispensable (Papp et al. 2003b). From the above perspective this is a surprising result; if the remaining duplicate (after knockout) is not expressed at the right time and place, then it obviously cannot provide backup. At least in yeast, most duplicate-associated backups involve genes that on average are not strongly coexpressed, do not share many similar 5′ motifs (that bind particular transcription factors), and diverged from each other a long time ago (Kafri et al. 2005). How can it be that such genes are redundant but not coexpressed?

The above logic has one obvious flaw. What is important for backup is not whether two genes are coexpressed under normal conditions, which is usually examined using coexpression data from wild-type cells. Rather, what is relevant is whether, when one of the two genes is knocked out, the remaining paralog is expressed in a compensatory manner. If the knockout itself initiates upregulation of the remaining paralog, then backup could still be

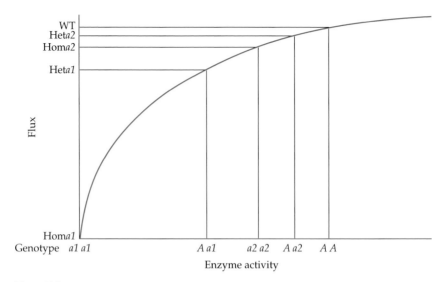

Figure 7.6 A mechanistic model for dominance: flux through a metabolic pathway as a function of activity of an enzyme. Consider a metabolic pathway that starts with chemical S. This is converted to intermediary product P1 by enzyme E1 (the reaction is reversible). P1 in turn is converted to P2 by enzyme E2. Finally, the pathway produces its end product, PN. The rate of production of this final product is the flux of the metabolic pathway. The curve shows the flux as a function of the activity of any one of these enzymes (E1, E2, etc.). It is assumed that the curve follows the law of diminishing returns; in other words, for every doubling in dose of the enzyme, there is less than a doubling of the flux. If we consider a normal wild-type individual (*AA*), the enzyme exists at a concentration consistent with flux WT. Now imagine two different mutant alleles. The first, *a1*, is a deletion of the gene and thus has a dramatic effect on the activity level of the enzyme. In the heterozygous condition the flux goes down, but not by all that much (flux Het*a1*). However, in the homozygous condition the flux goes to zero (flux Hom*a1*). If we suppose some linear mapping of flux levels onto fitness (and/or phenotype), then the dominance of the wild type over the deletion follows as a simple consequence of the form of the curve; that is, the heterozygote has much more than half the fitness of the wild type. Now, by contrast, consider a different allele, *a2*, that slightly impairs the function of the enzyme. The homozygote now has only a slightly lower flux (flux Hom*a2*) than the wild type. If the gene activity of the heterozygote is midway between that of the two homozygotes, we can also see that, due to the form of the curve, the heterozygote has a flux (Het*a2*) nearly midway between the two homozygotes. This is thought to explain the finding that slightly deleterious genes tend to be only weakly recessive but heavily deleterious mutations tend to be fully recessive (Charlesworth 1979).

provided. An analysis of expression profiles in single-gene knockouts suggests that dispensability is not passive and often is associated with upregulation of the previously silent paralog (Kafri et al. 2005).

What might be the mechanism of upregulation following deletion? Kafri and coworkers (2005) examined the correlation between RNA levels of pairs of paralogs under different growth conditions and found that, while the mean correlation of expression is low for dispensable paralogs, the variance in the correlation of expression is high. This was interpreted as evidence that the expression of dispensable paralogs is highly correlated under a few conditions, but much less correlated under most other conditions. As expected from this, dispensable genes do tend to have some 5' motifs in common. This, in turn, is suggested to underpin the ability to reprogram the transcription of one of the genes following deletion of the other (Kafri et al. 2005).

As another possible mechanism of upregulation, Kafri and coworkers (2005) proposed a simple model involving a pair of isozymes, both of which can convert a given substrate into the same product. This model assumes that both of the genes encoding these isozymes can be regulated by a particular transcription factor with different affinities for the two genes, such that it upregulates one gene (the one normally transcribed in the wild type) at a much lower concentration than it does the other (the backup). In turn, the concentration of the transcription factor is regulated by the substrate. Removal of the high-affinity isozyme gene leads to higher levels of the substrate, which in turn leads to increased levels of the transcription factor and hence upregulation of the nondeleted, low-affinity isozyme gene.

This model is inspired by the behavior of two yeast genes, *ACS1* and *ACS2*, each of which encodes an active acetyl-coenzyme A synthetase (van den Berg et al. 1996). Under glucose-limited conditions, *ACS1* is repressed, while *ACS2* is expressed. Notably, deletion of *ACS2* prevents complete glucose repression of *ACS1*, indicating that *ACS2* is indirectly involved in the transcriptional regulation of *ACS1*. It is currently unclear whether the reverse also holds—in other words, whether transcriptional reprogramming of *ACS2* is what compensates for deletion of *ACS1*.

The more general relevance of this sort of model remains undecided. Paralogs with relatively high similarity of regulatory motifs are less likely to be redundant than those with an intermediate level of overlap (Kafri et al. 2005). Moreover, He and Zhang (2006b), in reanalyzing the yeast data, argue that duplicate genes with low expression similarity appear to be highly dispensable because both expression similarity and gene dispensability are correlated with a third factor: the number of protein interactions per gene. There is, they conclude, little evidence supporting widespread functional compensation of divergently expressed duplicate genes by transcriptional reprogramming.

A similar rejection of widespread transcriptional reprogramming derives from analysis of synthetic lethals. It isn't necessary that the two genes (one the knockout, one the compensating gene) be sequence-related; they only need to be able to provide backup for one another. The key indicator of this more general compensatory relationship is a synthetic sick or lethal (SSL) interaction, in which removal of both genes yields a much more deleterious phenotype than either single mutation alone. In the case of synthetic lethals, both single knockouts will be viable while the double knockout will not be viable. Wong and Roth (2005) have used this as the basis of a general test for transcriptional upregulation in yeast. They started with a known set of gene pairs showing SSL interactions and an extensive set of expression profiles of single-gene knockout strains. For each transcriptionally profiled mutant, they then considered expression levels of genes known to be SSL interactors with the deleted gene, comparing expression before and after deletion. Importantly, SSL and non-SSL pairs show similar differences overall between wild-type and mutant expression, suggesting that transcriptional regulation of compensatory genes does not play a major role in dispensability. Moreover, there is no difference between SSL gene pairs that show some degree of homology (i.e., that are paralogs) and those that do not. These results appear to contradict the claim that upregulation for paralogs is common (Kafri et al. 2005). However, Wong and Roth (2005) do identify 13 instances where a significant level of upregulation is observed. In sum, while there is disagreement as to whether transcriptional compensation is rare (Wong and Roth 2005; He and Zhang 2006b) or not (Kafri et al. 2005), it does occur for at least some SSL pairs (see also Zhao et al. 1998; Lesage et al. 2004).

If upregulation of paralogs is rare, might there be alternative explanations for the association of dispensability and the presence of paralogs? For the most part it has been assumed that duplicates are associated with nonessential genes owing to some mechanism by which one protein can substitute for the other. But there could be at least two other explanations. First, what if genes that are dispensable when single copy are more likely to be successfully duplicated? What if, for example, highly condition-specific genes are easier to successfully duplicate? In this case we would see an abundance of nonessential genes (under laboratory conditions) with paralogs. Some evidence supports a relationship between dispensability and duplicability (He and Zhang 2006a). Alternatively, gene duplication could be followed by specialization of the two genes, with only one of the two adopting previously essential functions. The relevance of such a model is currently not clear.

Backup by paralogs or by distributed robustness: which is more important?

Given that there are at least two mechanisms for dispensability, it is valid to ask whether they are equally important. In a seminal study, Gu and colleagues showed that the fitness effect of single-gene deletion is generally lower for duplicates compared to single-copy genes (Gu et al. 2003). The authors concluded that compensation among duplicate genes accounts for at least a quarter of no-phenotype gene deletions in yeast. A similar observation was made in genome-wide RNAi experiments in *C. elegans* (Conant and Wagner 2004).

As noted above, the inference from these results—that gene duplicates provide backup—rests on the unjustified assumption that duplicate genes and single-copy genes encode equally important cellular processes. On the contrary, essential genes are less likely to undergo gene duplication (He and Zhang

2006a), possibly because gene duplication of an essential gene may cause serious dosage imbalance among protein complex subunits (Papp et al. 2003a).

To control for such a bias, we compared the proportions of experimentally verified essential genes that encode single-copy enzymes versus duplicated isoenzymes (Papp et al. 2004). Only genes predicted to encode essential reactions were considered. We found that very few essential enzymes are isoenzymes, strongly supporting previous claims that dispensability results partially from redundant gene duplicates (Gu et al. 2003). Two exceptions (failure of duplicates to compensate) were thioredoxin reductase, which has two isozymes (TRR1 and TRR2), and inorganic pyrophosphatase, also with two isozymes (IPP1, IPP2). Their failure to compensate might be due to lack of duplicate enzyme activity in the same subcellular compartment; in both cases, one isozyme is cytoplasmic, while the other is mitochondrial. Overall, we estimate that duplicates account for 14.6–27.8% of incidences of gene dispensability.

We next examined the effect of flux reorganization on in vivo gene dispensability. To avoid the complication of duplicate gene copies, we confined our analysis to single-copy, experimentally verified essential genes, comparing the proportions of these that encode essential versus dispensable (but non-zero flux) reactions. We hypothesized that essential and dispensable reactions should differ in the network's ability to compensate for loss. From our analysis we estimate that this mode of compensation can only explain 3.8–17% of gene dispensability.

In sum, of instances of nontrivial dispensability, there are possibly about two or three times as many examples of dispensability owing to duplication as to distributed robustness. These figures are in good agreement with ^{13}C-tracer experiments (Blank et al. 2005), in which flux is directly measured. These experiments suggest that, for 207 viable mutants of active reactions, network redundancy through duplicate genes is the major (75%) molecular mechanism, and alternative pathways the minor (25%) molecular mechanism of genetic network robustness in yeast.

The minor contribution of alternative pathways/distributed robustness may well be because the yeast metabolic network has difficulties tolerating extensive flux reorganization (Papp et al. 2004). However, it may be wise not to generalize too much from metabolic analyses, because in other systems (e.g., signaling systems) distributed robustness may be more common (Wagner 2005a).

Why Did Dispensability Evolve?

The most striking feature of knockout studies is just how many genes appear to be nonessential in laboratory conditions. Why is this? Could it be that the spread and retention of a duplicate (or alternative) pathway was favored because it provided backup against mutations? Or might backup simply be a fortuitous side product of a duplicate (or alternative) pathway that was retained for other reasons?

To address this question one can take either a general or a specific approach. In the general approach, we consider the conditions under which any mutation that reduces fitness could be buffered, while not specifying any detailed mechanism or considering any specific class of mutation (point mutation, deletion, etc.). In the specific approach, by contrast, we examine the known mechanisms of dispensability associated with knockouts (gene duplication, distributed robustness, condition specificity) and ask about the conditions under which selection might favor each one. For example, could selection oppose the deletion of one of a pair of genes because individuals with just one copy are more profoundly affected by deletion of the remaining gene?

General conditions for the evolution of redundancy

Mutation rate and population size are two key parameters that can determine whether selection can favor increased robustness against null mutations. As regards the influence of mutation rate, the logic is clear. If deleterious mutations are rare and have relatively little effect on fitness, then mutation is not going to provide the conditions under which compensatory mutations can evolve (e.g., Proulx and Phillips 2005). For example, Wilke and colleagues (2001) examined the evolution of populations of digital organisms exposed to high mutation rates. As the mutation rate was increased, competition began to favor the genotype with the lower replication rate. These slowly dividing genotypes, although at lower fitness peaks, were also located in flatter regions of the fitness surface—in other words, each mutation had less impact on fitness.

Deletion rates, at least of small deletions, are at least an order of magnitude lower than the point mutation rate (Ophir and Graur 1997). Hence, a priori, we would not expect selection to favor evolution of mechanisms to provide resilience to knockouts, just because they would be so rare. Similarly, Wright's major objection to Fisher's selectionist theory of dominance was that deleterious mutations held at mutation–selection equilibrium would be too rare to favor unlinked modifiers of dominance (for discussion see Bourguet 1999). It is notable that one of the few cases of empirical evidence for the evolution of robustness, at least to point mutations, comes from an organism with a very high mutation rate (see also Elena et al. 2006). Wagner and Stadler (1999), examining RNA viruses, looked at predicted RNA secondary structures to identify those that were conserved and those that were not. They then asked about the fate of point mutations in the two classes of RNA structures. Strikingly, they report that conserved structures are more robust to mutation than nonconserved ones.

Unfortunately, there is no general consensus on the impact of population size on the evolution of buffering/enhanced robustness against null mutations. The reason for this is that population size has different effects on the spread of deleterious mutants and modifier alleles. One group of models argue that, because deleterious mutations held at mutation–selection equilibrium are rare, selection on modifiers is generally very weak, unless population size is very high (Wagner 1999). On the other hand, because genetic drift dominates in small populations, many mutants will accumulate and the population will be far from maximizing fitness. In this latter case, it may be especially advantageous to minimize the impact of deleterious mutations (Krakauer and Plotkin 2002).

The latter model has some evidence to support it. In contrast to predictions of Wagner's model, recent empirical studies suggest that some viruses (of large population size) are not especially robust to mutations. For example, Burch and Chao (2004) demonstrated that, in contrast to classical observations in higher organisms, mutant φ6 viruses are *less* sensitive to mutations and to environmental perturbations than the wild type. Likewise, Krakauer and Plotkin's model may explain why intracellular bacteria (with very small effective population sizes) may buffer point mutations by expressing high levels of chaperones (Fares et al. 2002) to force proteins to adopt particular configurations. However, it is far from clear whether either Wagner's or Krakauer and Plotkin's model can fully explain the comparative biology of knockouts.

Krakauer and Plotkin support their model with anecdotal evidence that dispensability is more common for species with small population sizes. There is, however, little empirical support for this, aside from one in silico evolution study (Elena et al. 2007). In mice, which have a small population size, about 20–30 percent of genes (Hurst and Smith 1999; Wilson et al. 2005), perhaps even 50–60 percent (Liao and Zhang 2007), are essential. We would expect from Krakauer and Plotkin's model a higher proportion of essential genes in species with larger population sizes like yeast, worms, and most bacteria. However, if anything, we find the opposite: 20 percent are essential in yeast (Steinmetz et al. 2002) and 10 percent in worms, which have a genome rich in duplicates (Kamath et al. 2003). In bacteria, figures over 30 percent are found in only one genus, namely *Mycoplasma* (see Figure 7.1A). If anything, then, aside from *Mycoplasma*, rodents have the highest proportion (20–30%) of essential genes. However, such findings come with numerous caveats: for example, these studies used different methods to generate knockouts (or nulls) and assess viability (the 30% figure for rodents includes infertility). Nonetheless, the claim (Krakauer and Plotkin 2002) of more redundancy in small populations is far from self-evident.

The bacterial data do, however, strongly support a different pattern, namely, a higher proportion of essential genes in prokaryotic genomes with fewer genes (Spearman rank correlation, rho = -0.86, $P = 0.0001$; see Figure 7.1A). Consequently, the absolute number of genes thought to be essential in different bacteria is relatively invariant, averaging about 450–500 genes (see Figure 7.1B). Indeed, a simple model supposing there to be 450 essential genes in each bacterial genome provides a surprisingly good predictor of the proportion of genes that are essential in any given genome (see Figure 7.1A). This accords with the high proportion (78%) of genes that are essential in *Mycobacterium tuberculosis* and in the highly reduced genome of

Mycobacterium leprae (Sassetti et al. 2003). The most obvious explanation for this result is that as genomes shrink they are less likely to lose essential genes and, thus, the genomes become enriched in such genes (Pál et al. 2006b). A higher number of duplicate genes in larger genomes—duplication being a major means of increasing gene number—may provide a parallel explanation. However, the rarity of dispensable genes in the small genomes of *Mycoplasma* species may not just reflect the rarity of duplicates in small genomes, but also a higher incidence in larger genomes of genes required only under specific environmental conditions (Papp et al. 2004).

Further indirect support for the idea that organisms with small populations might be more robust comes from the observation that negative epistasis (mutations having larger effects together than expected) is especially prevalent in complex multicellular organisms (Sanjuan and Elena 2006). This pattern, it has been suggested, may be due to increased mutational robustness in the larger genomes of multicellular species (Sanjuan and Elena 2006). However, the trend toward negative epistasis in multicellular organisms may not be robust and should be treated with caution (Kouyos et al. 2007).

Specific models for dispensability support the "side-consequence" hypothesis

Specific models for dispensability have two distinct advantages over the more general models. First, they permit formulation of more precise questions. Second, they permit detailed consideration of alternative, "side-consequence" hypotheses. Consider, for example, the provision of redundancy by duplicates. Here we can ask: When is it likely that a newly created duplicate is favored by natural selection because it provides protection against disabling mutations of either gene? Another way to approach the same problem is to suppose that the duplication has reached fixation. We can then ask about the conditions under which a deletion mutation is opposed by selection because individuals with the deletion are more sensitive to loss of the paralog. For the same questions, we can frame alternative hypotheses. For example, duplicates may be maintained because a high gene dose may be selectively favored (see also Sugino and Innan 2006). An incidental side consequence of the retention of paralogs may be dispensability of the paralogs in single-gene knockouts.

Population genetical models largely agree with the mechanistically vaguer models: selection is weak if the mutation rate is low (Nowak et al. 1997). More generally, it is much more likely that a duplication deterministically spreads in a population because it provides a direct advantage, rather than because it confers buffering (Clark 1994). This conclusion is supported by empirical evidence. Flux balance analysis of the yeast metabolic network has shown that essential reactions are *not* more likely than nonessential reactions to be catalyzed by isoenzymes (Papp et al. 2004). Instead, isozymes (i.e., duplicated enzymes) appear at positions in the network where a high flux is needed (Figure 7.7). This suggests that duplicates were retained to

(A)

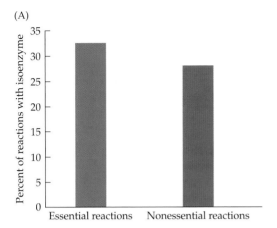

Figure 7.7 Metabolic network analysis of reactions catalyzed by single-copy enzymes and isoenzymes (duplicates that catalyze the same enzymatic reactions). Predictions on the reaction essentiality and optimal enzymatic fluxes across the network are derived by computational flux balance analysis (A) Essential reactions (i.e., those that are crucial for the production of key metabolites) are not especially likely to be catalyzed by isoenzymes compared to nonessential reactions (χ^2 test, $P > 0.5$). (B) In sharp contrast, reactions catalyzed by isoenzymes have larger fluxes than those catalyzed by single-copy enzymes (χ^2 test, $P < 0.0001$). (Data from Papp et al. 2004.)

(B)

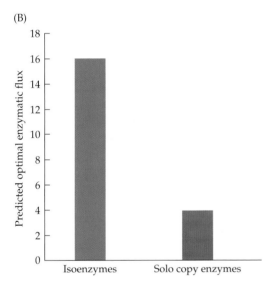

permit a selectively advantageous increase in flux rates, a secondary consequence of which can be buffering (see also Sugino and Innan 2006). Also, isozymes are often differentially regulated, indicating complementary roles during nutrient starvation or other forms of environmental stress (e.g., Kuepfer et al. 2005; Carlson 2007).

There have been relatively few attempts to consider how networks might evolve to permit distributed robustness. The evolution of this mechanism presents a particular challenge as it seems hard to explain how such a mechanism might evolve gradually (the stairs cannot function as an alternative to the lift if the staircase only goes part way; but see Wagner 2005b). An exception is provided by an analysis of hub architecture in metabolic networks (Pfeiffer et al. 2005). In these simulations of the evolution of metabolic networks, hubs and scale-free structures emerged simply under selection for growth rate (Pfeiffer et al. 2005). These results again support the notion that dispensability is just a side consequence.

Another way to ask about the evolution of distributed robustness in networks is to ask about the evolution of synthetic lethal pairs that are not sequence-related. Again we can ask whether selection is likely to favor the retention of both genes to provide such compensation or likely to produce these interactions as a by-product. There are some hints that support the second possibility. Consistent with earlier findings on bacteria and viruses showing that genetic interactions depend on environmental conditions (You and Yin 2002; Remold and Lenski 2004), at least 51 percent of synthetic lethal interactions are restricted to particular environmental conditions—that is, a large fraction of the genes involved in synthetic lethal interactions are essential under other conditions (Harrison et al. 2007).

These results are compatible with the following model (Figure 7.8). Two genes involved in a synthetic lethal interaction (e.g., A and B) sit on inde-

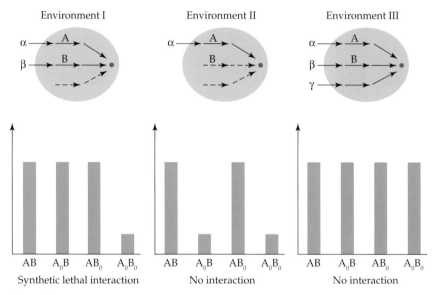

Figure 7.8 A model to explain plasticity of genetic interactions across the environment. Differences in the availability of external nutrients (α, β, and γ) across three hypothetical environments have a large effect on the activity of parallel metabolic pathways (see diagrams) and hence the impact of single (A_0B, AB_0) and double gene deletant (A_0B_0) genotypes (bar graphs). A key metabolite (red circle) can be synthesized through three different routes. Genes A and B, sitting on parallel pathways, show synthetic lethal interactions in environment I, where starting nutrients (α and β) of both pathways are present in the environment. However, B is unable to compensate for deletion of A in environment II, where only α is available. The double mutant (A_0B_0) is rescued in environment III, where a third starting nutrient is present (γ).

pendent pathways that are responsible for the utilization of two different nutrients (e.g., α and β, respectively). Under restrictive environmental conditions (e.g., when only α or β is present), the gene utilizing this nutrient (A or B) will be required for growth. Under more permissive laboratory conditions (e.g., when both α and β are present), the two genes may compensate for null mutations in each other. The retention of both genes is required by direct selection for the presence of each gene under different conditions as the organism keeps switching conditions. Hence, selection has favored retention of both genes, as they are required under different environmental conditions.

More generally, given (1) that deletion rates are low, and (2) that coherent "side-consequence" models receive more theoretical and empirical support than selectionist models, we are inclined to suppose that the high rate of dispensability of knockouts is not a reflection of selection for dispensability. This seems little different from the conclusion that dominance of wild-type alleles of metabolic genes is for the most part a side consequence of the way metabolism works (Wright 1934; Kacser and Burns 1981). This conclusion is further supported by an inability to detect the evolution of buffering in laboratory studies. Elena and Lenski (2001), for example, examined interactions between random insertion mutations and genetic background in *E. coli*. Each of the mutations was transduced into two genetic backgrounds, one ancestral and the other having evolved in, and adapted to, a laboratory environment for 10,000 generations. Importantly, the mutations were no less harmful in the derived background, suggesting that the derived bacteria had not evolved buffering mechanisms against the harmful effects of mutations.

Because of these considerations, we favor the "dispensability as side consequence" models, but it does not follow that evolution has not favored buffering of fluctuations in gene dosage (of which knockout is an extreme example). There are important perturbations other than mutation, such as developmental noise and environmental variance, which can alter the amount of functional gene product. These may potentially have more profound effects, as their rate and magnitude may be much higher than the mutation rate (Pál and Hurst 2000; Proulx and Phillips 2005). Buffering of these sorts of fluctuations might have side consequences for buffering of some genetic changes (not necessarily knockouts). This viewpoint is supported by computational studies. Ancel and Fontana, for example, found that RNA shapes that are robust against environmental perturbations are also robust against mutational perturbations (Ancel and Fontana 2000). Similarly, *E. coli* is capable of chemotaxis over a wide range of chemoattractant concentrations owing to an integral intracellular feedback system (Barkai and Leibler 1997; Alon et al. 1999). As a by-product, the system appears to be robust to changes in dosage of receptor proteins. Similar observations have been made for morphogen pattern formation (Eldar et al. 2004), segment polarity (von Dassow et al. 2000), and the cell-cycle network (Li et al. 2004).

Discussion

From the above considerations, we may conclude that dispensable genes are not really dispensable. In many cases they have essential functions that we have failed to identify owing to the nature of laboratory assays. Alternatively, there may be a paralog that can step in to perform the same function or there may be a work-around solution. Whatever the mechanism, proteins are typically under purifying selection. There is very little, if any, evidence that genes evolve dispensability owing to selection favoring robustness to mutation, except in cases where the mutation rate is exceptionally high. The alternative—that resilience to gene deletion is just a fortuitous side consequence—appears for the most part to be the more parsimonious model. Perhaps, in retrospect, this might be understood by combining theories. If both Krakauer and Plotkin (2002) and Wagner (1999) are partially correct, then the domain for the evolution of redundancy may well be small: Small populations benefit most from evolving mechanisms of redundancy (Krakauer and Plotkin 2002), but the population must also be big enough for efficient selection of genes providing robustness against mutations in the population (Wagner 1999). Hence, only at an intermediary population size might we expect selection favoring redundancy.

There are, however, many open issues. Empirical tests of the side consequence versus evolved adaptation hypotheses are relatively few. Likewise, strong comparative estimates of the rates of dispensability between genomes should be available in the not too distant future. For example, is it really true that mice have a higher proportion of essential genes than flies or free-living bacteria? If so, why? We can also ask whether gene essentiality tends to be a conserved feature. In eukaryotes it appears that it is. For example, orthologs of genes which are essential for early zebrafish development have a strong tendency to be essential for viability in yeast and for embryonic development in the worm (Amsterdam et al. 2004). In prokaryotes the emerging story seems to be different. For example, of 187 well-described essential genes in *H. pylori*, only 20 are essential in *E. coli, H. influenzae*, and *M. tuberculosis* (Salama et al. 2004; see also Sassetti et al. 2003; Zalacain et al. 2003). With such comparative data we can then also ask how genes become essential or, conversely, move from being essential to being dispensable.

The considerations above prompt numerous hypotheses about transitions between states. Consider, for example, the move from being dispensable to being essential. A more specialized lifestyle may be one factor driving this change. For example, an organism may have two energy sources, utilized by two independent pathways. In this case loss of one pathway need not be fatal. However, if the organism switches to having only one source of energy, the relevant pathway becomes essential. Alternatively, a genome may hold two copies of a gene, each providing backup for the other. Deletion of one may be neutral, in which case the population can easily evolve to a state in which only one gene remains, this one now being essential. In yet another situation, a given gene may perform more than one function.

Duplication may permit a separation of roles (subfunctionalization), leaving two possibly essential genes where before there was only one. A possible incidence of the creation of a novel essential gene via gene duplication is seen in Drosophila. K81 is a Drosophila paternal effect gene, necessary for zygote viability (Loppin et al. 2005). The distribution of K81 is restricted to just nine species comprising the melanogaster subgroup, suggesting a recent origin. The gene from which K81 was derived (probably by retroposition) is ubiquitously expressed, but it is unknown if it too had a necessary paternal effect prior to the duplication.

It is becoming increasingly clear that a large proportion of many genomes is transcribed but does not code for proteins (e.g. Semon and Duret 2004; Havilio et al. 2005). The noncoding RNAs present special challenges for the understanding of dispensability. First, they have largely been ignored in systematic screens and only in a few cases has a knockout been analyzed (e.g., H19; Leighton et al. 1995; Forne et al. 1997). Perhaps more importantly, recent analysis of such transcripts suggests that we need to reconsider what we mean when we say a "gene" is necessary. Typically, with protein-coding genes we have supposed that it is the protein product that is important. However, in yeast, some noncoding RNAs appear to be important, not because the transcript per se does anything, but rather because the process of transcription itself is important. Transcription of *SER3*, for example, is tightly repressed during growth in rich medium. The regulatory region of this gene is highly transcribed under these conditions and produces a noncoding RNA (*SRG1*). Repression of the protein-coding portion of the gene occurs by a transcription-interference mechanism in which *SRG1* transcription across the *SER3* promoter interferes with the binding of activators (Martens et al. 2004). Prevention of transcription of *SRG1* by mutation of its TATA box hence results in derepression of *SER3*. So, in what sense is *SRG1* necessary? Not in the same sense as protein-coding genes are necessary—that is, it is not the product per se that is important, it is the process of the making of the product.

Note, too, that the particular method used to create the null would influence observations. Notably, in the case of *SER3*, RNAi analysis might have had no phenotype, as it was never the transcript that was important.

CHAPTER **8**

Genome Defense

Christopher B. Schaefer, Mary Grace Goll, and Timothy H. Bestor

Introduction

TRANSPOSABLE ELEMENTS WERE FIRST IDENTIFIED IN MAIZE and were initially believed to be limited to plants. However, it is now known that almost all taxa contain transposons, and approximately 50 percent of the human genome is made up of transposon sequence.

The presence of transposons within the genome is not advantageous to the host but is in fact harmful. Active transposition threatens the stability and transcriptional regulation of the genome, and the overall fitness of the organism. For this reason, all organisms have evolved diverse mechanisms to protect themselves against the harmful effects of transposons.

Although organisms employ different mechanisms of genome defense, DNA methylation, RNA-mediated suppression, and enforced mutation of both DNA and RNA are common themes. The interferon response renders dsRNA-based silencing pathways less prominent in vertebrates. Genomes contain remnants of self-replicating DNAs that have been constrained by genomic defense systems, providing evidence of an ancient conflict between parasitic elements and their hosts that continues even today.

Transposable elements

The first evidence of mobile genetic elements was identified in now classic experiments performed by Barbara McClintock, in which she studied "jumping genes" in maize (McClintock 1946). These "jumping genes" contributed to changes in the color of maize kernels over generations of controlled crosses that could not be explained by standard Mendelian genetics—results that led to the hypothesis that at least some genes could move within genomes. Additional experimental work revealed the nature of these "jumping genes" as parasitic elements embedded in the host genome; these are

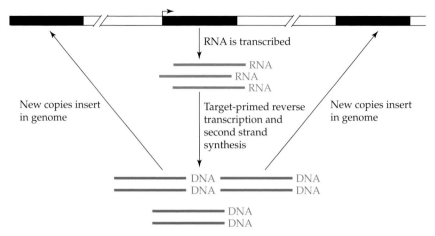

Figure 8.1 Retrotransposons replicate by an RNA intermediate. The first step in retrotransposon replication is transcription. The RNA is converted into DNA by target-primed reverse transcription and cDNA synthesis that leads to new dsDNA copies of the retrotransposons. These copies can integrate into new locations in the genome.

now referred to as transposable elements or transposons (Doolittle and Sapienza 1980; Orgel and Crick 1980).

Transposable elements are classified into two different groups (for review see Bestor 2005). Class I elements (retrotransposons or retroposons) move by means of an RNA intermediate (Figure 8.1). Integrated retrotransposon coding sequence is transcribed by host transcription factors. The RNA is then reverse transcribed into a new DNA copy of the transposon that can insert anywhere in the host genome. In mammals, transposable elements are almost exclusively retrotransposons; the largest classes of transposons in mammals include LINE-1 and SINE elements (Smit and Riggs 1996). LINE-1 elements appear to provide the sole source of transposon-derived reverse transcriptase in mammals (Deragon et al. 1990).

A second class of mobile genetic elements—Class II elements or DNA transposons—move as DNA segments without an RNA intermediate (Figure 8.2). These elements move by a cut-and-paste mechanism that requires an additional protein, termed a transposase, for transposition. This protein may be encoded within the mobile element itself or may be hijacked from other elements within the genome. The transposase binds to both inverted repeats that flank the DNA transposon, as well as to a target site. DNA at the target site is cut in an offset manner that generates short overhangs. Integration occurs via reversal of the excision reaction and a double strand break is left at the original site of the transposon. Although DNA transposons are widespread in many eukaryotes, they do not contribute significantly to mammalian genomes.

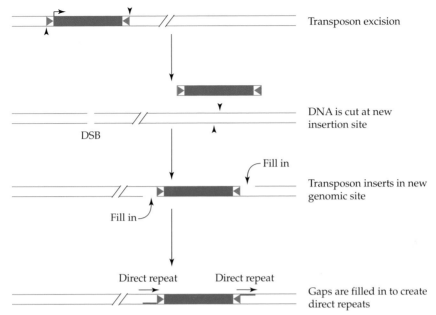

Figure 8.2 DNA transposons move by a cut-and-paste mechanism. A transposase, either self-encoded or produced by other transposons, recognizes the inverted repeats (red arrow heads) flanking the transposon and a new integration site (black arrow heads). The DNA is cleaved at the target site to generate staggered ends and the transposon is integrated by reversal of the excision reaction. A double strand break (DSB) is created at the site of excision.

Transposons compromise genomic stability

Although it is widely believed that the presence of transposable elements in host genomes proves that they provide some benefit to the host, there is little evidence to support such a hypothesis. While the sheer number of transposons in host genomes means that some will occasionally result in events that confer an advantage to the host, most transposons will harm the host, and the net effect of transposons on host fitness is negative (Hickey 1992; Bestor 2003). Transposons may actually retard evolution rather than accelerate it.

The most obvious way transposons can cause disruption of genomes is by insertion into functional genes (Figure 8.3). Transposon insertion can result in premature truncation of RNA transcripts, as well as chimeric RNAs that fuse gene and transposon sequences. Because transposons contain their own functional promoters, transposons which insert near functional genes can cause transcriptional deregulation of those genes or drive the production of antisense RNA which may interfere with mRNAs at a post-transcriptional level. In addition, multiple copies of transposable elements distributed throughout the genome may result in illegitimate pairing and recombination

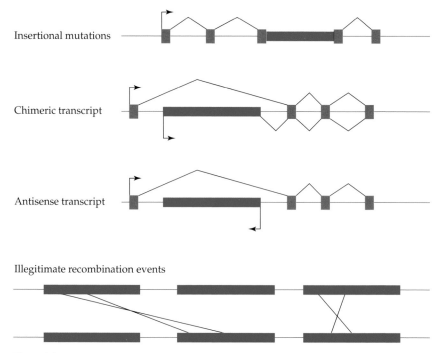

Insertional mutations

Chimeric transcript

Antisense transcript

Illegitimate recombination events

Figure 8.3 Transposons mediate the disruption of genomes. Insertional mutagenesis, chimeric transcript production, antisense effects, and illegitimate recombination are common mechanisms by which transposon activity can affect the overall fitness of its host.

between elements on nonhomologous chromosomes, resulting in deletions, insertions, and translocations (Yoder et al. 1997b).

Transposable elements are a greater threat to sexual hosts

Sexual organisms are at the highest risk from the detrimental effects of transposition (Hickey 1992). In asexual organisms, the fitness of the transposon and the host are equal, and any harm done to the host is felt to the same degree by the transposon. This is not the case in sexual organisms, where the fitness of the transposon is actually greater than that of the host organism. In sexual organisms, transposons can reproduce at a rate up to twice that of the host organism (Figure 8.4). A given host heterozygous for a transposon can transmit that transposon to all offspring. As a result, a transposon can go to fixation if it reduces host fitness by any value less than one-half, and harmful transposons can proliferate even if they substantially decrease host fitness. Sexual populations are therefore under pressure to evolve mechanisms that destroy or neutralize transposons (Bestor 2003).

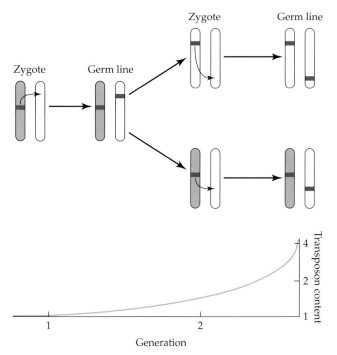

Figure 8.4 Exponential increase in transposon load in sexual populations. Transposons may be exposed to and spread to new, uninfected genomes after fertilization. At each successive generation the transposon load in the genome of a sexual host can increase exponentially.

Transposition in somatic cells may have detrimental effects, especially if transposition events result in deregulation of genes that control cell proliferation. However, the risk of transposition in somatic cells is minor compared to that of transposition in the germ line, where transposons have the opportunity to harm offspring via new insertions (Bestor 2003). In the nematode *Caenorhabditis elegans*, transposons are only silenced in the germ line and remain active in somatic cells (Emmons and Yesner 1984). In the flowering plant *Arabidopsis thaliana*, any cell has the potential to become a germ cell and *A. thaliana* has efficient transposon silencing systems that are active in all cells (Zilberman and Henikoff 2004).

Genome defense systems must continuously evolve

Genome defense systems are in a continuous evolutionary race against transposons. New transposons can arise by horizontal transfer or by retroviral infection that leads to the establishment of novel retrotransposons. Cellular sequences can acquire the ability to self-replicate and can become novel par-

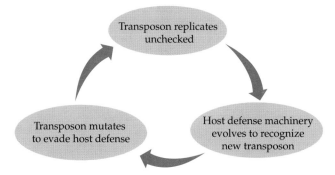

Figure 8.5 Transposons and host defense mechanisms are continually evolving. Recognition and silencing of transposons by the host puts selective pressure on transposons to evolve new mechanisms to evade the host defenses. The evolution of the transposon then requires the host to develop mechanisms to combat evasion by the transposon. This cycle will continue throughout the course of evolution.

asitic elements themselves. The primate-specific Alu retrotransposons are highly related to the 7SL RNA, and represent an example of a cellular sequence that has acquired the ability to self-replicate (Bestor 2003). In addition, the acquisition of new 5′ and 3′ sequences, recombination between transposons, and mutation of existing transposons may increase the efficiency of existing transposons. DNA sequence that becomes capable of self-replication by any mechanism can sweep to fixation in a sexual population, and genome defense mechanisms must perpetually adapt to respond to this constantly evolving threat (Figure 8.5).

Transposons and the expansion of genomes

Waves of transposon proliferation may explain what has been called the C-value paradox. The C-value paradox refers to the discrepancy between the size of an organism's genome and the number of genes encoded by that genome. As transposons proliferate and are then silenced within a host genome, evidence of their transposition remains in the form of inactive sequence remnants. The accumulation of silenced and inactive transposons results in expansion of the host genome (Pagel and Johnstone 1992; Rollins et al. 2006).

Although the reasons why one genome may harbor a larger percentage of transposons than another are complex, the extent of sexual outcrossing within a population and the efficiency of host systems devoted to transposon control both contribute substantially. Maize is a sexually outcrossing species and approximately 90 percent of the maize genome can be identified as transposable elements. In contrast, the genome of *C. elegans*, which has only rare sexual generations, is only about 12 percent transposons. No active transposons have been detected in the genome of the filamentous fungus *Neurospora crassa* (Galagan and Selker 2004), which employs one of the most aggressive systems of silencing DNA, while in *Drosophila melanogaster* one class of DNA transposons called P elements swept to fixation worldwide in only a century (Kidwell 1985).

Mechanisms of Genome Defense

The importance of double-stranded RNA (dsRNA) as a signal for the inactivation of gene expression at the transcriptional and post-transcriptional levels has only recently been fully appreciated. In mammals, dsRNA triggers an interferon response, which results in genome-wide inhibition of translation. Many other eukaryotes rely on the production of dsRNAs as a mechanism for specifically silencing homologous sequences. It has been known for some time that in many organisms the introduction of transgenes leads to the inactivation of both the transgene and the endogenous copy of the gene. Many models were created to explain these phenomena, but it wasn't until the discovery of small RNAs with homology to the transgene itself that the mechanism of RNA-mediated gene silencing began to be understood.

RNA interference

RNA interference (RNAi) was first observed when Jorgensen and colleagues attempted to make purple petunias darker by introducing an additional copy of the purple pigmentation gene that encodes chalcone synthase. Instead, the transgene-containing petunias were white, indicating that both the endogenous gene and the transgene were inactivated (Napoli et al. 1990). It is now known that the inactivation is not due to an interaction between the duplicated chalcone synthase genes themselves, as was originally thought, but is instead RNA based.

The link between RNAi and dsRNA was discovered in the nematode *C. elegans* by Andrew Fire and Craig Mello (Fire et al. 1998). While trying to interfere with the expression of a nonessential myofilament protein, UNC-22, they found that the introduction of dsRNA, but not either strand individually, led to potent and specific interference of their target gene. This discovery led others to utilize dsRNA to specifically inactivate their genes of interest in many other model organisms, such as Drosophila and plants, making RNAi a powerful molecular tool.

Double-stranded RNA is not typically produced within the cell and therefore serves as a perfect signal to alert the cell to a potentially deleterious event. The production of dsRNA can occur in at least three ways. First, many RNA viruses have a dsRNA intermediate as a part of their life cycle. Second, retrotransposons have strong constitutive promoters that can lead to the production of antisense transcripts and the formation of dsRNA. Third, the genomes of many organisms encode an RNA-dependent RNA polymerase (RdRP). It is believed that RdRPs generate dsRNA from aberrant single-stranded RNA by synthesizing a second strand de novo. Small RNAs, which are the targeting molecule of RNAi, are thought to prime the synthesis of the second RNA strand, resulting in amplification of the dsRNA signal.

Once researchers recognized that RNAi was mediated in both plants and *D. melanogaster* by small interfering RNAs (siRNAs) produced from long dsRNA, they began to search for the initiator protein of RNAi. The siRNAs appeared to have the hallmarks of dsRNA processed by an RNaseIII enzyme. This led to the identification of a class III RNaseIII protein in *D. melanogaster*

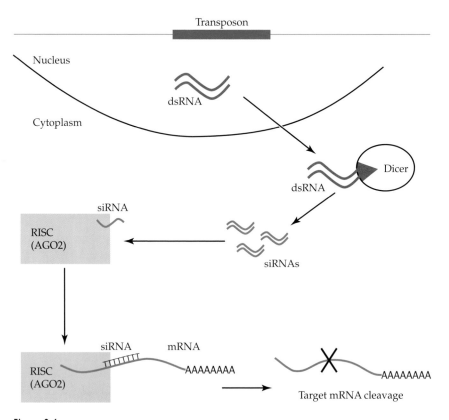

Figure 8.6 General mechanism of RNAi. Transposons produce dsRNA that is recognized and cleaved by the action of Dicer. This cleavage leads to the production of siRNAs, which then associate with Argonaute 2 (AGO2) of the RISC complex to mediate the destruction of complementary mRNA.

that was responsible for the production of siRNA (Bernstein et al. 2001). Due to its ability to cleave dsRNA into siRNAs, this protein was named Dicer. Double-stranded RNA first enters the RNAi pathway by its recognition and processing by Dicer. When Dicer binds to dsRNA, it cleaves the RNA into 21- to 23-nucleotide siRNAs that have two-nucleotide overhangs at the 5′ ends (Bernstein et al. 2001). The antisense strands of these siRNAs are loaded into a multisubunit protein complex known as the RNA-induced silencing complex (RISC) and are used to guide the sequence-specific degradation of homologous mRNA (Khvorova 2003; Schwarz et al. 2003) (Figure 8.6).

In addition to its role in processing dsRNA into siRNAs, Dicer is also required for the maturation of microRNAs (miRNAs; Grishok et al. 2001; Hutvagner et al. 2001; Ketting et al. 2001). miRNAs block gene expression by binding to partially complementary sequences located within the 3′ UTR of target mRNA (Doench and Sharp 2004). This interaction typically does not lead to cleavage of the mRNA but to blockage of translation through an unknown mechanism. The primary role of miRNAs appears to be in the regulation of gene expression during development, although miRNAs have also been identified that either enhance or inhibit viral replication (Jopling et al. 2005; Lecellier et al. 2005).

The RISC complex is the effector of the RNAi pathway. Once associated with an siRNA, a RISC-mediated surveillance mechanism is activated to scan for mRNA with sequences complementary to those of the siRNA. Once an mRNA has been recognized, it is cleaved by a "slicer" activity contained within the RISC complex, which leads to degradation of the mRNA (Tuschl et al. 1999; Hammond et al. 2000).

Characterization of the RISC complex by immunoprecipitation experiments in *D. melanogaster* led to the identification of Argonaute 2 (AGO2; Hammond et al. 2001). Argonaute proteins are a gene family conserved in most eukaryotic and several prokaryotic genomes. Argonaute proteins are characterized by two domains, the PAZ domain and the PIWI domain. AGO2 is the slicer component of RISC and has an RNaseH fold located within the PIWI domain that catalyzes the endonucleolytic cleavage of the targeted mRNA (Parker et al. 2004; Rivas et al. 2005). Additional components of RISC have been identified, but their role in RISC function or in RNAi is unknown.

Many organisms appear to use RNAi as a means to prevent or limit transposition at a post-transcriptional level, transcriptional level or both. The first natural function attributed to RNAi was the silencing of the Tc1 transposon in the germ line of *C. elegans* (Ketting et al. 1999). Characterization of miRNAs from the teleost fish *Danio rerio*, as well as *D. melanogaster*, *A. thaliana,* and the protozoan *Trypanosome brucei,* have led to the discovery of siRNAs derived from repetitive elements, which suggests an evolutionarily conserved role for RNAi in the regulation of transposon activity (Ullu and Tschudi 1984; Djikeng et al. 2001; Hamilton et al. 2002; Aravin et al. 2003; Chen and Meister 2005).

Recently a new class of small RNAs, called piwi-interacting RNAs or piRNAs, was identified in mammalian germ cells (Aravin et al. 2006; Girard et al. 2006; Grivna et al. 2006; Lau et al. 2006; Saito et al. 2006). While the function of piRNAs expressed at or following the pachytene stage of meiosis is not clear, piRNAs expressed prior to pachytene seem to be enriched for retrotransposon sequences (Aravin et al. 2007). The identification of small RNAs that appear to be derived from retrotransposons suggests that an RNA-dependent mechanism for transposon control may also be conserved in mammals.

Cosuppression

Cosuppression is a phenomenon in which the introduction of a transgene leads to the silencing of not only the transgene itself but also to endogenous genes with shared regions of homology. Cosuppression has been described in *D. melanogaster*, *C. elegans*, and plants (Pal-Bhadra et al. 1997; Dernburg et al. 2000; Ketting and Plasterk 2000). It has been suggested that cosuppression is a mechanism that allows an organism to monitor the copy number of a gene either by sensing overexpression of the gene or the production of aberrant transcripts from one or more copies of the gene; in addition, this system may have evolved as a way of repressing transposons present in multiple copies in these organisms.

Experiments with transgene expression in *D. melanogaster* demonstrated that two unpaired copies of the same transgene did not result in twice the level of gene expression, and the introduction of additional copies of the transgene led to further suppression of gene expression (Pal-Bhadra et al. 1997). This suppression of gene expression indicated that both the transgene and the endogenous copy were affected. Cosuppression in *D. melanogaster* appears to be relieved by mutations in the Polycomb complex of repressive chromatin proteins, which indicates that silencing is at the level of transcription (Pal-Bhadra et al. 1997).

Cosuppression of transgenes at the post-transcriptional level also occurs in *D. melanogaster*. The introduction of an increasing number of the alcohol dehydrogenase (*Adh*) gene causes a linear increase in *Adh* expression for up to five copies, after which the introduction of more copies leads to a decrease in expression (Pal-Bhadra et al. 1999). Results from nuclear run-on experiments indicated that the suppression was not at the level of transcription. siRNAs originating from the *Adh* gene were observed, indicating that an RNA-based mechanism acting at the post-transcriptional level was responsible for this type of cosuppression (Pal-Bhadra et al. 1999).

Quelling

Quelling is a gene silencing mechanism first described in the ascomycete fungus *N. crassa* (Romano and Macino 1992). Quelling was originally described as reversible inactivation of gene expression by transformation with repeated homologous sequences. Quelling takes place during vegetative growth and affects both endogenous genes and transgenes, as does cosuppression in plants and *D. melanogaster*. Screens to isolate quelling deficient (*qde*) mutants identified proteins homologous to those already shown to be involved in RNAi. The first *qde* gene identified (*qde-1*) was an RNA-dependent RNA polymerase (RdRP; Forrest et al. 2004). RdRPs have been shown to be essential for RNAi-like processes in the fission yeast *Schizosaccharomyces pombe*, *A. thaliana*, and *C. elegans*. The second *qde* gene identified (*qde-2*) was a member of the PPD/Argonaute family of proteins. It is probable that during vegetative growth repetitive sequences in the genome are transcribed and create aberrant RNA. This aberrant RNA is amplified and converted to dsRNA by the activity of RdRP, and is cleaved by DCL-2 (the *N. crassa* Dicer homolog) to produce siRNAs. These are then incorporated into a RISC-like complex containing QDE-2, which leads to the search for and destruction of mRNA containing complementary sequences (Catalanotto et al. 2004).

Meiotic silencing by unpaired DNA

Meiotic silencing by unpaired DNA (MSUD) is a recently identified RNA-based mechanism first found in *N. crassa* (Shiu et al. 2001). MSUD was discovered while studying meiotic transvection of the *Asm-1* gene. During vegetative growth, *N. crassa* is haploid; however, it has a transient diploid stage that results from the fusion of two haploid nuclei of opposite mating types, giving rise to a zygote. The zygote undergoes meiosis, leading to pairing of homologous chromosomes. MSUD inactivates genes that are not paired during meiosis. This inactivation is not limited to the unpaired gene but includes all copies of the gene present in the genome, whether the additional copies are paired or unpaired. This suggests that a signal that works in *trans* is responsible for the inactivation, as in the case of quelling. A screen for suppressor mutations that affected MSUD identified the *Sad-1* gene (Shiu et al. 2001). This gene was found to encode an RdRP similar to QDE-1, which suggests that MSUD and quelling are mechanistically similar phenomena. Two additional mutants in MSUD, *Sms-2* and *Sms-3* (suppressor of meiotic silencing 2 and 3), encode a QDE-2 argonaute/PPD family–like protein and a DCL-2 Dicer–like protein respectively (Shiu et al. 2001). While both quelling and MSUD use similar proteins to carry out their functions, it appears that they represent independent RNAi-like pathways in *N. crassa*.

DNA elimination

The protozoan *Tetrahymena thermophila* exhibits an RNAi-like mechanism that uses siRNA-like scan RNAs (scnRNAs). These are produced from non-genic, bidirectional, micronuclear transcripts as a means to direct genome rearrangements that lead to the elimination of specific DNA sequences (Mochizuki and Gorovsky 2004).

T. thermophila shows nuclear dimorphism, with each cell containing a germ line micronucleus and a somatic macronucleus. Although the macronucleus is derived from the micronucleus, the macronucleus lacks 15 percent of the micronuclear genome due to sequence elimination that occurs during macronuclear development. Approximately 6000 internal eliminated sequences (IES) are excised from the macronucleus. IES in *T. thermophila* vary in size from 0.5 to 20 kb and consist primarily of repetitive DNA, including transposable elements. Therefore, DNA elimination in *T. thermophila* represents a novel adaptation of RNAi for the control of transposable elements.

Twi1p, a member of the PPD/Argonaute protein family, is specifically expressed during conjugation and is required for genome rearrangement (Mochizuki et al. 2002). Twi1p interacts with and is required for the accumulation of conjugation-specific small RNAs. The current model of DNA elimination in *T. thermophila* proposes that the micronuclear genome is transcribed in a bidirectional manner early in conjugation, which results in the formation of dsRNAs. These dsRNAs are processed into scnRNAs in a Dicer-like process. The scnRNAs accumulate in the parental macronucleus, where those having homology to macronuclear DNA sequences are degraded. As a result, only scnRNAs homologous to micronucleus-specific sequences remain in the old macronucleus. These scnRNAs then move to the developing new macronucleus, where sequences homologous to scnRNAs are identified and targeted for degradation.

Transcriptional gene silencing

In some RNAi-competent organisms, siRNAs not only prevent gene expression at the post-transcriptional level but also act at the transcriptional level. RNA-dependent transcriptional gene silencing (TGS) has been observed in *S. pombe*, *D. melanogaster*, *A. thaliana*, and in the soma of *C. elegans* (Mette et al. 2000; Pal-Bhadra et al. 2002; Volpe et al. 2003; Grishok et al. 2005).

The nuclear effects of RNAi are best understood in *S. pombe*. Database searches revealed that *S. pombe* had a Dicer homolog, which argued for the existence of an RNAi-type mechanism. Cloning of siRNAs from *S. pombe* revealed that a large component of the siRNA population consisted of centromeric repeat sequences (Reinhart and Bartel 2002). Dense regions of chromosome structure termed heterochromatin are present at the centromeres, the silent mating-type locus, and telomeres in *S. pombe*. This suggested that the RNAi pathway might play a role in formation of heterochromatin at these chromosomal regions. Subsequent experiments support this idea. Mutations in components of the RNAi pathway lead to loss of heterochromatin at centromeres and to production of both sense and antisense transcripts originating from centromeres (Volpe et al. 2002). Although both sense and antisense transcripts are produced, no siRNAs are detectable when the RNA-dependent RNA polymerase gene *rpd1* is deleted (Motamedi et al. 2004). This indicates the essential role of Rpd-1 in the production of siRNAs.

The complex responsible for directing heterochromatin formation at centromeric sequences in *S. pombe* is termed the RNA-induced initiation of transcriptional gene silencing (RITS) complex (Verdel et al. 2004). In a manner similar to RISC, siRNAs associate with an argonaute homolog in the RITS complex, but instead of directing the destruction of mRNA, RITS targets regions of the genome for alterations in chromatin structure (Verdel et al. 2004).

Di- and trimethylation of histone H3 lysine 9 (H3K9) is a common hallmark of silenced sequences. In *S. pombe*, H3K9 methylation is carried out by the activity of the histone methyltransferase Clr4. It is believed that the siRNA-containing RITS complex recruits Clr4 to a sequence homologous to the siRNA (Volpe et al. 2002). Methylation of histone H3K9 leads to recruitment of the heterochromatic protein, Swi6, and leads to stable association of RITS with its target sequence (Noma et al. 2004). The binding of RITS and Rdp-1 to chromatin directly couples the generation of siRNAs to chromatin modification and allows both transcriptional and post-transcriptional silencing to be maintained in a *cis*-acting manner.

Heterochromatin formation at transposons and repetitive elements in other organisms also depends on components of the RNAi pathway, which suggests that a conserved mechanism silences the expression of deleterious elements. TGS is widely observed in the flowering plant *A. thaliana*. In the case of *A. thaliana*, it appears that siRNAs direct the methylation of cytosine residues, rather than modification of histones, through a process termed RNA-directed DNA methylation (RdDM; Aufsatz et al. 2002). The phenomena of RdDM will be discussed more extensively in the context of cytosine methylation later in this chapter.

Mammals most likely do not utilize a siRNA-mediated transcriptional silencing mechanism in somatic cells. First, mammalian cells activate a nonspecific response known as the interferon response when dsRNA is present in the cell (reviewed in Katze et al. 2002). This response is unique to mammals and results in the genome-wide inhibition of translation in response to the presence of dsRNA. The presence of dsRNAs is detected by a dsRNA-dependent protein kinase (PKR) that binds to dsRNA in the cytoplasm. Activation of PKR kinase by binding of dsRNA results in phosphorylation of the elongation and initiation factor (eIF)-2α protein and blocks translation of both viral and cellular proteins. The presence of generalized silencing in response to dsRNA not only makes siRNA-mediated TGS unnecessary, but also renders additional dsRNA-dependent mechanisms of silencing hazardous to the cell.

While the RNAi components in mammalian somatic cells may be utilized exclusively for miRNA production under normal conditions, recent data suggests a role for small RNAs in DNA methylation in mammalian germ cells. Null mutations of the piRNA-interacting proteins, Mili and Miwi2, result in demethylation of retrotransposon sequences and lead to their derepression in germ cells (Aravin et al 2007; Carmell et al. 2007). While it is clear that Mili and Miwi2 both play a role in methylating retrotransposon DNA in mammalian germ cells, it is not clear whether they are responsible for the de novo methylation of these sequences or the maintenance of DNA methylation.

Systemic silencing

A fascinating feature of RNAi observed in plants and *C. elegans* is systemic acquired silencing. Systemic acquired silencing allows for the RNAi pathway to be activated throughout an organism despite initiation in an individual cell or tissue. Systemic silencing appears to have developed as an antiviral defense mechanism that operates at the post-transcriptional level.

Systemic silencing was first observed in transgenic tobacco plants (Palauqui et al. 1997; Palauqui and Balzergue 1999). Spontaneous inactivation of the endogenous nitrate reductase gene (*Nia*) in transgenic plants carrying an additional copy of the *Nia* gene led to a reduction in the amount of available nitrogen and resulted in chlorosis, a yellowing of leaf tissue due to lack of chlorophyll. The silencing initiated in small foci and over time spread throughout most of the leaf and eventually into new growth. The signal for systemic silencing moves from cell to cell through plasmodesmata and over greater distances through the vascular system. Systemic silencing in tobacco can be triggered in new plants by introduction of sense, antisense, or promoterless DNA, or total RNA from systemically silenced plants. Therefore, it appears that any type of nucleic acid can lead to systemic silencing in plants, provided it shares homology with genomic sequences. Two size classes of siRNAs have been identified in plants. One class is 20–21 nucleotides in length and mediates RNAi in plants. The other class is 23–25 nucleotides in length and is responsible for systemic silencing (Hamilton et al. 2002).

Experiments in *C. elegans* showed that when dsRNA was accidentally injected into the body cavity instead of the gonads, this still led to silencing of the target gene in germ cells (Fire et al. 1998). This spreading results in the inheritance of the RNAi effect in the F1 progeny of the injected worms. The signal responsible for systemic silencing in *C. elegans* appears to be dsRNAs and not siRNAs. Injection of siRNAs into gonads is efficient in inducing RNAi against germ line–specific genes, but is not capable of silencing genes in other tissues. Moreover, siRNAs against a gene required for proper embryonic development lead to the death of half of the embryos when injected into one of the two *C. elegans* gonads, as would be expected in the absence of siRNA spreading (Tijsterman et al. 2002).

A screen for systemic silencing mutants identified the systemic RNAi defective 1 (*sid-1*) gene, which encodes a transmembrane protein (Winston et al. 2002). SID-1 functions in a cell autonomous manner and is expressed in cells that are in direct contact with the environment. Expression of *sid-1* in *D. melanogaster* allows the uptake of dsRNA, and the efficiency of uptake increases with longer dsRNA molecules (Feinberg and Hunter 2003).

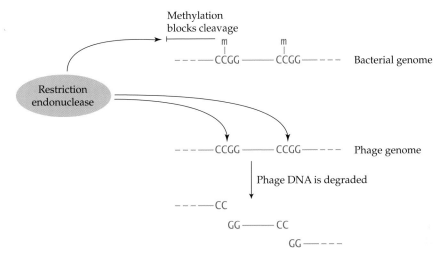

Figure 8.7 The bacterial restriction modification system. Bacteria express restriction endonucleases that recognize and cleave specific sequences in DNA. Methylation of these sequences by bacterial methyltransferase enzymes prevents cleavage of host DNA. Phage DNA is unmethylated and will be destroyed.

Cytosine Methylation

Restriction modification in bacteria

In bacteria, DNA methylation is utilized as part of the restriction modification system that protects genomes against invading bacteriophage (phage) DNA (Arber and Linn 1969). Phages are obligate intracellular parasites that invade bacteria—often integrating into the host genome—and utilize the host machinery for their own replication. In order to recognize invading phage DNA, bacteria utilize modified bases. The bacterial genome is methylated by endogenous DNA methyltransferases at the C5 or N4 positions of cytosine residues or at the N6 position of adenine residues in specific sequence contexts. Host-expressed restriction enzymes cleave unmethylated DNA at specific sequences identical to those methylated by host methyltransferases, but are unable to cleave methylated DNA at these sites (Figure 8.7) (Roberts 1985). Methylation thereby protects the host DNA from restriction, and provides a mechanism for the recognition and elimination of foreign DNA.

5-Methylcytosine and heritable silencing in eukaryotes

The modified base 5-methylcytosine is produced by the addition of a methyl group to the 5 position of the cytosine ring (Figure 8.8). 5-Methylcytosine is present in DNA of plants, vertebrates, some fungal species, and a subset of

Figure 8.8 The modified base 5-methylcytosine. 5-Methylcytosine is produced by the addition of a methyl group (CH_3) to the 5 position of the cytosine ring. The methyl group is shown in bold.

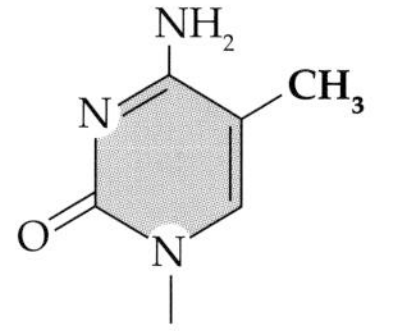

5-Methylcytosine

protostome invertebrates (Goll and Bestor 2005). DNA 5-methylcytosine provides for very strong transcriptional silencing when it is present at promoter regions. Because of the repressive capabilities of DNA 5-methylcytosine, this modification was first proposed to operate as a means of regulating gene expression. However, 5-methylcytosine represents a chemically stable mark and no confirmed demethylases have been reported. DNA 5-methylcytosine is not observed in the genomes of some popular model organisms, including *S. cerevisiae, S. pombe, D. melanogaster,* and *C. elegans* (Goll and Bestor 2005). In no case has regulation of methylation been shown to act as a reversible regulator of gene expression in any organism. Rather, 5-methylcytosine is suited to the long-term silencing of particular sequences within the genome, and its primary role in eukaryotes is likely as a mechanism for silencing of components of the genome that require transcriptional silencing over the entire life of the organism.

The bulk of DNA cytosine methylation in the genomes of plants and mammals, and almost all of that in the fungus *N. crassa*, lies within repetitive DNA (Goll and Bestor 2005). Much of this 5-methylcytosine is located in transposable elements, and loss of cytosine methylation leads to their reactivation in both plants and mammals (Li et al. 1992; Walsh et al. 1998; Okano et al. 1999; Kato et al. 2003). The fact that loss of 5-methylcytosine alone is enough to trigger the reactivation of transposable elements in eukaryotic organisms indicates that where 5-methylcytosine is present, it is the major mechanism of transposon silencing. The contributions of RNAi-related pathways or of other factors, such as histone modification, may be small. To date, only demethylation has been shown to reanimate transposons in the mammalian genome.

In addition to its ability to silence transcription, methylation of cytosine residues is mutagenic. Deamination of methylated cytosine converts it to thymidine (Figure 8.9) (Schorderet and Gartler 1992). These conversion events become fixed in the genome after replication. Mutagenesis via 5-methylcytosine results in genomes that are depleted in cytosine residues, especially in the context of CpG dinucleotides, where methylation is most prevalent. This CpG depletion is especially observed at highly methylated sequences within genomes, such as transposable elements (Schorderet and Gartler 1992). Over time, the accumulation of C-to-T mutations in promoter and coding sequences will result in the inactivation of transposable elements.

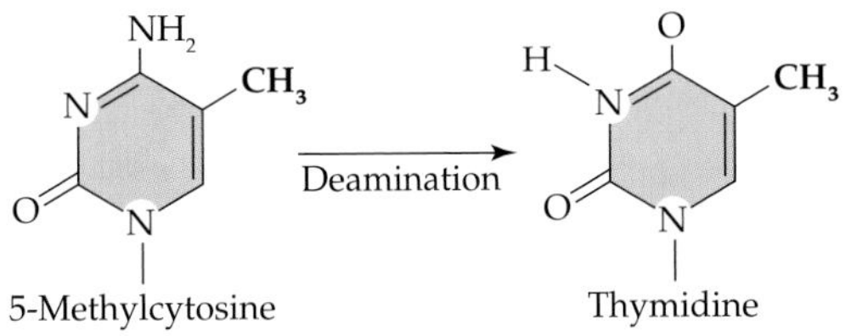

Figure 8.9 5-methylcytosine is mutagenic. Deamination of 5-methylcytosine produces thymidine. Over long periods of time, deamination of 5-methylcytosine results in accumulation of thymidine residues in transposon sequences. These mutations will eventually inactivate the transposon.

DNA methylation drives the permanent inactivation of transposons and the accumulation of inactive copies that cause the genome to expand.

DNA methylation in vertebrates

In vertebrates, 5-methylcytosine predominates at the CpG dinucleotide. The symmetrical nature of these sequences allows for clonal inheritance of methylation patterns during DNA replication. After DNA has undergone replication, the new DNA duplex is hemimethylated. Hemimethylated sites recruit the DNA methyltransferase Dnmt1, which has a preference for hemimethylated CpG dinucleotides (Figure 8.10) (Yoder et al. 1997a). Dnmt1 represents the bulk of methyltransferase activity in the cell and is localized to replication forks during S-phase of the cell cycle (Leonhardt et al. 1992). As a result of this activity, methylation in the context of CpG dinucleotides

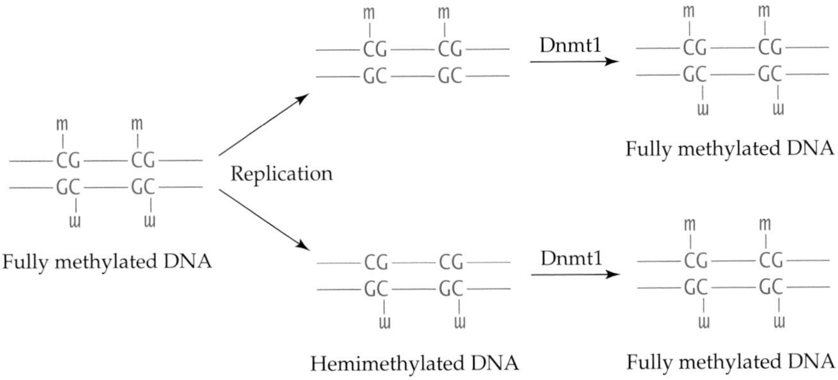

Figure 8.10 Methylation marks at CpG dinucleotides are passively maintained by proteins from the Dnmt1 family of DNA methyltransferases. Following replication of methylated DNA, two hemimethylated DNA molecules are produced. These hemimethylated sites are recognized and fully methylated by the action of Dnmt1, which leads to the faithful maintenance of the methylation state of the genome.

is propagated in an extremely stable manner: DNA methylation marks have been shown to be maintained with high fidelity for more than 80 generations in cultured cells (Schubeler et al. 2000).

Mechanisms responsible for the initial establishment of methylation patterns in vertebrate organisms are not well understood. Dnmt3 methyltransferases are required for methylation of newly integrated retroviral sequences in embryonic stem (ES) cells, and are thought to contribute to establishment of genomic methylation patterns (Okano et al. 1999). However, Dnmt1 accounts for the bulk of DNA methyltransferase activity in vertebrate organisms and has strong activity on unmethylated as well as hemimethylated sites (Yoder et al. 1997a). Therefore, it is likely that Dnmt1 may also make contributions to the establishment of methylation in some contexts.

Although DNA methylation is essential for the silencing of transposons in mammals, it is not entirely clear what signals are responsible for targeting DNA methylation to transposons. In addition to active DNA methyltransferases, the only other gene that has been demonstrated to be essential for methylation of transposable elements is *Dnmt3L* (Bourc'his and Bestor 2004). Dnmt3L proteins appear to be present exclusively in mammalian genomes, and, although Dnmt3L lacks key catalytic motifs essential for methyltransferase activity, it shows homology to Dnmt3 methyltransferases in regions outside of the catalytic motifs. Dnmt3L is essential for methylation of transposons in the male germ line of mammals. Loss of methylation from these elements in the male germ line leads not only to reanimation of transposable elements, but also results in male sterility. In the absence of Dnmt3L, developing spermatocytes exhibit large numbers of incorrectly synapsed chromosomes during meiosis (Bourc'his and Bestor 2004). Errors in synapsis are likely to result from illegitimate pairing of unmethylated transposons. Dnmt3L has been shown to interact with the DNA cytosine methyltransferase Dnmt3a, and is likely to be important for directing Dnmt3a to transposable elements in the male germ line. Dnmt3L is evolving at a much faster rate than the Dnmt3a proteins, as would be expected for an enzyme involved in recognition of transposable elements. Dnmt3L may serve as an adaptor protein that can rapidly evolve in response to changes in the nature of parasitic elements, without the constraints imposed by the retention of enzymatic activity that apply to the methyltransferase Dnmt3a (Goll and Bestor 2005).

Additional mechanisms are also likely to be involved in targeting DNA methylation to transposable elements. Dnmt3L is only important in the male germ line and no mechanisms of directing methylation to transposable elements in the female germ line or in somatic cells have been identified. Several recent papers have suggested that Polycomb group proteins, known for their repressor activities in *D. melanogaster*, may also interact with DNA methyltransferases to target DNA methylation in vertebrates. In addition, some evidence suggests that duplication of sequences itself may be a trigger for DNA methylation.

Methylation in insects

Based on studies in *D. melanogaster*, which lacks DNA cytosine methylation, it had become a common misconception that all insects lack DNA methylation. However, cytosine methylation has been reported in many other insect species, and sequence coding for Dnmt1 homologs has been identified in *Bombyx mori*, *Apis mellifera*, and *Tribolium castaneum*, members of the orders Lepidoptera, Hymenoptera, and Coleoptera, respectively. A Dnmt3 homolog has been identified in *A. mellifera* and it is likely that homologs will also be detected in the genomes of other insects as their genomic sequences become more complete. The Diptera are probably unusual in that they lack Dnmt1 and Dnmt3 homologs, and therefore lack DNA methylation. These data suggest that the majority of insects may utilize methylation systems similar to those of vertebrates in defense against genomic parasites (Goll and Bestor 2005).

Methylation in plants

As is the case in mammals, the bulk of DNA cytosine methylation in plants is mediated by DNA methyltransferases of the Dnmt1 and Dnmt3 families (Goll and Bestor 2005). The large majority of 5-methylcytosine in plants is also found in transposons and pericentric repeats, and much of this methylation is found at CpG dinucleotides (Goll and Bestor 2005). This methylation can be passively maintained through DNA replication, as is the case in vertebrates.

Unlike vertebrates, plants show significant methylation of cytosine residues in non-CpG contexts. Methylation of CpNpG sites is prevalent in transposable elements and is mediated by plant-specific methyltransferases termed chromomethylases (Lindroth et al. 2001). CpG and CpNpG methylation at transposable elements appear to serve overlapping functions in transposon silencing, as loss of either type of methylation alone results in lower levels of transposon reactivation than the combined loss of methylation in both contexts (Kato et al. 2003). Evolution of coding sequence devoid of CpG dinucleotides could represent one mechanism by which transposable elements could evade silencing by DNA methylation. Methylation at CpNpG sites may serve to block this strategy for transposon evasion of the methyltransferase machinery. Unlike CpG methylation in plants and vertebrates, methylation at CpNpG sites is entirely dependent on the modification of histones (Jackson et al. 2002). In the absence of methylation at histone H3K9, methylation at CpNpG sites is lost from plant genomes and transposable elements become partially reactivated.

Plants also exhibit methylation at all cytosine residues when this methylation is catalyzed by DRM methyltransferases. These proteins are members of the Dnmt3 family of DNA cytosine methyltransferases, but, unlike their vertebrate counterparts, appear to be directed to gene promoters and transposable elements by small RNAs. This process is termed RNA-directed

DNA methylation, or RdDM, and is initiated through an RNAi-like mechanism (Cao and Jacobsen 2002). Nuclear dsRNA is processed by a Dicer-like protein (DCL3) into siRNAs. If dsRNA from promoter sequences is produced, and processed into siRNAs, these siRNAs lead to the methylation of homologous DNA promoter sequences. Because RdDM occurs in nonsymmetrical sequences, it cannot be passively maintained in the absence of ongoing signal and requires a constant RNA stimulus for propagation through multiple rounds of cell division. Although it is tempting to speculate that small RNAs could represent a general mechanism for targeting DNA methylation, to date compelling evidence of RdDM has been observed only in flowering plants.

Methylation in N. crassa

All detectable cytosine methylation in the filamentous fungus *N. crassa* resides in repetitive elements that are silenced by a premeiotic scanning process called RIP (Repeat Induced Point mutation). RIP was the first genomic defense system identified in eukaryotes and appears to be unique to a subset of fungal species (Cambareri et al. 1989). All duplicated sequences in the *N. crassa* genome accumulate a large number of inactivating C-to-T transition mutations. Mutation only occurs in the region of homology, and all duplications that share more than 80 percent identity are detected in a single passage through the sexual cycle. Although the mechanism by which RIP generates mutations is not entirely clear, it has been shown that the DNA methyltransferase homolog RIP defective (*rid*) is required for this process (Freitag et al. 2002). RIP-mediated mutagenesis is expected to utilize the mutagenic properties of cytosine residues to trigger the same C-to-T transition events observed in other organisms, but on an accelerated time scale. It is possible that RID methylates sequences that are then subjected to deamination by a yet-to-be identified enzyme. Alternatively, RID may participate directly in deamination of cytosine residues to uracil residues, which are then fixed as thymine during replication.

Although DIM-2 catalyzed DNA methylation is dispensable for RIP itself, sequences subjected to RIP are methylated at all nonmutated cytosines in response to an increased content of TpA dinucleotides generated through RIP. This methylation can be attributed to the DNA cytosine methyltransferase DIM-2, which is a divergent member of the Dnmt1 family. In *N. crassa*, DIM-2 is responsible for methylation of cytosine residues in all sequence contexts (Kouzminova and Selker 2001). This methylation requires trimethylation of histone H3K9 (Tamaru and Selker 2001). A similar process, termed methylation induced premeiotically (MIP), in which repeated sequences are densely methylated but not mutated, occurs in a number of other fungal species and is mediated by a methyltransferase related to DIM-2 (Rhounim et al. 1992).

Silencing in *N. crassa* is extremely efficient, and the *N. crassa* genome shows a complete absence of mobile genetic elements. However, such efficient protection against genomic elements comes at a high evolutionary cost. Because silencing is so efficient, absolutely no genome duplication is tolerated. As a result, *N. crassa* cannot evolve by gene duplication and divergence, and most if not all paralogs in the *N. crassa* genome are thought to have duplicated and diverged before the emergence of RIP (Galagan and Selker 2004).

Editing as a Form of Defensive Mutagenesis

Restriction of retroviruses by G-to-A hypermutation

Recently, a novel mechanism of genome defense against parasitic DNA was identified which actively blocks retroviral replication. In primates, following retroviral infection, the host protein APOBEC3G is incorporated into the viral particle. During first strand synthesis, APOBEC3G acts to deaminate cytosine residues, which results in the accumulation of uracil residues in the cDNA (Mangeat et al. 2003). During replication, uracil is recognized as thymidine, causing adenine to be inserted into the newly synthesized DNA strand in the place of guanine (Figure 8.11). By this mechanism up to 25 percent of the guanine residues within the retroviral cDNA may be mutated to render the invading retrovirus inactive. In addition to deamination of retroviruses, APOBEC3G has also been shown to inhibit retrotransposition of Long Terminal Repeat (LTR)-based retrotransposons, which are closely related to exogenous retroviruses (Dutko et al. 2005). APOBEC3G does not appear to inhibit non-LTR retrotransposons, such as LINE-1 elements, probably because LINE-1 transposition is nuclear and APOBEC proteins are cytoplasmic (Turelli et al. 2004).

Evolution of cytosine deaminases as an antiretroviral defense appears to be restricted to mammals. Members of the APOBEC family, including the cytosine deaminase AID, which is essential for somatic hypermutation in adaptive immune response, are apparent in all vertebrates. However, APOBEC proteins outside of mammals do not appear to have antiviral properties. The mouse genome encodes one homolog of APOBEC3G called CEM15 and this protein is capable of inhibiting retroviral replication (Zhang et al. 2003). APOBECs appear to have expanded considerably in primates. Humans show evidence of nine APOBEC3 homologs. Although it is not clear that all of these APOBEC3 proteins are involved in inhibition of retroviral replication, evidence has been provided that APOBEC3B and APOBEC3F also share retroviral cDNA editing capabilities (Bishop et al. 2004). Retroviral diseases are less common in humans than in most other mammals.

It appears that some retroviruses have already begun to develop mechanisms for evading deamination by APOBEC proteins. The retrovirus HIV-1,

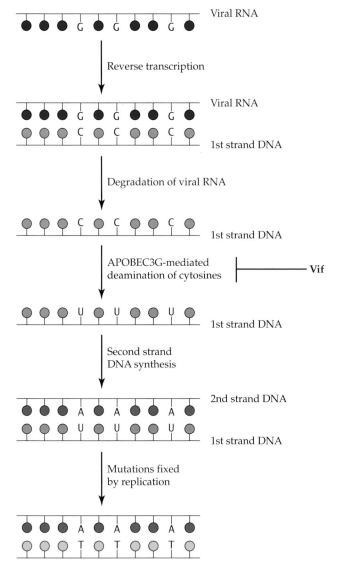

Figure 8.11 APOBEC3G-mediated mutagenesis of retroviral genomes. Upon infection, the RNA genome of the retrovirus is reverse transcribed, leading to synthesis of the first DNA strand. When incorporated into the viral particle, APOBEC3G deaminates cytosine residues in the first strand, which converts them to uracils. Following replication, these uracil residues become fixed in the viral genome through replication. The HIV-1 retrovirus combats APOBEC3G-mediated mutagenesis by production of Vif, which inhibits APOBEC3G function.

responsible for Acquired Immune Deficiency Syndrome (AIDS) in humans, encodes the Vif protein, which is an inhibitor of APOBEC3G (Mariani et al. 2003). HIV-1 strains lacking Vif cannot replicate in human cell lines, unless these lines also lack APOBEC3G. Vif provides a clear example of the ever-escalating battle between parasite and host.

A-to-I editing of SINE elements

More than 10 percent of the human genome is represented by the primate-specific SINE elements of the Alu family. Early in primate evolution these elements amplified at a tremendous rate, with approximately one new insertion fixed per generation. The rate of Alu transposition has now decreased one hundred–fold. Comparison to known genomic sequence revealed a high number of mismatches consistent with adenine (A) to inosine (I) substitutions in mRNA transcripts containing inverted Alu repeats within introns (Athanasiadis et al. 2004; Kim et al. 2004; Levanon et al. 2004). A-to-I editing is mediated by a family of adenosine deaminases (ADARs) that act on regions of dsRNA, such as those which result from inverted repeats. It is likely that the A-to-I editing trigger is not the Alu elements per se, but rather regions of dsDNA formed by the inverted repeats. The proliferation of Alu elements in primates has led to a high frequency of their occurrence in inverted repeats. Significantly lower levels of RNA editing are detected in other vertebrate organisms that also have significantly fewer Alu elements and, consequently, fewer inverted repeats present in their genomes (Kim et al. 2004).

Translation and splicing machinery interpret the nucleoside inosine as guanine and, as a result, A-to-I editing can influence the coding capacity of a given sequence. It may also serve to eliminate the potential for aberrant splice sites within Alu elements, which could be potentially mutagenic. In addition, A-to-I editing represents a mechanism by which organisms can protect regions of dsRNA in coding transcripts from being subjected to RNAi and interferon responses. It has been shown that hyperedited inosine-containing RNAs are retained in the nucleus (Zhang and Carmichael 2001), where they may be prevented from participating in dsRNA-based genome defense mechanisms in the cytoplasm. A connection between A-to-I editing and the RNAi pathway has been shown in *C. elegans*, where disruption of the editing machinery leads to silencing of transgenes. Components of the RNAi pathway have been identified as suppressors of ADAR mutations in *C. elegans*, which suggests that in the absence of ADARs some endogenous RNAs are accidentally fed into the RNAi pathway. It is likely that in mammals the lethality observed in ADAR-deficient mice is in part due to a host-directed interferon response to unedited regions of dsRNA present in RNA transcripts (Hartner et al. 2004; Wang et al. 2004).

Conclusion

The most ancestral form of immunity is that of genomic immunity. Mechanisms that protect the host genome from insults resulting from insertion of parasitic DNA are universal among sexual taxa. Genome defense mecha-

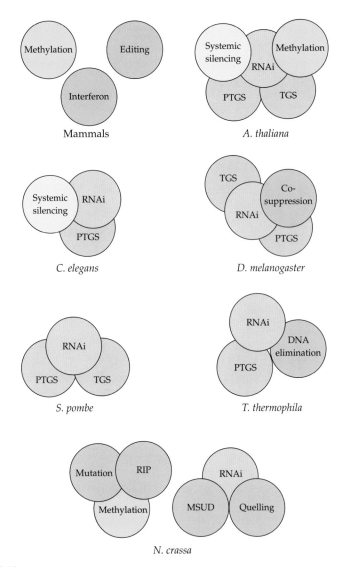

Figure 8.12 Interrelationship of host defense mechanisms in various organisms. Many organisms utilize RNAi-based mechanisms for control of parasitic elements. Mammals instead rely on a combination of the interferon response, cytosine methylation, and nucleic acid editing. *A. thaliana* has partially incorporated methylation into RNA-based mechanisms of silencing, while *N. crassa* has developed the additional system of RIP to control duplicated sequences.

nisms can be found in organisms ranging from protozoa to mammals, indicating not only that the chronic problem of parasitic DNA has been present throughout evolution, but also the tremendous danger these elements pose to genomes.

While all organisms are plagued by parasitic DNA elements, various mechanisms of genomic defense are employed to varying extents (Figure 8.12). DNA methylation, DNA elimination, mutation, and silencing of gene expression at the transcriptional and post-transcriptional levels are the most common. While some organisms use unique mechanisms to protect their genome, most organisms share a common mechanistic framework that has been modified during the course of evolution. These modifications also suggest that changes in strategies of genome defense are driven by the evolution of parasitic elements as they diverge in order to evade host genome defenses.

CHAPTER **9**

Sex-Biased Genomic Expression

Brian Oliver

Introduction

IT HAS LONG BEEN RECOGNIZED THAT FEMALES AND MALES have different morphology and reproductive roles. With the advent of the theory of evolution, it became clear that these sexual differences were important contributors to the reproductive success of each sex (Darwin 1871). Mate selection between the sexes and competition within a sex for mates are good examples of sexual selection. Some early recognized examples of the results of reproductive evolutionary pressures include brightly colored male peacocks and guppies, and outrageous appendages, such as antlers in ungulates. These traits are best seen as a consequence of the balance between the advantage they confer in competition between members of the same sex and the cost they impose on individual survival.

The rapidity of change in sexually dimorphic traits in the course of evolution has been repeatedly noted for well over a century (Darwin 1871). In a set of closely related species, reproductive differences are often noteworthy; however, systematic data on the genetic underpinning of these has been lacking. The opening of the genomic era is already providing us with important direct information on the influence of sex on the genome. We are beginning to ask some simple but very important questions about the sex-biased use and evolution of the genome. A very informative line of study has been to compare the patterns of sex-biased expression between the sex chromosomes and autosomes. Through the simple analysis of these patterns, in light of what we know about the transmission and structure of sex chromosomes, we are beginning to see full-genome–scale evidence for how copy number, tissue-biased expression, female–male competition, and the biology of chromosome structure can affect rates of change (Vicoso and Charlesworth 2006).

This is, of course, just the beginning. Genomics holds great promise for transforming the study of evolution as an extension of both natural history and experimental genetics.

One Species, Two Optimal Genomes

Phenotypic traits and the underlying genes persist because of successful transmission from one generation to the next. Gene survival depends on how well a particular organism is adapted to the environment and the stability of that particular environment. While the environment is typically thought of as a locale, such as a certain valley on the island of Hawaii or a lake in eastern Africa, interactions between and within the sexes are essential for the successful transmission of genes from one generation to the next (Figure 9.1).

Selection for alleles required for general life functions, like metabolism or immunity against infection, occurs in both females and males. The sexes share most or all of the underlying genes involved in generating these traits and those genes are subject to the same selective environment. However, given that the genome must be differentially deployed in females and males in order to generate female- or male-specific somatic structures and gametes, the optimum genome for females and males must be different (Rice 1984).

The importance of sex in evolution depends critically on the underlying magnitude of sex-biased expression. If there are only a few genes that are used differently in females and males, then evolution towards an optimized genome creates little conflict. If these differences are great, and especially if genes required for one sex are detrimental to the other, then conflict is greater. Sex-biased expression in humans and some of the major genetic model systems appears to be quite extensive; how much depends greatly on where one draws the line between nonbiased and biased expression, but at least 15 percent of the genome shows strong sex-biased expression in mammals (Rinn and Snyder 2005), fruit flies (Parisi et al. 2004), and nema-

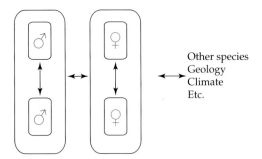

Figure 9.1 Interactions between the species and the environment, between the sexes, and among members of the same sex are all important forces in natural selection.

todes (Jiang et al. 2001). Many or most of these differences are directly related to gamete production. That the same genome is used by females and males in these species in different ways is de facto evidence that many genes have a greater importance for one of the sexes. This raises the odds that expression of some of those genes will be good for one sex and bad for the other.

Because such a large fraction of the genome shows sex-biased expression, the optimum genomes for females and males are likely to be significantly different. Because the compromise genome transits through females and males with a more-or-less equal probability, the genome is constantly being pulled in a female- or male-biased direction by selection and counterselection. The exceptions to this rule are the sex chromosomes (Charlesworth and Charlesworth 2005; Vicoso and Charlesworth 2006).

The transmission and heterogeneity of sex chromosomes provides an outstanding and simple "controlled experiment" where we can draw inference about how sexual selection modifies the genome. A long history of genomic analyses of Y chromosomes (Charlesworth et al. 2005) supports a masculinizing effect on a chromosome which has transmission solely through males. Additionally, Y chromosomes show clear evidence of degeneration, a predicted outcome from the lack of a recombination partner during meiosis. Theoretical treatments of X chromosome evolution are not as numerous and the results of genomic analyses reveal surprising differences between species. X chromosomes are masculinized in mammals (Khil et al. 2004), but not in *Drosophila melanogaster* (Parisi et al. 2003) or *C. elegans* (Reinke et al. 2004). There are some reports of rapidly evolving X chromosomes in Drosophila (Parisi et al. 2003) and others that suggest that X chromosomes and autosomes evolve at similar rates (Thornton et al. 2006). Why this is so, is not entirely clear. What is clear is that we now have the tools to address these problems systematically.

Y Chromosomes

While this chapter focuses on the X chromosome, a number of the important guiding principles are well-illustrated by the more extensively studied Y chromosomes (or the thematically similar W chromosomes of birds and butterflies; see Chapter 10).

Sex is determined by either environment or by sex chromosomes (Figure 9.2). In an environmental sex-determination system, temperature or a cue (such as the abundance of a given sex in the local population) is the determinant of female or male differentiation. One simple way that this could be encoded is by having a temperature-sensitive sex-determination gene. In such a system there is no need for sex chromosomes as the temperature-sensitive allele could be homozygous in both sexes.

However, a mutant gene that abolished temperature sensitivity would set in motion a long series of changes that could result in sex chromosome differentiation from the autosomal progenitor. Assuming that the non–temperature-sensitive allele was dominant over the wild-type and dictated male

Figure 9.2 Y chromosomes can arise from an autosomal precursor (top) and progress through a series of developmental steps. Alleles with male-biased expression (blue) are indicated on the sex chromosomes. Newly arriving/arising genes (blue arrows) and inversions (black arrows) are also indicated. Red X indicates loss-of-function mutation.

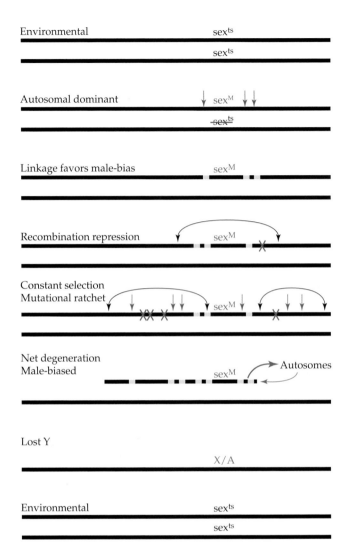

development, then it would be considered a neo–Y chromosome. Its presence would select for the loss of the recessive temperature-sensitive wild-type allele, as this allele would no longer be required for generating males within the population. This would favor specialization of the chromosome that lost the temperature-sensitive sex-determination locus as the neo–X chromosome. In this type of system, XY individuals are males and XX individuals are female.

The presence of a dominant male-determining allele results in a region of the Y chromosome that is favorable to alleles with male-biased function.

On the ancestral autosome, alleles are equally subject to selection in both females and males. Now, with the dominant sex-determining gene being passed (by definition) exclusively through males, selection can mold the genome for optimum male performance in the local region in *cis* to the dominant male-determining allele. The smaller the recombinational distance between the dominant male-determining allele and a given nearby gene on the neo–Y chromosome, the more time an allele spends in males. Mutations that are advantageous to males in those alleles most closely linked to the male-determining allele accumulate in the population even if they are detrimental to females, in part because they spend such a disproportionate amount of time in males.

Spontaneous alterations in the organization of the genome can greatly accelerate the evolution of the Y chromosome. Inversions on the neo-Y, in which a block of the chromosome is in reversed orientation, are of special interest because they effectively prevent recombination with the neo-X, creating a block of Y-linked genes that pass exclusively through males. Within the nonrecombining block, male-biased function flanking the dominant sex determining allele can flourish. These alleles are never available for selection in females. Further inversions provide traps or incubators for yet more alleles with male-biased function, eventually leading to the whole or nearly all of the Y chromosome being inherited as a masculinized block.

There is no free lunch for the Y chromosome and males. While the Y chromosome environment is good for the creation of genes with male-biased function, loss of recombination also abrogates Y chromosome repair. On autosomes and the X chromosome, recombination separates alleles at different loci, such that the fate of a deleterious allele at one locus and an advantageous one at another locus can be separated. The consequences of losing this ability are clearly seen in young to middle-aged Y chromosomes. There is no way to purge the chromosome of transposons, retroviruses, and deleterious mutations. The chromosome is selected for or against as a unit. This creates pressure to migrate Y chromosome–encoded functions to other chromosomes. While new genes do arise on the Y chromosome through amplification and translocation (Skaletsky et al. 2003), the net trend is clearly towards loss of functionality.

At some point the Y chromosome may lose enough functionality that there is no longer selection to maintain it (Graves 2006). Theoretically, this could set the stage for another round of sex chromosome evolution. Indeed, there is experimental genetic evidence for just such a transition. In *C. elegans*, temperature-sensitive alleles in a major sex-determination gene switch the sex-determination system from chromosomal to environmental (Hodgkin 2002).

Important determinants of Y chromosome evolution can be briefly summarized as residency, dominance, and mechanics. The Y chromosome shows exclusive residency in males, where it is under constant male-specific selection. There can be little doubt that this leads to a shift towards an optimal male-biased expression of the Y-linked genome. A Y chromosome allele does not necessarily have a corresponding X chromosome allele. Once a Y-linked allele has diverged sufficiently from the X-linked equivalent, it effectively becomes hemizygous. All new mutations in such alleles are dominant and under immediate selection. This is important. New recessive alleles on autosomes must become abundant enough in the population to give rise to a homozygote. This is a serious hurdle in a large outbreeding population. The first male to receive such a Y chromosome allele from his father will be affected by that mutation; an allele with a similar magnitude of effect on an autosome might be eliminated by drift before a homozygote could undergo selection. The lack of Y chromosome recombination during meiosis provides the mechanics favorable for acquisition of male-biased functions that are protected from ever being selected in females.

X Chromosomes

What about X chromosomes? Some simple thought experiments, and more serious theoretical treatments, suggest that the sex-biased nature of the X chromosome should also differ from the autosomes (Rice 1984). Unlike the case of the Y chromosome, the scales are not tipped as extremely, and a tipping point favoring either sex can be envisioned.

An autosomal allele with a strongly sex-biased function can spread in the population even if it is detrimental to the other sex, but only if the net effect of such accumulation is beneficial. The equal residency time of the allele in females and males means that there is equal opportunity for positive or negative selection in each sex. Y-chromosomes spend all their time in males, and are thus skewed towards male-biased functions. X chromosome residency is female-biased: two-thirds in females, one-third in males (assuming a 1:1 population sex ratio). This means that X chromosome alleles have more opportunity for selection in females. While this potentially favors a female-optimized X chromosome, there is a catch (Figure 9.3). Because there are two X chromosomes in females, a fully recessive allele is only under selection in homozygotes. Therefore, like an autosomal allele with female-biased function, an X chromosome allele with female-biased function must become abundant enough in the population to generate a homozygote. In contrast, an X chromosome allele favoring males is immediately available for selection, because there is no second allele to "cover" the recessive allele. Furthermore, a female homozygous for a recessive allele favoring females and disfavoring males must have a father that bore the allele. This should make it far easier to select for X chromosome alleles that favor males.

The strength of this hemizygosity effect depends critically on the degree of genetic dominance. The other copy of the X chromosome in females masks a new, fully recessive allele, but if that allele is at least partially dominant, then it would be subject to immediate selection in females (Figure 9.4). Because of the female-biased residency of the X chromosome, dominance can give the edge to female-biased genes. Geneticists studying the genes

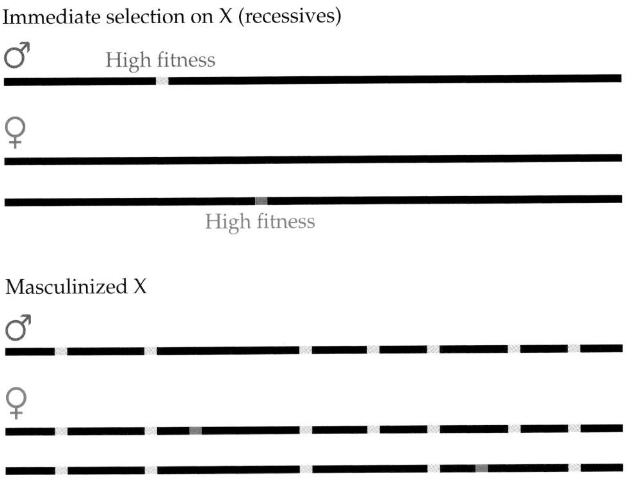

Immediate selection on X (recessives)

♂   High fitness

♀

High fitness

Masculinized X

♂

♀

Figure 9.3 Recessive mutations leading to increased fitness in either males or females are differentially selected on the X chromosome. Chromosomes (black) and alleles favoring males (blue) or females (red) are shown. The allele favoring males is immediately subjected to selection. The allele favoring females does not result in a phenotype because of the second allele with a more sex-neutral activity masks it. The result is a masculinized X chromosome with more alleles showing male-biased function. These alleles with male-biased function can become fixed in the population, while the alleles with female-biased function have a lesser chance of fixation.

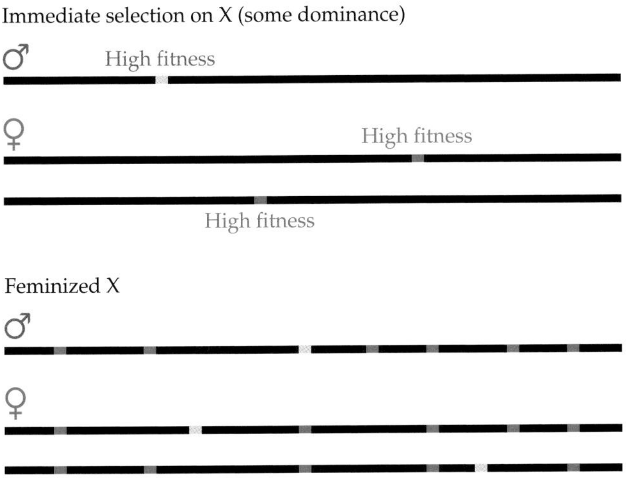

Immediate selection on X (some dominance)

♂   High fitness

♀   High fitness

High fitness

Feminized X

♂

♀

Figure 9.4 Dominant mutations leading to increased fitness in either males or females are differentially selected on the X chromosome due to residency. Dominant alleles favoring males or females are immediately subjected to selection. The alleles are selected two times in females for every time they are subjected to selection in males. The result is a feminized X chromosome with more alleles showing female-biased function.

required for development or viability of model organisms typically uncover recessive mutations, where a defective allele is covered by wild-type alleles. Very few loci are overtly haplo-insufficient and mutations giving rise to new functions are similarly rare. What is much less clear is how many alleles occupy the gray zone of partial dominance. For example, it is clear that there are isozymes and isoalleles that are both expressed and have slightly different optima for activity. These must show a degree of dominance. Selection over evolutionary timescales is a far better measure of an allele's contribution to fitness and subtle degrees of dominance, than the phenotypic characterizations performed by experimentalists; but there are some hints from experimentation suggesting that dominance may be more extensive than anticipated from the simple analyses of individual loci. Studies of oligogenic traits by quantitative genetics, screens for modifiers of existing phenotypes, and dosage compensation all suggest that many or most genes have some degree of dominance (Zhang and Oliver 2007). If a single locus is made less effective, the self-correcting nature of biological networks can compensate for the insult without generating an obvious phenotype, but minor insults to several neighbors in the gene network can result in phenotypic changes (Figure 9.5). These oligogenic phenotypic changes are under selection, and if there are enough of them, there could be a net feminization of the X chromosome.

Sex-biased expression patterns can now easily be tested. If one assumes that sex-biased expression is related to sex-biased function, then these data allow us to test ideas on X chromosome evolution. The results are exciting, as strong sex-biases in gene expression from the X chromosome have been observed in mammals, flies, and nematodes. The results are also a bit con-

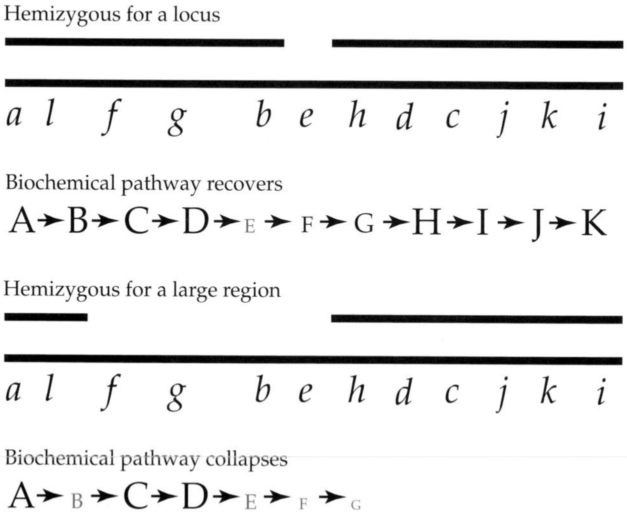

Figure 9.5 Heterozygosity for single genes are corrected by the robust nature of biological systems (top), but multiple dose changes can collapse a biological pathway (bottom). This suggests that many or most genes show some dominance. Chromosomes (thick lines), deletions (gaps), genes (lowercase italic), and proteins (uppercase roman) are shown. Direct lesions in the protein network (red) recover (increasing font size) or collapse (decreasing font size) in subsequent steps.

fusing, as there is strong evidence for a female-biased X chromosome in flies and nematode, and a male-biased X chromosome in mammals (Wang et al. 2001; Kelly et al. 2002; Lercher et al. 2003; Parisi et al. 2003; Ranz et al. 2003; Khil et al. 2004; Reinke et al. 2004;). While we are still too early on in our study of the X chromosome to know with certainty, it seems likely that similar forces are acting in these organisms. If this is true, then selection is tipped towards females or males by some differences in the biology of those organisms. There is much debate and thought going into what might mediate these tipping points, but as yet no clear resolution (Hurst and Randerson 1999; Rogers et al. 2003; Wu and Xu 2003; Oliver and Parisi 2004; Vallender and Lahn 2004; Khil et al. 2005).

Comparing Mammalian, Drosophila, and *C. elegans* X Chromosomes

It has long been suggested that the mammalian X chromosome is enriched for genes involved in reproduction and brain function. An extensive study of sex-biased gene expression in mammals shows that genes with male-biased expression are generally enriched on the X chromosome (Lercher et al. 2003; Khil et al. 2004). This result suggests that immediate selection for genes with male-biased function has been a strong force in the evolution of the mammalian X chromosome. This relatively rapid selection for genes favoring males, even if detrimental for females, is most consistent with theories of sexually antagonistic selection under the assumption of recessive alleles in females (Rice 1984).

It is also true that alleles favorable to both males and females should be selected more readily on the X chromosome via immediate selection in males. After all, not all genes that are advantageous to males would be expected to be disadvantageous to females! For example, the explosive increase in brain size in recent human evolution could be due to females choosing more intelligent mates, or to a non–sex-biased selection for more intelligent individuals. Recent work showing that the X chromosome is overexpressed in brain tissues in both female and male humans (Nguyen and Disteche 2006), might be more consistent with the latter.

The exception to the rule that genes with male-biased expression accumulate on the mammalian X, is the late male germ line (Khil et al. 2004). This is an important exception, as gametogenesis is associated with the most dramatic sexually dimorphic expression in many species (Parisi et al. 2004; Reinke et al. 2004; Rinn and Snyder 2005). It seems clear that the paucity of genes showing male-biased expression on the X chromosome during late spermatogenesis is secondary to requirements for chromosome segregation during meiosis (Wu and Xu 2003; Khil et al. 2004). High fidelity meiosis in mammals requires homolog pairing and recombination, but the X is homologous to only a short region of the Y chromosome. It is thought that precocious condensation of the X and Y chromosome during meiosis assists in

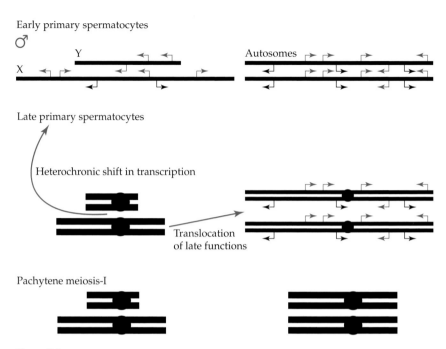

Figure 9.6 X inactivation in the male germ line is expected to alter the distribution of genes with male-biased expression in late spermatogenesis. Early (top), middle, and late (bottom) stages in development of primary spermatocytes. Genes (arrows) with male-biased (blue) or nonbiased (black) expression are shown. Transcription of male-biased and non–sex-biased functions are hindered by a repressive chromatin state during early inactivation of the X chromosome relative to autosomes (see middle stage). Those functions need to be either expressed from the X chromosome earlier (and be regulated post-transcriptionally), or moved to an autosome. This movement can be through the physical change in the linkage of a gene by transposition or through the gradual usurping of functionality by a different gene.

pairing (Figure 9.6). Given that condensed chromatin is generally repressive for transcription, this would be a strong force for moving late spermatogenesis functions to autosomes, or for heterochronic alterations in expression, such that X chromosome genes required for spermatogenesis are transcribed prior to shutdown. This is precisely what appears to be occurring. Late spermatogenesis is characterized by a deficit in X chromosome gene expression, while early spermatocytes express an excess of genes with male-biased expression (Wang et al. 2001; Khil et al. 2004).

While there may be a contribution of later reactivation in most meiotic spermatocytes, X inactivation could be a powerful factor in driving functions required for spermatogenesis to other chromosomes. It has been suggested that some of this movement could be physical displacement of genes and indeed, there is greater-than-expected traffic of retroposed genes to and from the X chromosome (Bertran et al. 2002; Khil et al. 2005). It seems likely that, in the trade-off between better chromosome transmission and completion of the spermatogenesis transcription program, there are multiple mechanisms whereby the functions of inactivated genes are taken up by genes elsewhere in the genome.

Briefly, the preliminary functional genomic data for mammals suggests that there is an overall bias towards an optimized male genome for the X chromosome. However, it is important to note that this is a general trend applying to the population of several thousand X chromosome genes, and that there could be many genes on the X chromosome that are in fact being driven by female advantage. Global expression analysis simply suggests that there are more genes where immediate selection has a greater influence than residency.

The X chromosome of Drosophila has fewer genes with male-biased expression than expected (Parisi et al. 2003; Ranz et al. 2003). There is no recombination in Drosophila males during meiosis; so, proper segregation of chromosomes in this species does not rely on recombination of nonhomologous X and Y chromosomes. Given the strong effect of X inactivation in the male germ line of mammals, it is tempting to suggest that the loss of male functions from the X chromosome in Drosophila is also due to a secondary effect of meiosis. Additional work on X chromosome inactivation in Drosophila is sorely needed, but there is weak evidence for precocious X inactivation in the Drosophila male germ line (Lifschytz and Lindsley 1972), and it also appears that there are fewer X-linked genes with late functions in spermatogenesis (Wu and Xu 2003). However, global analysis of whole testis shows no overall reduction in X chromosome transcription (Gupta et al. 2006). Most importantly, if X inactivation is the main force in decreasing the number of X chromosome genes with male-biased expression, then only genes expressed in the germ line should be affected. While sex-biased expression in the soma of Drosophila is more modest than in the germ line, there are also fewer genes with male-biased somatic expression on the X chromosome than expected (Oliver and Parisi 2004). Interestingly, the many genes that encode components of the seminal fluid are not located on the *D. melanogaster* X chromosome. In the interesting case of a translocation of an autosomal arm to the X chromosome in *D. pseudoobscura*, all four of the genes encoding these proteins on the ancestral autosome are either completely lost or relocated away from the the neo-X chromosome arm (Wagstaff and Begun 2005). This is the opposite of what is observed in mammals, where there are more genes with male-biased expression on the X chromosome. These data suggest that in Drosophila the population of X chromosome genes is more influenced by residency than by immediate selection of recessive mutations. Why this should be so is unclear, but perhaps the more streamlined genome of Drosophila has a more precariously balanced gene expression network, and thus has a higher portion of alleles exhibiting partial dominance.

The *C. elegans* X chromosome is arguably the most ancient. There are even fewer X-linked genes with male-biased expression in the germ line than is observed in Drosophila (Reinke et al. 2004). In *C. elegans* the sex is either male or hermaphrodite (basically females that produce a few sperm). Like in mammals, it is quite clear that the X chromosome is largely heterochromatic in the XO male germ line, and also in the XX hermaphrodite germ line (Bean et al. 2004). Just as immediate selection of recessives on the X can lead to the fixation of alleles benefiting males or all individuals of either sex, X inactivation drives male-biased and non–sex-biased functions off the X chromosome. This is quite evident in *C. elegans*. There is an underrepresentation of genes with sex-biased expression on the X chromosome, such that it is becoming a soma-biased chromosome (Reinke et al. 2004).

X inactivation in *C. elegans* might be a mechanism to promote meiosis, but the logic is less clear than for mammals. Disjunction of the X in the male germ line is irrelevant, as the X has no pairing partner. Further, why would X inactivation occur in the hermaphrodite germ line which has two Xs and so each X has a homolog available for pairing? Regardless of the ultimate cause of X inactivation in the germ line, it is clear that general repression of gene expression from the X chromosome would make this a disfavored location for genes required for spermatogenesis.

Inactivation may not be the only force acting on the *C. elegans* X chromosome. Like in Drosophila, the lower-than-expected frequency of X-linked genes with male-biased expression is not limited to the *C. elegans* germ line. Hermaphrodite soma-biased expression is enriched on the X chromosome (Reinke et al. 2004). As with Drosophila, this may mean that the increased residency in hermaphrodites is a more important factor than immediate selection on recessive alleles. Indeed, because hermaphrodites are self-fertile (but prefer outbreeding to males), residency of X chromosomes in hermaphrodites is more than two-thirds.

Moving Forward

In the three genomes under consideration, the X chromosome has an unusual collection of genes with male-biased expression, but three well-studied cases is hardly a large sampling. As more genomes are analyzed, we can look forward to reading more about the distribution of sex-biased expression in those genomes. While the next set of examples will probably include a lot more mammalian, insect, and nematode genomes, work on a wider range of species with different sex determination mechanisms and different reproductive strategies will be important for uncovering universal truths about the evolution of sex and sex chromosomes. Indeed, recent work on ZW systems is especially welcome (Malone et al. 2006; Storchova and Divina 2006) as the effect of maleness and femaleness can be parsed from the effects of hemizygosity and residency (which are reversed with respect to XY systems).

There is also much follow-up work needed on the major model organisms. For example, the idea that X chromosome inactivation in the male germ line is a major force in X chromosome evolution seems clear from the study of mammals and *C. elegans*, but there has been little work on understanding the role of X inactivation in the male germ line of Drosophila. The underlying reason for X inactivation is also unclear. The model for promotion of pairing with the Y makes a lot of sense for mammals, but does a poorer job describing the need for inactivation in Drosophila and *C. elegans* (Figure 9.7). Cause and effect are particularly difficult to sort out. For example, X inactivation in the male germ line might be a defense against the expression of genes with female-biased function and only secondarily a mechanism for promoting chromosome pairing and segregation.

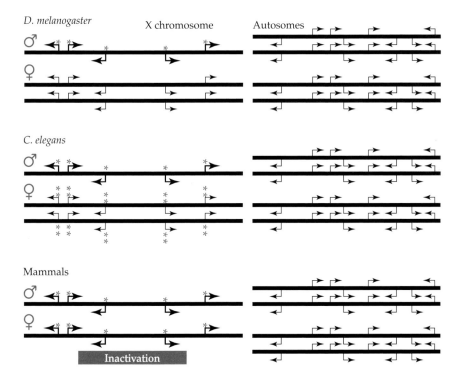

Figure 9.7 Dosage compensation restores the balance in gene expression between X chromosomes (left) and autosomes (right). Genes are represented by arrows. Increased expression (green) and/or decreased expression (red) regulates dosage compensation relative to an autosomal standard. Mammalian dosage compensation is particularly interesting because one X is inactivated in females. This monoallelic expression could allow for immediate selection of alleles affecting female fitness, but the residency time of active X chromosomes is now equal.

There are also peculiar features of mammals that could yield important general insights. For example, the mechanism of X chromosome dosage compensation in mammals provides a greater opportunity for immediate selection in females. During the course of sex chromosome evolution, there is a gradual loss of genes from the Y chromosome. This results in genomic imbalance in males, which become monosomic for a large fraction of the genome. Such imbalances are not tolerated for any other chromosomes. Having a single copy of any major chromosome is incompatible with organismal viability. Two recent studies suggest that this imbalance is rectified by increasing X chromosome expression in males (Gupta et al. 2006; Nguyen and Disteche 2006). The classic inactivation of one of the X chromosomes in mammalian females might be a mechanism to reduce overexpression of X-linked genes. Because of X inactivation, mammalian females are a mosaic of cells with one or the other X inactivated (Heard et al. 1997). Immediate selection of recessive alleles should be occurring in these females and, interestingly, this also alters effective residency. While it is true that two-thirds of X chromosomes reside in females, residency of active X chromosomes is equilibrated. This could be an important contributor to X chromosome evolution in mammals, but is complicated by a number of factors, including clonal selection within individual females due to random X inactivation. Additionally, some X-linked genes subject to imprinting are particularly interesting cases for examining the effects of functional hemizygosity (Seymour and Pomiankowski 2005). Thus, despite the added complexity of mammalian genomes, genome-scale analysis will be an important tool in this ongoing exploration.

This chapter was written in a personal capacity and does not represent the opinions of the NIH, DHHS, or the United States Federal Government.

CHAPTER **10**

Sex Chromosome Origins and Evolution

D. Charlesworth

Introduction: The Diversity of Sex Chromosomes

SEX CHROMOSOMES ARE INTRIGUING GENOME REGIONS. There is good evidence that sex chromosomes evolved from ordinary chromosome pairs, yet the "classical" Y chromosomes of mammals (Figure 10.1) and *Drosophila* species (Figure 10.2) are cytologically much smaller than the X chromosomes, and have lost most genes that must originally have been present and are still present on the X (Muller 1914; Skaletsky et al. 2003). Gene numbers per megabase are thus much reduced compared with other genome regions— a process called *genetic degeneration*. Recent developments in genetics and molecular genetics, together with genomic sequences of species with sex chromosomes (or portions of them), are giving new understanding and providing ways to test some current theories about how nonrecombining genome regions will evolve. A large amount of information is now available, and this chapter introduces the main concepts relating to each of the important questions evolutionary biologists have asked about the evolution of sex chromosomes.

Sex chromosomes evolved independently in animals and dioecious plants and they are quite diverse (see Figures 10.1–10.3; reviewed by Bull 1983). The heteromorphic sex chromosomes in model organisms (e.g., humans, mice, and *Drosophila melanogaster*; see Figures 10.1 and 10.3) consist of large genome regions that do not undergo genetic crossing-over between the Y and the X chromosomes, and small *pseudo-autosomal* regions that do recombine. Similarly, most of the large Y chromosome of the white campion, *Silene latifolia* (one of the few flowering plants with separate sexes), does not pair with the X in meiosis (Westergaard 1958). In contrast, as illustrated in Figure 10.1, the sex-determining genes of some organisms are located in small genome

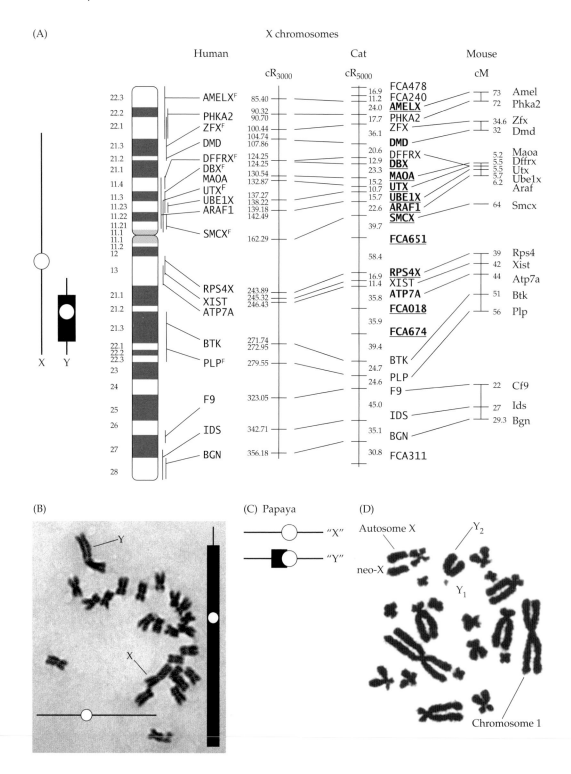

◀ **Figure 10.1** The diversity of sex chromosomes in various species. In the schematic diagrams, recombining regions are shown as narrow lines, and the nonrecombining, male-specific region of the Y chromosomes as black boxes. (A) A schematic diagram of the human X and Y chromosomes (left) shows the small size of the Y chromosome. To the right, the human X chromosome is compared with those of the mouse and cat, showing the similar gene content and order of genes; other chromosomes, including the Y chromosome, have undergone many rearrangements (Bourque et al. 2004). (B) *Silene latifolia*, a dioecious plant with sex chromosomes, in which the Y chromosome is the largest. The X and Y chromosome pair only at one end, but the size of the pseudoautosomal region is not accurately known. (C) The sex chromosomes of the plant *Carica papaya*, which are largely pseudoautosomal, with a small male-specific region. (D) The bat, *Carollia brevicauda*, in which the X chromosome is fused to an autosome, creating a new large Y_2 chromosome, in addition to the original small Y_1 chromosome. (A, from Murphy et al. 1999; B, courtesy of author; D, from Parish et al. 2002.)

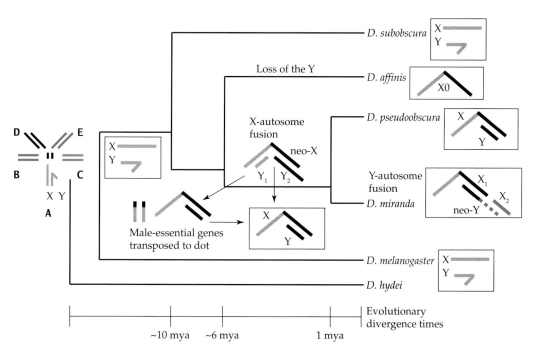

Figure 10.2 Sex chromosomes in the genus *Drosophila*. The diagram at the left shows the ancestral set of six chromosomes, five rod and one "dot." The rest of the figure shows the sex chromosome constitutions in males of several species, illustrating examples of a species with an X0 sex-determining system, and of fusions with autosomes, forming neo–sex chromosomes. (I thank A. Larracuente, A. G. Clark, and M. Noor for their unpublished information about *D. pseudoobscura*.)

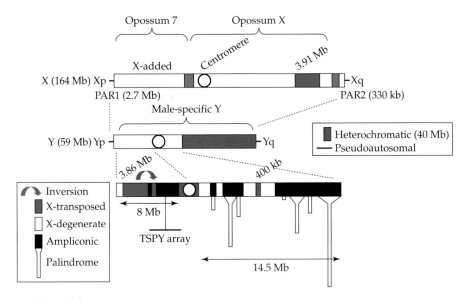

Figure 10.3 The human sex chromosome regions in detail. The similarity of the X chromosome (top) to the neo–sex chromosomes in *Drosophila* is evident. The "old" X region (carrying many genes homologous to the opossum X chromosome) consists of the Xq arm and a small part of the p arm near the centromere (the pseudoautosomal PAR2 region was recently added at the Xq tip); the X-added region (largely carrying genes homologous to the opossum chromosomes 4 and 7; see Mikkelsen et al. 2007) forms most of the p arm. The pseudoautosomal regions, PAR1 and PAR2, contain 25 genes and 5 genes, respectively. Although gene density in the rest of the X is lower than that of most human chromosomes (except two regions with homology to chicken chromosomes [green]), the total of more than 1,000 X-linked genes is roughly equally distributed along the p and q arms (see Carrel et al. 1999). The Y chromosome (middle diagram) is more than one third the size of the X chromosome, but much of it is heterochromatin. The bottom diagram indicates (not to scale) the Y regions with the different categories of MSY genes, including the large palindromic regions. The figure and Table 10.1 show clearly that the Y chromosome is largely a degenerated and rearranged form of the X chromosome. (After Skaletsky et al. 2003; Ross et al. 2005; Bhowmick et al. 2007.)

regions, surrounded by autosomal regions, and there is no cytological heteromorphism; examples include the threespine stickleback (Peichel et al. 2004) and, among plants, papaya (Ma et al. 2004; McDaniel et al. 2007; see Figure 10.1). Some of these sex-determining regions have now been shown to resemble sex chromosomes in that they lack genetic recombination, and these *male-specific Y* (or MSY) regions are clearly evolving some properties like those of the classical sex chromosomes. These nonclassical systems are receiving increasing attention, because they may have evolved recently. This remains to be tested (see later discussion in this chapter) but, if it is correct, these species may be used to study the first steps in sex chromosome evolution.

Sex chromosomes are not confined to diploids. Some bryophytes have X and Y chromosomes (e. g. Yamato et al. 2007). In these plants, the dominant life stage is the haploid gametophyte, which can produce either female gametes (X gametophytes) or male gametes (Y gametophytes), and the brief diploid stage is XY. Some organisms with heteromorphic sex chromosomes have female heterogamety (ZW females and ZZ males); these include the Lepidoptera, birds (McQueen et al. 2001), a number of reptiles (Matsubara et al. 2006), some amphibia, and a few dioecious plants (though most of these have male heterogamety; Westergaard 1958). Female heterogamety probably evolves rarely, and may often be derived from male heterogamety. ZW/ZZ systems are less frequent than XY/XX systems, and, in teleost fish and amphibia, they appear to be derived from XY species, which are often found among related species, sometimes very close relatives (Miura et al. 1998; Mank et al. 2006).

Even in the era of genome sequencing, classical Y chromosomes and MSY regions are difficult to study. Their gene density is low (Table 10.1 shows the data from humans), and their intergenic regions and introns contain large amounts of repetitive sequence, which imposes several technical difficulties. Furthermore, they have rearrangements, so that one homolog cannot be used to help align the other, unlike the autosomes. The X chromosomes are not greatly rearranged among different species, so it is inferred that the Y chromosome has undergone many rearrangements and contains repeats

Table 10.1   Human sex-linked genes

Region and stratum	Functional genes (campliconic genes)	Pseudogenes	Genes with male function[a]	Genes with ubiquitous expression
X and Y copies (X-degenerate)				
1 (old X)	5 (4)	0	4 (3 campliconic)	1 (non-campliconic)
Regions added to X ≈ 120 mya (p arm)				
Stratum 2	3 (1)	0	1	2
Stratum 3	7 (1)	3	1 (campliconic)	6
Strata 4 and 5	6 (2)	7	1 (campliconic)	2
Recently X-transposed genes				
——	3 (0)	0	1	0
Other genes on the Y but not the X				
——	3 (3)	0	3	0

[a] Estimated number (showing the predominance of such genes on the Y chromosome).

Sources: Skaletsky et al. 2003; Ross et al. 2005; Bhowmick et al. 2007.

(see Figure 10.3). To avoid confounding repeats in a single Y chromosome with polymorphisms between different individuals, the highly repetitive Y chromosome can be sequenced from a single individual, but this does not solve all difficulties. Assembly of highly repetitive genomes is evidently very difficult—it requires large sequenced regions, such as BAC clones, but these may be difficult to sequence if they contain repetitive sequences (which are sometimes unstable when cloned, and so cannot be sequenced, or may compete in PCR reactions, so that some copies fail to amplify and thus remain unknown). If the repetitive sequences are AT-rich, poor strand separation may impede sequencing reactions. Even with a sequence in hand, low gene density makes finding genes very difficult. The *D. melanogaster* and mosquito (*Anopheles gambiae*) genomes illustrate the difficulties (Carvalho 2002; Krzywinski et al. 2006). Nevertheless, the gene content of these chromosomes and the unusual evolutionary processes affecting them are starting to be understood.

Before discussing sex chromosome sequences, we will first consider the evolution of separate sexes and explain why this leads to nonrecombining regions on a *proto–sex chromosome*. Sex chromosomes have evolved independently many times, in different animal and plant taxa, which is a favorable situation for testing evolutionary hypotheses. This situation may also be helpful for understanding differences between the systems, for example, why different taxa have different degrees of gene loss and genetic degeneration.

After describing the reasons for recombination suppression in sex chromosomes, we will consider why the nonrecombining regions of sex chromosomes are predicted to lose genes, why they undergo rearrangements, and finally, how much they adapt after becoming restricted to one sex. Such adaptation may involve changes in genes due to selection acting on the genes themselves—for example, to increase male fertility—but some changes may be responses to degeneration of genes driven by selective processes occurring in other, linked genes. (As will be explained later in this chapter, "hitchhiking" processes, in which selection on one gene interferes with selection on linked genes, are much more important for sex chromosomes than for genome regions where recombination occurs.) Changes in the gene content of the chromosomes may also occur as they evolve into sex chromosomes, and sometimes regions from other chromosomes are translocated to the sex chromosomes.

All these processes also happen in nonrecombining genome regions other than sex chromosomes. Why and how nonrecombining genome regions evolve is an important question in its own right, and sex chromosomes are not the only such regions (see later text). Indeed, the diversity of such situations should help us to discern their common features. The similarity between sex chromosomes and the large, nonrecombining regions of some fungal genomes in which the incompatibility genes are located is so striking that fungal geneticists sometimes refer to these as sex chromosomes

(Fraser and Heitman 2004), despite the fact that fungal incompatibility does not involve any sex differences. In this chapter, however, the term *sex-determining region* is used only for species in which individuals are divided into those producing large gametes (female gametes: the eggs of animals or egg cells in the ovules of plants) and those producing small gametes (male gametes: sperm in animals and sperm cells of plants, which in flowering plants are reduced to one of the nuclei within the pollen). Production of large and small gametes (anisogamy) preceded the evolution of different male and female body types in animals (and male and female flowers in plants), and sex chromosomes evolved after the sex-determining systems themselves evolved. Understanding the evolution of sex chromosomes is thus also connected to understanding why separate sexes (dioecy) evolved; here the term *sex chromosome* will be used only for dioecious species. However, we shall employ the term sex chromosome for both classical, highly heteromorphic sex chromosomes and also the nonrecombining, sex-determining regions found in many dioecious species without chromosome heteromorphism.

The Origins of Sex-Determining Loci and Recombination Suppression around the Sex-Determining Locus Region

Evolution of an initial sex-determining region

The first step in the evolution of sex chromosomes is the evolution of a sex-determining locus; that is, a region of genome in which, in species with genetic sex-determination, genes are located that control whether the individual develops as a male or female. In many animals, including many fish (Mank et al. 2006), and some plants, sex-determination is environmental, depending on temperature or other growth conditions (Bull 1983). In most mammals, including humans, and in the plant *Silene latifolia*, primary sex-determining genes on the Y chromosome determine male development, and absence of a Y chromosome leads to female development. The Y chromosomes of all these species also carry genes required for male fertility; deletion of these genes leads to male sterility. (In most genetically well-studied organisms, male and female fertility functions involve genes in many different genome regions, and those genes on the sex chromosomes are only a subset of all fertility genes.) In *D. melanogaster* the Y chromosome does not carry sex-determining genes; sex development depends on a mechanism that counts the numbers of X chromosomes and autosomes (e.g., Schutt and Nothiger 2000). The *D. melanogaster* Y chromosome appears to carry only male-fertility function genes, all of which appear to have originated from elsewhere in the genome, since they do not have X-linked homologs (Carvalho 2002). In humans, about half of the genes on the Y chromosome have male-fertility functions, but the rest have various other functions, and, unlike *D. melanogaster*, several X-Y homologous gene pairs are known (see Table 10.1). Two of these, like the *D. melanogaster* Y-linked genes, have been duplicated onto the Y chromosome, and have no X-linked homologs. (DAZ genes have an autosomal origin and were transposed onto the Y chromosome, and this may also be the origin of other genes, including a set named the CDY genes [see Bhowmick et al. 2007]. But, for other genes that have autosomal copies, the latter may have originated by movement from the sex chromosomes; the origins of the BPY and PRY genes are unknown.)

Recombination suppression

It is often stated that recombination suppression evolves around a sex-determining locus, but it is important to understand that here the term sex-determining locus is used to mean a region in which the sex-determining genes are located, not a single gene. Selection for recombination suppression requires that some combinations of alleles at two different loci are disfavored, so that recombination is disadvantageous. Thus no such selection will occur with just a single sex-determining gene.

In the case of sex chromosomes, it has long been understood that the disadvantage of recombination is that the X and Y chromosome probably carry genes with advantages in one of the two sexes, whose expression in the other sex is disadvantageous. Specifically, in the simplest case of two genes (which probably arises in the early stages of sex chromosome evolution, as will be explained shortly), females (and thus the evolving X chromosome) must carry alleles causing male sterility, while males (and thus the Y chromosome) must have female-sterility alleles (Figure 10.4A). Ignoring, for the moment, how the two sexes evolve, it is evident that recombination will be disfavored, because, as the figure shows, it will produce chromosomes with both sterility alleles (i.e., sterile, neuter individuals) (Nei 1969).

This explanation can account for why recombination between sex-determining genes is disfavored, but it does not explain how these genes became linked in the first place. To understand this, we need to understand how the two sexes evolve. Consider a nondioecious ancestral population (with no sex chromosomes). The evolution of genetic control of sex must have involved at least two changes at nuclear genes, perhaps first producing females (by a male-sterility mutation) and then males (the two mutations in Figure 10.4). Male sterility could be advantageous, allowing such a mutation to spread in the population, despite the selective disadvantage of loss of male fertility. There are two kinds of potential advantages: (1) avoidance of self-fertilization, and (2) an increase in female fertility due to abolishing investment of reproductive resources in male functions (Lloyd 1974, 1975).

Evolution of separate sexes and proto–sex chromosomes

In an ancestral population of self-compatible hermaphrodites (or monoecious plants, with individuals bearing separate male and female flowers), avoiding self-fertilization could be strongly advantageous, since inbreeding depression is often severe. If the ancestral population had environmen-

(A)

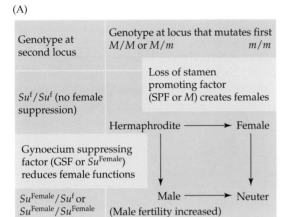

(B)

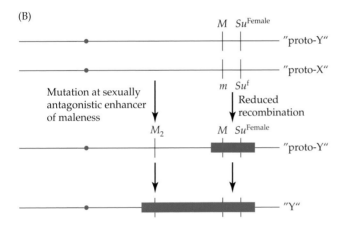

Figure 10.4 (A) The diagram shows how two successive mutations can each be advantageous, but may interact in such a way that the combination of both mutations is detrimental. One mutation (causing male sterility) creates females, which can be selectively advantageous in an initial population of hermaphrodites (see text). Another mutation (from the original allele in females, Su^f, to Su^{Female}) causes female sterility, creating males or male-biased hermaphrodites that may have higher fitness than the initial hermaphrodites (see text). This mutation is very disadvantageous if expressed in females— thus the mutations are sexually antagonistic. (B) Such situations lead to the evolution of reduced recombination (shown by the blue box around the first two genes, corresponding to the MSY regions in Figure 10.1). The lower diagram also shows how a similar interaction with a subsequent mutation might occur if a tightly linked region behaving as a single genetic sex-determining locus has evolved, and then a sexually antagonistic mutation, M_2, occurs at another locus. A similar situation arises if a single gene has taken control of sex-determination, creating a new sex-determining region, and a sexually antagonistic mutation occurs at another locus.

tal sex-determination, individuals would already have been either male or female, so this advantage would be absent. In many cases, there would be the additional advantage of resource reallocation and, if this is large enough, genetically determined females can invade an outcrossing population.

Once females are established in the population at intermediate frequencies, it can be advantageous for the hermaphroditic or monoecious individuals (often termed *cosexual*) to become more male in function. For example, in plants, male and female flower functions must often compete for resources; thus mutations increasing male fertility may often reduce female fertility; that is, they are partial female-sterility mutations. This is an example of *sexual antagonism*. The presence of females can make it advantageous to allocate more resources to male functions than in the ancestral population, even if partial female sterility results (Charlesworth and Charlesworth 1978). Male fertility must be increased more than female fertility decreases, and—a critically important point—the increase in male fertility necessary to allow such a mutation to spread in the population depends on the linkage to the locus with the male-sterility mutation that created the females. Linkage is therefore needed for the initial establishment of populations with males and females controlled by alleles at two loci.

Since male and female reproductive functions are complex, involving many genes, sterility mutations can occur at many loci. Therefore, it is not implausible that a male-sterility mutation at one locus (creating females) could be followed by a female-sterility mutation at a locus not too distant on a homologous chromosome without the male-sterility mutation, forming a *proto–Y chromosome* (Figure 10.4B). Since mutations with suitable effects on male fertility at linked loci will be rare events, the evolution of such a sex-determining region will probably take a long time. Once this has happened, the selective force for tighter linkage between the two loci, described above, will then arise (see Figure 10.4B), perhaps reducing recombination further. Further changes towards maleness are predicted, if the "males" still have some remaining female function (Charlesworth and Charlesworth 1978).

This hypothesis predicts that a chromosome initially evolves into a proto–sex chromosome without gene movements or chromosome rearrangements, and that the rearrangements common in Y (and W) chromosomes occur in the later stages of sex chromosome evolution (later in this chapter we will see some reasons why rearrangements may occur once sex chromosomes have partially evolved). The two evolving sex chromosomes will thus initially have similar gene content; this is supported by the finding that the X and Y chromosomes carry pairs of homologous sequences spread across large regions of the X genetic or physical map. Examples include the sex chromosomes in humans (see Table 10.1), birds (Lawson-Handley et al. 2004), and snakes (Matsubara et al. 2006). In the plant genus *Silene*, in which dioecy evolved within the past 10 million years (discussed later in this chapter), four genes spread over a large genetic map distance of the *S. latifolia* X chromosome, and also present on the Y chromosome, have been mapped in the nondioecious species *S. vulgaris* to a single chromosome, strongly suggesting that the sex chromosomes inherited a similar gene content from a single ancestral chromosome (Filatov 2005).

Evolution in Young and Proto–Sex Chromosomes: Sexually Antagonistic Genes

Once a sex-determining region has evolved on a chromosome, and there are two sexes in the population, a new type of sexually antagonistic allele may appear, involving further loci in addition to the sex-determining genes in the model just outlined. Once two sexes exist, mutations can arise that are advantageous in one sex, but reduce fitness in the other (see Figure 10.4; Rice 1987b). These may include mutations involved in adaptation and specialization of the newly evolved males, making them better males. In many dioecious plants, the males are hermaphrodites with an allocation bias towards male functions, but still capable of producing some seeds ("inconstant males"; Lloyd and Webb 1986). These hermaphrodites may evolve greater maleness, either by mutations at unlinked loci (which can invade populations if their effects are not sexually antagonistic, or if the effects are male-limited), or by mutations at loci on the evolving proto–Y chromosome (Charlesworth and Charlesworth 1978). Sexually antagonistic mutations will arise if, in the ancestral hermaphroditic or monoecious population, alleles at some loci have evolved to balance the selective advantages of male and female functions. Once separate sexes evolve in the population, genes on the proto–Y chromosome are free to respond and become male-biased. Tighter linkage is then advantageous, as explained earlier.

Another distinct possibility that could explain the loss of recombination between X and Y chromosomes involves a single sex-determining locus. Sexually antagonistic mutations arising at linked loci can again spread in the population only if they are closely enough linked. Such single sex-determining loci could arise in a species with a genetic sex-determining system, either by a mutation at a different locus which takes control of sex-determination (see Bull 1983), or when a sex-determining locus moves to a new genomic location (Traut and Willhoeft 1990). Some organisms that have a small sex-determining region on one chromosome may be cases of such "secondary" sex-determining regions, for example, in some fish whose related species are dioecious, such as medaka (Kondo et al. 2006), and perhaps in threespine sticklebacks (Peichel et al. 2004). In medaka, the sex-determining locus is a duplicate of an autosomal copy, and there is no copy on the homologous ("X") chromosome (Kondo et al. 2006); consistent with this, the sex-determining region in a related species is on a chromosome not homologous to the medaka X/Y pair, and there is no apparent homolog of the medaka sex-determining locus (Kikuchi et al. 2007). Although only a single sex-determining gene would initially be present, the chromosome could evolve into a new nonrecombining sex chromosome through accumulation of sexually antagonistic mutations arising at linked loci.

Determining the Age of Sex Chromosomes

Even though, as we have just seen, a small nonrecombining MSY region is predicted to exist in the early stages of sex chromosome evolution, independent evidence is necessary to conclude that a sex chromosome system evolved recently. An alternative to recent origin is that the sex-determining region simply has not become extended across a large part of the chromosome. To know whether this second possibility is plausible, we need approaches to date the origins of sex chromosomes.

Phylogenetic evidence

The mammalian case illustrates one source of information allowing the age of sex chromosomes to be determined. The mammalian sex chromosomes are known to be old, because many of the X-linked loci are on this chromosome in all mammals tested thus far, including non-Eutherian species, whose common ancestors lived more than 300 million years ago (mya) (see Figures 10.3 and 10.9A). Figures 10.3 and 10.9A also illustrate the important finding that mammalian sex chromosomes are neo–sex chromosomes (created by translocation of a new arm onto older sex chromosomes) like those that have arisen several times later in mammals (see Figure 10.1D) and in the genus *Drosophila*, by fusion of autosomes to the X or Y chromosome (see Figure 10.2). The "X-added" region (XAR) is large and carries many genes (Carrel et al. 1999; Ross et al. 2005), and this region must have been added to both the X and Y chromosome. In both the old X chromosome and the XAR, homologous X- and Y-linked gene pairs can be recognized (Skaletsky et al. 2003). The evolution of neo–sex chromosomes is discussed further later in this chapter.

Divergence between X- and Y-linked sequences

With genomic data, homologous X- and Y-linked sequences can be compared to estimate the divergence times of specified regions (i.e., the times since individual Y sequences stopped recombining with the homologous X sequences, which may not be the same in all regions of an XY pair). These estimates require information about the amounts of time corresponding to different sequence divergence values (a *molecular clock*). Ideally, sequences of genes (as opposed to intergenic regions) should be analyzed, to give reliable alignments and to provide estimates based on synonymous and intron sites, which are often only slightly affected by natural selection. The currently known gene content of the human MSY, summarized in Table 10.1, illustrates the contrast with the X chromosome, which carries more than one thousand loci (Ross et al. 2005). As one would expect, few genes in the ancient X region have retained homologs in the human Y chromosome (most have disappeared from the Y chromosome), and divergence estimates between the X and Y sequences of such *X-degenerate* genes are extremely large, consistent with this representing an ancient sex chromosome (Skaletsky et al. 2003).

There are more remaining Y homologs of genes in the XAR, and about half of the recognizable MSY homologs of genes in this region appear to be functional. The other half are pseudogenes (see Table 10.1). As discussed later in this chapter (see Figure 10.9A), divergence of these genes is lower than for the anciently X-linked genes (because they ceased recombining later in their evolution). At least 10 of the X-degenerate pairs in humans have no known Y-linked copy in mice. Some of these genes might have become lost in the rodent lineage, but the mouse Y chromosome sequence is not yet complete, so these genes may be found in the future. This difference between species is consistent with gene degeneration and loss being at least partially random, and potentially different in independent lineages.

Apart from mammals, there are only a few species in which we can compare X-Y gene pairs with copies on the Y chromosome, and estimate the times since recombination stopped. Bird W chromosomes are estimated to be have originated between 100 and 170 mya, based on two genes that seem to have stopped recombining with their Z-linked homologs before evolutionary radiation of the birds (Lawson-Handley et al. 2004). For the plant *S. latifolia*, the MSY region of the Y chromosome originated much more recently (estimated at about 10 mya, based on sequence divergence transformed into years using a molecular clock; Nicolas et al. 2005).

In *D. melanogaster*, this approach is not possible, because there are no X-Y gene pairs. All that is known is that mosquito genome sequences show that their X chromosomes are not homologous; that is, the same X chromosome has not been maintained since these species diverged (estimated to be about 250 mya).

Without gene sequences from the X and Y chromosomes, it is difficult to determine the age of a sex chromosome system, though data on related species can sometimes be helpful. If a dioecious species has nondioecious relatives known from sequence divergence to be closely related, then dioecy must have evolved recently—at least if one can exclude the possibility that nondioecious species have reverted from dioecious ones (reversion is a common event in plants). If, on the other hand, all the related species are dioecious, as in papaya, the highest divergence among the species can give a minimum estimate of the age of the system. If this minimum age is not recent, the sex chromosomes may not be young, even if they are nonheteromorphic (assuming that dioecy did not evolve several times independently in the species group). In papaya, low X-Y divergence for several genes supports a young sex chromosome system (R. Ming, pers. comm.).

Degeneration and Loss of Adaptation of Genes on the Nonrecombining Sex Chromosome

Many Y (or W) chromosomes have degenerated genetically (genes have been lost or gene functions have deteriorated, compared with the X or Z chromosome). In many species, an X chromosome is required for viability, and YY genotypes are inviable (including some plants; reviewed in Westergaard 1958). This is not due to a single lethal Y-linked mutation, because it is a general property of these chromosomes, not just some Y genotypes. In dioecious plant populations, sex ratios are often biased towards females, and it has been suggested that this is because Y chromosome–bearing pollen grow more slowly than X chromosome–bearing pollen, and fertilize less than half the ovules. Consistent with this, the sex ratio bias is indeed present in seeds of *Rumex nivalis* (Stehlik and Barrett 2005).

In the present age of genome sequences, sequence data can be used to test for the presence of functional Y-linked homologs of X-linked genes, as opposed to no homolog at all, or just a pseudogene (see Table 10.1). Even when a Y-linked copy remains, it may still exhibit degeneration; indeed, such pairs of alleles can be used in tests based on analyses of nonsynonymous and synonymous substitutions, which can reveal more than the mere fact of degeneration. In comparisons using single sequences from different mammalian species, an excess of nonsynonymous substitutions is detectable in primate Y-linked genes, but not in X genes, despite the Y genes being maintained over a long evolutionary time and having higher estimated numbers of synonymous substitutions. In all three genes studied, the results suggest that a strikingly high proportion of Y-linked substitutions behave as neutral, again suggesting that selection is unable to act effectively on the sequences of Y-linked genes (Gerrard and Filatov 2005). This suggests that some Y-linked loci are degenerating, despite selection to maintain their functions (selective constraint). Later in this chapter, we discuss further results from *Drosophila miranda* suggesting that selection is unable to prevent degenerative changes in genes on a neo–Y chromosome.

Recombination suppression around the sex-determining locus region—not just sheltering of recessive mutations kept heterozygous—is clearly the critical factor for degeneration. Modelling of the genetics shows that recessive deleterious mutations cannot reach high frequencies or become fixed on a Y chromosome if it recombines even rarely with its homolog. However, in the absence of recombination, deleterious mutations can fix even if they are not fully recessive, if the population has a low effective size (Nei 1970). Moreover, there is evidence for degeneration, including loss of a few genes, in non-recombining haploid situations other than sex chromosomes, such as the mating-type regions of *Chlamydomonas reinhardtii* and fungi, where sheltering of recessive mutations is not possible (e.g., Ferris et al. 2002; Lengeler et al. 2002).

A common property of several processes that are likely to affect evolving sex chromosomes is that they cause low effective population sizes. The ability of natural selection to eliminate deleterious mutations, or to fix advantageous ones, depends on the effective population size as well as on selection coefficients. Thus Y chromosomes are predicted to accumulate deleterious mutations that would not increase if they were X-linked, and they will also fail to benefit from advantageous mutations that would spread through a population's X chromosomes (Orr and Kim 1998). The different processes affecting nonrecombining genome regions are reviewed in Charlesworth and Charlesworth (2000), and are briefly outlined here.

Degeneration due to deleterious mutations

The first process proposed to explain degeneration involves deleterious mutations. When a region evolves to become nonrecombining, there will be a distribution of the numbers of mutations carried by individual chromosomes in the population, ranging from mutation-free chromosomes to chromosomes with mutations in several loci. Theory predicts a Poisson distribution of mutation numbers, whose mean depends on the mutation rate and selection coefficient against mutant alleles. (Most models simplify the biological reality and assume that mutations always have the same selection coefficient, whereas mutation effects actually range from small to drastic.) In a finite population, this can lead to *Muller's ratchet*—the chance loss of rare mutation-free chromosomes (Figure 10.5), which cannot reappear,

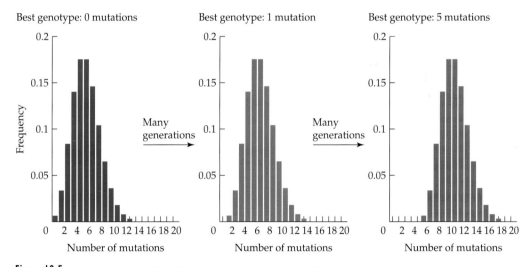

Figure 10.5 Muller's ratchet. The figure shows the expected distribution of numbers of deleterious mutations in a nonrecombining region, assuming a mean value of five mildly deleterious mutations. In finite populations, the low number of mutant-free chromosomes among the parents of each generation may allow this class to disappear from their progeny. The chromosomes with one mutation then have the highest fitness. Over the generations, the number of mutations in the best class of chromosomes will increase from time to time, successively lowering the mean fitness—the numbers of mutations do not come to an equilibrium. Because all future generations trace their ancestry back to such a best class of chromosomes in an ancestral generation, the population is effectively passing through successive bottlenecks and its effective size is reduced.

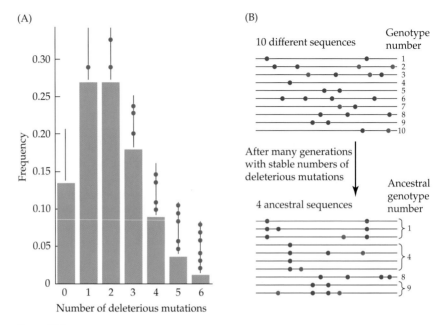

Figure 10.6 Background selection. Each chromosome carries genetic variants (blue dots), some of which may be strongly deleterious mutations (red dots). (A) The equilibrium distribution of numbers of deleterious mutations in such a region, assuming a mean value of two such mutations. (B) The selective elimination of the deleterious variants lowers the effective population size—roughly speaking, the number of lineages ancestral to the chromosomes in the population. As mutant-free sequence types in the ancestral population replace sequences carrying deleterious mutations, the descendant population accumulates multiple copies descended from only part of the ancestral population. The diagram shows how this leads to lower diversity; this effect is less than with Muller's ratchet. In turn, the lower effective population size reduces the ability of selection to fix advantageous mutations or eliminate deleterious ones.

because there is no recombination. Thus their loss could contribute to irreversible Y chromosome degeneration (Charlesworth 1978; Gordo and Charlesworth 2000). Over time, this process leads to fixation of mutations (Charlesworth and Charlesworth 1997).

Another process driven by deleterious mutations is *background selection* (Figure 10.6). Mutations with deleterious effects large enough to prevent their fixation are removed from the population by natural selection. Unless recombination happens, only mutation-free chromosomes in any generation can become ancestors of chromosomes in future generations (chromosomes carrying mutations in any of their genes have no long-term descendants). Thus, background selection, acting in a genome region with many loci, reduces the effective population size for those genes to the number of

mutant-free chromosomes; this number could be small for a large nonrecombining genome region containing many genes with a plausible mutation rate per locus. Background selection is one form of *genetic hitchhiking*,
as neutral variants in the genome region will also be removed from the population due to their linkage to the deleterious mutations being removed.
Thus, this process also reduces neutral diversity, as shown in Figure 10.6B.

Interference with the spread of favorable Y-linked mutations

Yet another process involved in degeneration is the *Hill-Robertson* effect of
deleterious mutations too weakly selected to be eliminated quickly by natural selection (for instance, synonymous changes in coding sequences, and
some conservative changes in proteins' amino acid sequences). Under muta-
tion-selection balance and genetic drift, such weakly selected mutations will
be present in populations at a range of frequencies, including frequencies
high enough for back mutation to be important. After a region stops recombining, mutations will build up until most chromosomes carry some mutations. These mutations will interfere with selection in the region—eventually stopping any further accumulation—because selection that lowers the
frequency of one mutation will increase the frequency of another one. Y
chromosomes should thus reach a low level of fitness that does not decrease
further. The presence of deleterious mutations on all individuals' Y chromosomes would then hinder the spread of favorable Y-linked mutations.

Selective sweeps

Selection for advantageous mutations can also lead to Y chromosome genes
accumulating deleterious mutations through selective sweeps, another form
of hitchhiking (Rice 1987a). Whenever a favorable mutation arises on a Y
chromosome in a population and spreads to fixation, this also causes the
fixation of any disadvantageous mutations carried on the chromosome on
which the favorable mutation originally occurred (Figure 10.7A). Equiva-
lently, a selective sweep reduces the effective population size for linked
genes, since, if only one allele remains, this causes a bottleneck in their population. In addition, advantageous mutations will need to have large enough
beneficial effects to outweigh selection against any deleterious mutations
on the same (nonrecombining) chromosome. Consequently, some advantageous mutations on Y chromosomes will fail to increase in frequency (Figure 10.7B), even though a mutation with the same selection coefficient could
have increased if it had occurred in an autosomal gene.

Evidence for degeneration processes

In addition to the abundant evidence that deleterious mutations constantly
occur, there is indirect evidence that these processes are collectively impor-

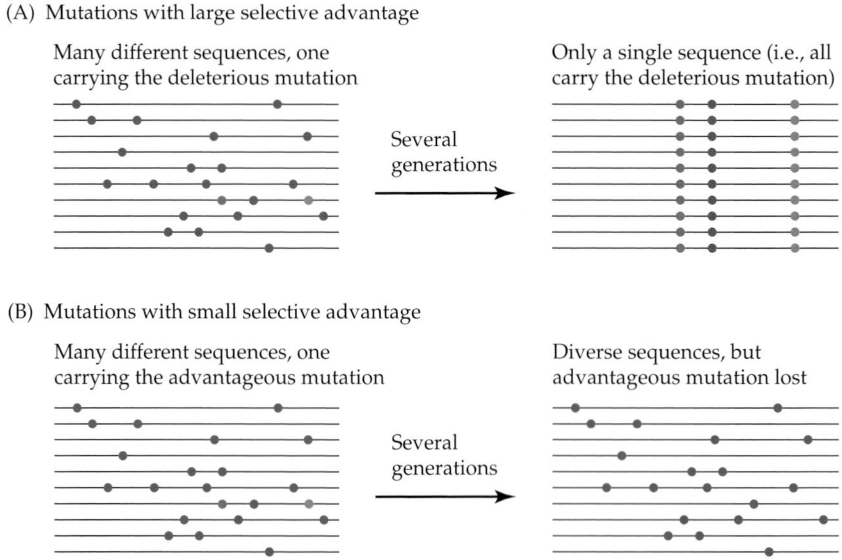

(A) Mutations with large selective advantage

Many different sequences, one
carrying the deleterious mutation

Several
generations

Only a single sequence (i.e., all
carry the deleterious mutation)

(B) Mutations with small selective advantage

Many different sequences, one
carrying the advantageous mutation

Several
generations

Diverse sequences, but
advantageous mutation lost

Figure 10.7 Reduced adaptation and diversity in a nonrecombining genome region
in which advantageous mutations (red dots) can occur, and in which there are
genetic variants (blue dots). (A) The mutation's selection coefficient is large
enough that it spreads through the population in a selective sweep. Fixation of the
initially rare, favorable mutation reduces variation at both selected and neutral
sites, and is equivalent to a bottleneck in the population; that is, to a severely
reduced effective population size. If any of the preexisting variants on the fixed
chromosome are deleterious (green dots), they may also be fixed. (B) The mutation cannot spread through the population because it is not selectively advantageous enough to outweigh the fitness effects of the deleterious alleles carried by
the chromosome (green dot). Thus adaptation is slowed down.

tant, and are active in some Y chromosomes (though it is difficult to distinguish between them and assess their relative importance). This evidence
comes from studies of DNA sequence diversity, showing that some process
or processes are reducing the effective population size of Y chromosomes. The
evidence relies on homologous pairs of X- and Y-linked genes. The first type
of evidence is that nonsynonymous substitutions are accumulating faster in
Y-linked alleles than their X-linked homologs in the recently evolved
Drosophila miranda neo–Y chromosome. This will be explained later in this
chapter, when this particularly informative chromosome is described in detail.
    The second type of evidence comes from studies of sequence diversity.
Since nucleotide diversity depends on the effective population size (assuming similar mutation rates per nucleotide), the degeneration models predict
reduced diversity of Y-linked genes, compared with other parts of genomes
(Gordo and Charlesworth 2000), as illustrated for two of the models in Fig-

ures 10.6B and 10.7A. Even taking account of the fact that Y-linked genes should have lower effective population sizes than other loci due to being present only in one copy per male, they have unusually low diversity in several species. This has been found in *D. miranda* (see Bartolomé and Charlesworth 2006), and in the plant *S. latifolia* (Filatov et al. 2000; Filatov et al. 2001; Laporte et al. 2005), in which Y chromosome mutation rates are, if anything, higher than for the X chromosome (Filatov and Charlesworth 2002). Low diversity is also estimated for bird W chromosomes (Montell et al. 2001; Berlin and Elle-gren 2004). This observation could be due to the commonly observed lower female than male mutation rate (which is also observed in birds; Carmichael et al. 1999), but low W chromosome diversity persists even after making a correction for this effect. More loci must, however, be studied in order to compare Y- (or W-) linked genes with autosomal genes, to ensure that the X-Y (or Z-W) differences are not due to unusually high X (or Z) diversity.

Nucleotide diversity estimates for human Y-linked genes are not greatly reduced compared with other loci (Sachidanandam et. al 2001), after taking into account the Y chromosome's effective population size. This does not disprove the theory that low Y chromosome sequence diversity reflects the activity of processes causing Y genetic degeneration, because degeneration is probably no longer happening in the ancient human Y chromosome (which now has few active genes that could drive hitchhiking processes), whereas it is probably still ongoing in *S. latifolia*, whose Y chromosome evolved recently (see previous discussion). The modest lowering of diversity of mammalian Y-linked genes could be due to a higher variance in reproductive success of males versus females. This might be caused by sexual selection with males mating with multiple females, which further reduces the Y chromosome effective population size (Handley et al. 2006).

Selective sweeps are expected not only to reduce diversity, but also to lead to a strong excess of rare variants. Such events within the species' moderately recent evolutionary past may thus be detectable by comparing variant frequencies with the neutral expectation. There is evidence for such an event in the recently evolved *D. miranda* neo–Y chromosome (Bachtrog 2004). A selective sweep on a nonrecombining chromosome will, however, obscure any signs of earlier hitchhiking events, so this does not demonstrate that degeneration of this species' Y chromosome has been driven exclusively or largely by spread of advantageous mutations; other processes may also have contributed to genetic degeneration.

Do Y Chromosomes Always Degenerate, and What Determines Which Genes Are Retained?

Even in ancient Y chromosomes, X-Y gene pairs often remain. For example, for genes on the human Y chromosome, X-Y gene pairs are found mostly in the region that was added onto the older sex chromosomes, but the oldest part of the Y chromosome also contains some such gene pairs (see Table 10.1). The categories most likely to be retained are genes involved in male functions, or with ubiquitous expression, particularly when large amounts of gene product are important (i.e., the gene is haplo-insufficient). Human Y genes are mainly in these categories. Surprisingly, several X-Y gene pairs known from the human genome are found to have become nonfunctional in the chimpanzee sequence (Hughes et al. 2005). As explained above, the selective processes that cause degeneration will be slow, because only a few genes remain. Therefore, degeneration in such old sex chromosomes is unexpected, and the reason for its recent occurrence is unknown. One possibility is a lower Y chromosome effective population size due to sexual selection (Hughes et al. 2005), though this is not supported by data on diversity, since chimpanzee Y-linked sequences have higher diversity than in humans (Stone et al. 2002). Possibly, genome rearrangements may affect gene expression and cause these events.

In dioecious plants, degeneration of Y-linked genes may be prevented because the haploid stage of the life cycle is extensive (Haldane 1933); many genes are expressed in the vegetative nucleus of pollen (Honys and Twell 2004), whereas animal sperm do not transcribe their genes. At present, too few sex-linked loci have been studied in plants to compare the frequencies of degenerated genes with those in animals whose sex chromosome regions stopped recombining at similar times; however, sex-linked gene sequences are becoming increasingly available from plants.

In haploid plants, such as bryophytes, there is no equivalent to the XX genotype of diploid plants and animals; all diploids are XY. Thus the effective population sizes are the same for both sex chromosomes, and equal degeneration of both the X and Y is expected (Bull 1983). The bryophyte *Marchantia polymorpha* Y chromosome contains about 10 Mb of DNA and the X chromosome is almost twice as large. This heteromorphism might be due to additions to one chromosome, or deletions from the other. Recent results from the Y chromosome (Yamato et al. 2007) have found two distinct regions. In region Y1 (4 Mb), 8 out of 9 genes (89%) are found in female (haploid) gametophytes as well as haploid males, and are thus presumably homologous to genes on the X chromosome. The more gene-rich Y2 region (6 Mb) has an estimated 76 percent of genes homologous with X-linked genes, based on 55 genes. These results suggest some degeneration (though the failure to amplify homologs of Y genes by PCR could be due to high divergence, and data from five X-Y gene pairs suggests very ancient divergence), but, consistent with the prediction above, most genes are clearly not degenerated. Of the total of 14 Y-linked genes detected in males only, several probably have male functions (Yamato et al. 2007). It will be interesting when the X chromosome gene content is also known, thus making further analyses possible.

Y chromosomes cannot degenerate completely, with loss of all genes, unless the sex-determining genes are replaced by another gene or genes controlling sex-determination. This has presumably happened in the ancestors of species, such as some species in the genus *Drosophila*, whose Y chromosomes do not carry the male-determining factor (some *Drosophila* species have even evolved X0 sex-determination, as in many Orthopteran insects; see Figure 10.2). Such a change has probably occurred in the fly *Megaselia scalaris* (Traut et al. 1999), and in the medaka fish, whose male sex-determining region is only 258 kb and contains only one functional gene (Kondo et al. 2006). A related fish species has, again, a small male sex-determining region, but in a different genome location (Kikuchi et al. 2007).

Transposable element accumulation

Transposable element insertions are often found at unusual abundances in nonrecombining regions of genomes, including Y and W chromosomes, for example, in the mosquito *Anopheles gambiae* Y chromosome (Krzywinski et al. 2004), the W chromosome of the silkworm, *Bombyx mori* (Sahara et al. 2003), and the Y chromosome of *M. polymorpha* (Yamato et al. 2007). The introns of the few known *D. melanogaster* Y genes are extremely large, partly due to the presence of multiple transposable elements (Carvalho 2002).

Transposable element insertions can lead to expansion of the region's size and thus to low gene density, even in species with quite small MSY regions, for example, in papaya (Liu et al. 2004). Consistent with these results, plant Y chromosomes are often larger than the X chromosomes (Westergaard 1958; Parker 1990), as in *S. latifolia*, suggesting that accumulation of repetitive sequences may be an early effect of recombination suppression. This is also suggested by data from the recently added arm of the *D. miranda* neo–Y chromosome (Bachtrog 2003). Because the neo–Y chromosome can be compared with the homologous autosome (indicating the ancestral state), the findings in *D. miranda* demonstrate that the higher abundance of transposable elements in this pair is due to an increase on the neo–Y chromosome.

Low gene density of MSY regions

The expectation that repetitive sequence should accumulate means that low gene density of an MSY region does not necessarily imply genetic degeneration. It must be shown that the low density is not merely due to an increase in the content of transposable and other repetitive elements. Transposable element insertions may, however, contribute to genetic degeneration, because they may insert into genes, or into functionally important regions whose disruption may affect gene expression levels. They may also cause chromosome rearrangements, including duplications. In mammals, the Y chromosomes of different species are highly rearranged (for example, the human Y chromosome contains regions in a quite different order from the X chromosome; see Figure 10.3), whereas genome sequences and genetic mapping of distantly related mammals shows that the set of X genes, and their arrangements, are highly conserved (e.g., Pecon-Slattery et al. 2000; Bourque et al. 2004; see Figure 10.1A).

How Is Recombination Suppressed?

Chromosome inversions

The fact that Y chromosomes have undergone rearrangements raises the possibility that these rearrangements include inversions that were selected in order to suppress recombination. However, there is no direct evidence for the involvement of inversions, and it is also possible that recombination was suppressed in a more gradual manner, with no sudden loss of recombination. In this case, the inversions in the human and other sex chromosomes might merely be a consequence of high repetitive-sequence content leading to Y chromosome rearrangements. Certainly, inversions have occurred in regions of Y chromosomes where recombination was already suppressed (Schwartz et al. 1998). Recently evolved sex chromosomes will be helpful in testing the role of inversions in suppressing recombination. For example, in the threespine stickleback, the region carrying the sex-determining locus is not inverted, yet the region with the male-determining allele has lower recombination than the homologous chromosome (Peichel et al. 2004). On the other hand, in papaya part of the MSY region is inverted (R. Ming, pers. comm.).

Progressive recombination suppression

It is now known that recombination suppression between the X and Y chromosome of mammals, birds, and the plant *S. latifolia* has progressively absorbed previously pseudoautosomal regions. Sequence divergence between the members of different X-Y gene pairs varies too much to be explained if recombination stopped at the same time across the entire nonrecombining X and Y region (Lahn and Page 1999; Atanassov et al. 2001; Filatov et al. 2001). Moreover, in all these species, the least-diverged gene pairs are closest to the pseudo-autosomal end of the X (the human PAR1; PAR2 was recently added to the sex chromosomes, and is therefore not relevant to the events that stopped recombination).

In the human sex chromosomes, synonymous site divergence (K_s) between X-Y gene pairs, or pseudogene pairs, decreases towards the X chromosome PAR1 (Skaletsky et al. 2003). A major reason for this is the addition of autosomal material to the mammalian sex chromosomes (see Introduction in this chapter). As one would expect (Figure 10.8), X-added genes have lower X-Y divergence than those in the older X region, dating the addition to about 110 mya (Lahn and Page 1999; Ross et al. 2005). Even among the X-added genes, however, the divergence data suggest that recombination at the PAR1 end stopped more recently than in other parts of the Y chromo-

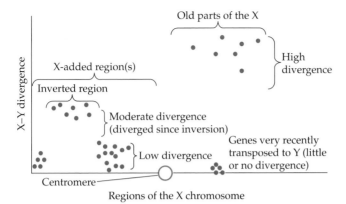

Figure 10.8 Hypothetical example showing possible divergence levels between X-Y gene pairs. Discontinuities in divergence values are expected when parts of the sex chromosomes are older than other regions (since divergence depends on the evolutionary time during which the X and Y genes have not recombined). Inversions will also create discontinuities, and the figure shows an X region containing genes with moderate divergence values from the Y homologs. The region of low divergence could have suppressed recombination due to a further inversion, more recent than the first one, or by some other mechanism (see text).

some (Lahn and Page 1999; Skaletsky et al. 2003; Figure 10.9A), and the same is true in other species whose sex chromosomes are not known to be neo–sex chromosomes (Lawson-Handley et al. 2004; Nicolas et al. 2005; Bergero et al. 2006; Figure 10.9B). Nucleotide divergence values have high variances, and different sequence regions diverge at different rates, making it difficult to determine whether different X-Y pairs differ in a discontinuous manner, forming truly distinct "strata" (rather than just a trend towards more divergence with distance from the PAR). Since mammalian sex chromosomes have few X-Y gene pairs, it may be impossible to test for involvement of inversions by this approach. If plant sex chromosomes have indeed only undergone partial degeneration, finer details may be revealed if enough X-Y pairs can be found. Moreover, if the Y chromosome can be physically mapped in sufficient detail, it may be possible to test whether inverted regions with respect to the X chromosome (confirmed by comparison with species without sex chromosomes, so that we can be sure that the Y chromosome has been inverted) correlate with the divergence values.

The only plant for which there is comparably detailed evidence is *S. latifolia*. In this case, it does not seem that the low-divergence gene pairs were transposed onto the sex chromosomes, since all X-linked gene homologs that have so far been mapped in the nondioecious species, *S. vulgaris*, are found on a single chromosome (Filatov 2005). The number of genes mapped so far is small,

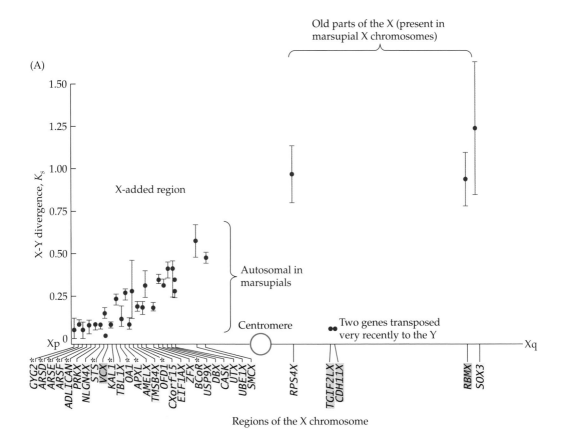

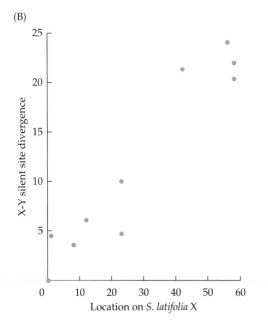

Figure 10.9 Observed divergence levels between X-Y gene pairs (A) in humans and (B) in the plant *S. latifolia*, for the few X-linked loci so far known in this species. (A, from Skaletsky et al. 2003.)

however, and we will need to investigate more genes to sure of this. One case has been found of an autosomal gene being transposed to the *S. latifolia* Y chromosome, creating a new gene, *SlAp3Y*, that has no X homolog and is expressed mainly in anthers (Matsunaga et al. 2003); the size of the region transposed is unknown (but could potentially be studied if enough genes on this species' sex chromosomes were known). Translocations involving the sex chromosomes are, however, well-documented in other plants, including the well-studied *Rumex acetosa*, in which both Y arms have accumulated repetitive sequences (Shibata et al. 2000; Mariotti et al. 2006); the phylogenetic evidence suggests that this is an older sex chromosome system than that in *Silene*.

Further Chromosome Rearrangements

High repetitive-sequence content is predicted to lead to duplications and deletions, and lack of recombination can allow such rearrangements to persist. Both kinds of rearrangements are documented in the human Y chromosome (see Figure 10.3), and the rearrangements have scrambled the Y gene order relative to that on the X chromosome. Deletions may contribute to loss of degenerated genes, and rearrangements may bring genes into regions in which their expression is impaired, potentially affecting their chances of degeneration. Rearrangements may occur at unusually high rates compared with other genome regions. Differences between the Y chromosomes of different individuals, with polymorphism for length and copy number of certain regions, are detectable in humans (Repping et al. 2006) and in the plant *R. acetosa* (Wilby and Parker 1986). As illustrated in Figure 10.3, the human (and chimpanzee) Y chromosomes contain *ampliconic* regions with large palindromes (mirror image repeats), between which gene conversion occurs (Rozen et al. 2003). Although these duplications may contribute to the high rate of deletions, the existence of these duplicated regions has the effect of preserving functional gene copies on the nonrecombining Y chromosome, even as degenerative processes occur.

Rearrangements are a likely consequence of repetitive sequence accumulation. For instance, inversions can be caused by pairing between transposable elements in different locations, and there is evidence for this in the human Y chromosome (Schwartz et al. 1998). Since their split about 6 mya, the human and chimpanzee Y chromosomes have each undergone a large inversion in the X-degenerate region, in which the genes are single-copy (Hughes et al. 2005). Clearly, this inversion has nothing to do with suppressing X-Y recombination, but is a consequence of the presence of repetitive sequences.

Neo–Sex Chromosome Formation

An important category of chromosome rearrangement includes translocations forming neo–sex chromosomes (Figure 10.10). These chromosomes offer an opportunity to test whether recent cessation of recombination leads to the predicted effects—specifically, whether lack of recombination leads to loss of

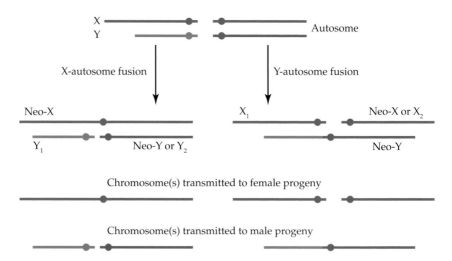

Figure 10.10 The formation of neo–sex chromosomes by chromosome fusions, and the exclusive transmission of neo–Y chromosomes through males, which implies that in species with no male recombination, the previously autosomal arm immediately stops recombining.

adaptation. Neo–sex chromosome formation initially creates a larger X or Y chromosome, one arm of which was previously autosomal. A Y-autosome fusion causes the previously autosomal neo-Y arm to be restricted to males. Such an X_1X_2Y system, in which X_2 is a neo-X, is found in the plant *R. acetosa*. Similarly, an X-autosome fusion creates an XY_1Y_2 system in which the neo-Y (Y_2) and the original Y chromosome both segregate from arms of the neo-X (as in the bat, shown in Figure 10.1D). If recombination occurs in male meiosis, the neo–sex chromosome arm can continue to recombine with its autosomal homolog. Sometimes recombination may stop with formation of a neo–sex chromosome, because one of the chromosomes involved is inverted; this has occurred in a mammal, the black muntjac deer (unpublished data of Wen Wang and colleagues, Chinese Academy of Sciences, Kunming).

Molecular evolutionary evidence for genetic degeneration in neo–sex chromosomes

In the genus *Drosophila*, which has no recombination in any chromosomes in males, either Y-autosome or X-autosome fusion leads to complete cessation of recombination on the former autosome that is restricted to males. Since *Drosophila* chromosome arms are large genome regions, carrying many hundreds of loci, such an event should lead to very strong operation of the processes expected to cause genetic degeneration, and therefore should strongly affect the new Y chromosome's evolution. Even though recombination stopped only about one million years ago, due to a Y-autosome fusion, the *D. miranda* neo–Y chromosome indeed shows evidence of the predicted degeneration (reviewed in Bartolomé and Charlesworth 2006). Based on comparisons of many pairs of loci still present on the neo-Y and neo-X, the neo-Y diversity is low for silent sites, neo-Y substitution of nonsynonymous variants is higher, and many neo-Y loci are pseudogenes (in the largest survey so far, ratios of K_a to K_s values for divergence from the closely related species *D. pseudoobscura*, in which the arm corresponding to the neo-Y is still autosomal, were estimated from 68 genes to be 0.13 for the neo-X, versus 0.47 for the neo-Y; Bachtrog 2005). The neo-Y has also accumulated transposable elements.

It is important to distinguish whether the higher substitution of nonsynonymous variants is due to advantageous changes in Y-linked genes or reduced adaptation of protein sequences. The latter would be evidence for degeneration despite the action of purifying selection. Indeed, sequences of a set of nonpseudogene loci show overall evidence of selective constraint (i.e., $K_a < K_s$). By using homologous sequences from a moderately distant outgroup species, *D. affinis*, it was possible to infer which substitutions occurred in the neo-Y lineage and which in the neo-X since these chromosomes stopped recombining (Figure 10.11). Numbers of substitutions in nine such loci were analyzed by McDonald–Kreitman tests, modified to use the

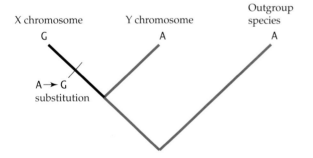

X chromosome nucleotide	G	A	G	T	T	C	C	T	C	T	C
Y chromosome nucleotide	A	G	G	C	C	T	C	C	G	A	T
Outgroup nucleotide	A	G	A	T	C	C	G	T	A	T	C
X or Y changed	X	X	–	Y	X	Y	–	Y	–	Y	Y

Figure 10.11 Using an outgroup species and parsimony to estimate sequence substitutions in the X- and Y-chromosome lineages (indicated by black and blue lines, respectively). The tree shows the states of the first site where the three sequences differ. At this site, one can deduce that the X sequence has changed, since it differs from the state in both the Y chromosome and outgroup sequences, which can be assumed to indicate the ancestral state. The table shows how, at many such sites, one can deduce which sequence has changed (the changes in the X and Y sequences are then indicated in the bottom row of the table). Some sites, however, do not provide clear information.

Table 10.2 Results of McDonald–Kreitman tests for substitutions in the same set of nine genes in the neo-Y and neo-X lineages of *Drosophila miranda*[a]

	Fixed differences		Polymorphisms	
	Silent	Amino acid replacement	Silent	Amino acid replacement
Neo-X alleles	27	16	67	21
Neo-Y alleles	47	64	1	7

[a]Although there are more substitutions of nonsynonymous (replacement) sites in the neo-Y (see Figure 10.11), there is no evidence that this is more than expected, given the higher numbers of nonsynonymous polymorphisms, compared with the X-linked alleles. This suggests that these substitutions occur because selection is ineffective for Y-linked alleles.

Source: Bartolomé and Charlesworth 2006.

inferred numbers of substitutions in the neo-X- and neo-Y-linked alleles. If these substitutions were driven by selection, their numbers relative to synonymous substitutions should be significantly higher than among polymorphisms on the neo-Y. However, Table 10.2 shows that this is not the case. There is thus no support for selection having driven the neo-Y substitutions, suggesting that failure of selection to eliminate these variants is responsible (Bartolomé and Charlesworth 2006). Expression of many neo-Y genes is also low, compared with the neo-X chromosome alleles, although this could be partially due to increased expression of the neo-X genes as dosage compensation is evolving, as will be discussed later in this chapter (Bachtrog 2006).

We have already mentioned the evidence that adaptive changes in one or more Y-linked genes have led to a selective sweep on this chromosome (Bachtrog 2004), but this does not, of course, imply that such events are a major cause of degeneration. Even with genomic data, it will be difficult to estimate the relative importance for Y chromosome degeneration of selective sweeps versus deleterious mutations. The findings in *D. miranda*, indicating rapid degeneration of neo-Y-linked genes, are consistent with the fact that neo–Y chromosomes of other *Drosophila* species are short-lived (see Figure 10.2).

Neo–sex chromosomes in species other than Drosophila

Evidence of degeneration, and low expression, has also been found for genes in the recently evolved neo–Y chromosome in the black muntjac deer (which is estimated to have arisen in the past half million years; unpublished data, Wen Wang and colleagues, Chinese Academy of Sciences, Kunming).

In mammals, the X-added region must also have been added to the Y chromosome (see Figure 10.3), and, as explained earlier, this may have been accompanied by an event that prevented the regions on the X and Y chromosome recombining. The neo-Y region recently added to the human Y chromosome (the X-transposed region) will presumably also have stopped recombining with its X homologous region, since it is transposed to a region that does not recombine with the X chromosome; however, it is probably too small to test for degeneration, as only three genes at most are included in the region. This Y region has the highest content of repetitive sequences and transposable elements—largely due to LINE elements (Skaletsky et al. 2003)—consistent with such accumulation being the first change after recombination stops. The evidence for accumulation of repetitive sequences in the small MSY regions of threespine sticklebacks and papaya (Yu et al. 2007) is also consistent with this.

Evolution and Adaptation of Ancient Sex Chromosomes: Evolution of Dosage Compensation and of Changed Gene Content

Box 10.1 summarizes the types of changes that occur during sex chromosome evolution. The evolution of sex chromosomes is, however, not a story only of degeneration of the Y chromosome, and the question of how much the genomes of Y and X chromosomes change in response to their new situation is a very interesting one. The Y chromosome is transmitted exclusively through males, and the X chromosome experiences a similar, though milder, female-biased transmission. The MSY genes never recombine with the X chromosome, while the homologous regions of the X chromosome recombine only in females (i.e., half as often as autosomes, assuming a 1:1 sex ratio). Both sex chromosomes probably undergo many changes as they evolve, and at present little is known about adaptation after new sex chromosomes evolve. As Y-linked genes degenerate due to lack of recombination, the activity of their gene products decreases relative to that of others in the genome. This may select for duplications to maintain activity levels, perhaps explaining some of the duplications of Y-linked genes encoding proteins, such as dyneins that are important in *Drosophila* for male functions (Carvalho 2002). There will also be selection for compensation mechanisms to restore correct expression levels relative to those of other genes.

Dosage compensation

The Y chromosome loses genes and functions gradually, so the situation is not the same as for an aneuploid mutant lacking a whole chromosome (monosomy). Once the Y copy has degenerated, each sex-linked gene that is required in two copies is indeed monosomic for the X copy, but this situation evolves gradually. As explained above, some genes may be prevented from degeneration unless new copies evolve on the sex chromosomes or elsewhere in the genome. At the other extreme, if loss of function is fully

BOX 10.1
Possible order of events in Y chromosome evolution

- Evolution of sex-determining region with linked male-sterility and female-sterility loci on an ordinary chromosome, creating proto–sex chromosomes

- Evolution of reduced recombination between these genes (and perhaps around them), leading to a nonrecombining region around the sex-determining genes

- Reduced recombination leads to

  Accumulation of repetitive sequences, including transposable elements, and expansion of introns, increasing the size of the Y chromosome

  Sexually antagonistic genes on the Y chromosome evolve towards more male-biased alleles

  —————— Selection for further recombination reduction

  Transposition onto the Y chromosome of genome regions carrying male function genes

  Genetic degeneration, with loss of gene functions nonessential for male fertility

- Rearrangements, including inversions, in nonrecombining regions can lead to loss of genes and other deletions, reducing the size of the Y chromosome

  Evolution of dosage compensation

---

recessive, degeneration is, of course, not opposed by selection, since most diploid individuals will have intact X-linked copies. Y-linked genes whose mutations are intermediate between these extremes may lose the struggle to survive and succumb to degenerative processes. Clearly, loss of gene functions is selectively disadvantageous, since dosage compensation mechanisms have evolved, and only a few *D. melanogaster* and human X-linked genes are not compensated (an estimated 15% in humans, most often genes in the X-added region; Carrel et al. 1999; Carrel and Willard 2005).

Expression of X-linked genes in somatic tissues of diploid males and females of dioecious animals (with X:AA males, where A stands for an autosome set, and XX:AA females) is indeed equalized, and correct X:A expression levels are thus maintained in both sexes. Species thus far studied include mammals, *Drosophila* species, *C. elegans* (reviewed in Gupta et al. 2006), and birds, though compensation may be incomplete in the latter (Itoh et al. 2007). One might expect compensation to involve increased expression from the single X in XY males, and this is how dosage compensation works in *D. melanogaster* (as explained shortly, different organisms have different systems). Control of expression might be exerted individually on each X-linked gene, as its Y-linked homolog degenerates, or might affect the X as a whole.

In either case, different selection pressures act in the two sexes, giving further situations with sexually antagonistic effects of different alleles. In species with the Drosophila-type of dosage compensation system, for example, unless X upregulation is stringently male-specific, it will also affect X-linked genes in females, causing too high dosage and/or imbalance with autosomal genes. If such a situation develops, selection may then favor a general downregulation of X genes in females, relative to autosomal ones (Charlesworth 1996). This may explain the puzzling mammalian system, in which one X chromosome is inactivated in females and its genes are not expressed, or in *C. elegans* hermaphrodites, whose two X chromosomes both express genes at about half the level of that in X0 males, equalizing expression of X-linked genes in both sexes (although this is not a dioecious species, it is descended from dioecious *Caenorhabditis* ancestors, from which this dosage control is presumably inherited).

Indeed, there is some evidence supporting this complex evolution and counter-evolution of X-linked genes' expression. The set of genes that are X-linked in all mammals, including marsupials, had probably degenerated and evolved dosage compensation before the new X arm was added. In marsupials, the female's paternal X chromosome is specifically inactivated by an imprinting mechanism, and this also occurs in early development of Eutherians; random inactivation is established slightly later (Nguyen and Disteche 2006). In somatic tissues of mice and humans, which have random X-inactivation, the increased expression of X-linked genes in both sexes, hypothesized above, is indeed detected (Nguyen and Disteche 2006). The ideal experiment would compare recently duplicated loci, one X-linked and one autosomal; the few tests with such genes support the conclusion of X chromosome overexpression. An alternative test is to use data averaged across large numbers of genes (to obscure effects of individual loci, which evidently vary widely in transcription levels). Estimated levels of transcript from X-linked loci (expressed from just one X in males or females) are indeed roughly equal to those from autosomal loci (each expressed from two gene copies), whereas one would predict half that level if expression of X-linked genes is determined purely by inactivation of one X in females or presence of just one copy in males. Furthermore, a distinctive X chromosome–specific methylation pattern has recently been discovered in cultured human cells, marking alleles of genes on both the single X chromosome in males and the active X chromosome in females (Hellman and Chess 2007). These results are all consistent with the hypothesis that raised transcription from the X chromosome is still detectable today, even though it evolved early on, as Y genes degenerated; it appears that the problem of overexpression in females was solved later by evolving X inactivation. In *Drosophila*, with no X inactivation, this hypothesis predicts that increased transcription from the X should be detectable, and there is evidence supporting this also (Gupta et al. 2006).

Discussion: Adaptation of Sex Chromosomes

As previously mentioned, the Y chromosome is predicted to both degenerate and evolve. Some Y-linked genes essential for male functions may be
preserved, and genes with sexually antagonistic effects may also evolve new
alleles with improved male functions, once they are confined exclusively to
males. Although little direct evidence is yet available documenting genes
with sexually antagonistic effects, several experiments suggest that such
variants exist in *D. melanogaster*. In these experiments, the transmission patterns of chromosomes are altered such that these genome regions are likely
to be selected for male functions alone, whereas they had previously been
subject to selection in both sexes. It is then predicted that male-related characters should be enhanced, and female-related characters should decline,
and this has been observed (Prasad et al. 2007).

There are several reasons to expect that the X chromosome must also
evolve. Whatever the mechanism of dosage compensation, it is evident that
the systems that recognize the X chromosomes and count their number must
interact with these chromosomes, and that fine-tuning of X-linked gene
expression levels must evolve. As dosage compensation evolves, X-linked
alleles must be replaced by new alleles, perhaps leading to selective sweeps
in the X chromosome region near the gene affected.

The gene content of X chromosomes also evolves. Male-benefit alleles are
expected to accumulate on the X chromosome as well as the Y chromosome,
because (like characteristics caused by rare X-linked alleles, such as color-
blindness, that are much more common in males than females) advantageous mutations at X-linked loci whose Y homolog is absent or nonfunctional will be expressed more often in males than in females (Charlesworth
1992). On the other hand, X-linked genes are not expressed in late spermatogenesis (meiotic sex chromosome inactivation), so genes are expected to
transpose to the autosomes before they can evolve a role in processes late
in spermatogenesis, as is indeed observed (Khil et al. 2005).

In the future, it may be possible to distinguish between different possible explanations for the changing gene content of sex chromosomes, and
to test how much genes that remain on these chromosomes adapt, versus
how much Y-linked alleles change through degenerative processes. There
is already some sequence-based evidence for continued adaptation in Y-
linked genes. The evidence for a recent selective sweep on the *D. miranda*
neo-Y has been mentioned previously, and the analysis of three primate Y-
linked gene sequences suggests that some nonsynonymous substitutions
may be driven by selection (Gerrard and Filatov 2005).

At present, the timecourse of degenerative and adaptive changes is
known only roughly. Clearly, the great age of the mammalian and *Drosophila*
sex chromosomes has allowed most genes to be lost from the chromosomes
ancestral to the MSY. In the case of neo–Y chromosomes, very fast degeneration is clearly evident, but (as explained earlier) these are extreme situ-

ations. The timescale of de novo sex chromosome evolution and degeneration is probably much slower, particularly if initial nonrecombining regions are small and contain few genes. At present, it is not clear whether accumulation of repetitive sequences, including transposable elements, occurs before gene sequences start to degenerate, nor whether such accumulation contributes to reducing gene expression. It will be interesting to find out whether expression of a given gene changes before, or along with, amino acid substitutions that may impair the encoded protein's functions, and whether chromosome-wide control of expression occurs. If lower expression is an evolved response to presence of individual damaged Y-linked genes, all members of a species with an evolving sex chromosome may be expected to have similar expression levels of particular genes. If, however, reduced expression of Y-linked genes is a direct effect of mutations in non-coding sequences of genes (e.g., their promoters), reduced expression may be seen only in some individuals, as fixation of such deleterious mutations will be slow.

CHAPTER **11**

Molecular Signatures of Adaptive Evolution

Alan Filipski, Sonja Prohaska, Sudhir Kumar

Introduction

DARWIN'S GREAT DISCOVERY OF NATURAL SELECTION was rightfully considered to be the fundamental principle of adaptive evolutionary change at both the phenotypic (Fisher 1930) and, later, the DNA sequence level. At the same time, suggestions were made that evolutionary changes could be random as well as adaptive. This was given a mathematical basis by Sewall Wright's description of random evolutionary change in populations in the form of what is now called *random genetic drift* (Wright 1931). The relative importance of genetic drift and natural selection on the evolution of protein sequences was much debated at that time. Because of the belief that most protein change had selective impact and that effective population sizes were so large that random effects would be smoothed out, the effect of genetic drift tended to be underappreciated.

In the 1960s, the genetic code was deciphered and more protein sequences became available; hence, it finally became possible to assess in a quantitative way the amount of genetic variation within and among species. In 1968, the tide began to turn as Motoo Kimura argued, on the basis of the limited sequence data then available, that most protein differences within and among species could be attributed to random genetic drift, and that the differences were not the result of natural selection only (Kimura 1968, 1983). This argument was based on calculations that demonstrated that the observed quantity of protein polymorphism in populations and the inferred rate of evolution between species were far greater than would be expected if selective mechanisms were the primary determinants (Kimura 1983).

Kimura made an early and compelling argument about the relative importance of natural selection and genetic drift; this argument exploited the degeneracy of the genetic code (Kimura 1977). Many DNA sequence changes in certain codon positions do not result in amino acid changes (i.e., they are synonymous changes), which allows them to escape natural selection and accumulate *neutrally* by random genetic drift. This stands in contrast to other codon positions where the vast majority of changes will alter the amino acid encoded (nonsynonymous changes). These changes will often be subject to positive selection, which speeds up the rate at which they proceed to fixation in a population, or to what is often called purifying selection, which removes them from the population. If the mutations are fixed by positive selection, then the rate of fixation of nonsynonymous changes will be greater than the rate of fixation (substitution) of synonymous changes in any codon.

In contrast, Kimura's neutral theory predicts that the evolutionary rate of synonymous DNA substitutions will be greater than that of nonsynonymous substitutions, because purifying selection is operating to remove the vast majority of nonsynonymous mutations. This prediction was confirmed (Kimura 1977, 1991). Over time, these observations were extensively supported by growing sequence data sets, which gradually led to the acceptance of the neutral theory.

In comparing the rate of synonymous and nonsynonymous substitutions in the same protein, Kimura implicitly assumed that the observed substitution rate of synonymous codon sites may be used as an estimate of mutation rate for the entire gene (Kimura 1977, 1983). Some form of this estimation method has been virtually universal since then in studies relating to adaptation at the DNA level (Miyata and Yasunaga 1980; Nei and Kumar 2000; Bustamante et al. 2005; Nielsen et al. 2005a). The common modus operandi in these tests of selection is to estimate synonymous divergence per synonymous site (denoted d_S or K_s), and nonsynonymous divergence per nonsynonymous site (d_N or K_a) for any protein-coding gene, and then to derive the selection ratio $\omega_S = d_N/d_S$. Many methods have been developed for estimating d_S and d_N based on somewhat different assumptions, but the principle remains the same (see overviews in Nei and Kumar 2000; Yang 2006). If ω_S is significantly greater than 1.0, then we attribute this excess to positive selection on the gene. If ω_S is significantly less than 1.0, then we attribute the difference to negative (purifying) selection.

For a vast majority of proteins, ω_S is found to be less than 1.0 when it is computed as an average over all the codons in a protein. This is because most of the nonsynonymous mutations will have negative fitness effects and will be eliminated by selection. Even in the presence of positive selection on some codons, ω_S will generally be less than 1.0 because of the purifying selection against most of the nonsynonymous mutations. For this reason, estimating ω_S for each codon separately is advocated (e.g., Nielsen and Yang 1998; Suzuki et al. 2001; Suzuki and Nei 2001; Zhang et al. 2005). A recent account by Nei (2005) provides an excellent overview of many historical and recent studies, where scientists have looked for genes and codons that have undergone positive selection (Darwinian evolution) at the molecular level.

Even though d_N/d_S and K_a/K_s have been used historically for referring to the extent of natural selection, it is most accurate to define the selection ratio (ω) as the ratio of the rate of nonsynonymous *substitutions* per site to the rate of nonsynonymous *mutations* per site. That is, $\omega = (d_N/2t)/\mu_N$, where μ_N is the rate of nonsynonymous mutations per nonsynonymous site, and t is the time of species divergence.

Of course, the rate of nonsynonymous mutation is difficult to determine directly, and synonymous divergence is often used as a proxy for the denominator. In this case, ω becomes identical to ω_S. A number of authors have also used sequence divergence in introns (d_I) and in pseudogenes (d_φ) in the denominator, which makes ω_I and ω_φ, respectively, equal to ω (Li and Tanimura 1987; Wolfe et al. 1989; Li and Graur 1991; Bergström et al. 1999; Keightley and Eyre-Walker 2000; Nachman and Crowell 2000b; Chen et al. 2001).

In the following, we discuss various factors that often invalidate the procedure of equating the rate of synonymous, intron, and pseudogene divergence with the rate of nonsynonymous mutations to estimate ω_S, ω_I, and ω_φ, respectively.

Codon Usage Bias and the Estimates of Selection Ratio (ω)

We begin with the assumption that the rate of synonymous mutations (μ_S) is identical to the rate of nonsynonymous mutations (μ_N). In this case, if all synonymous mutations are "strictly neutral," then we can directly use the synonymous divergence per synonymous site (d_S) in estimating the selection ratio ($\omega = \omega_S$). In the 1970s, it was thought that the synonymous mutations were all indeed strictly neutral, and, therefore, directly useable for estimating ω (Kimura 1977). However, recent comparisons of the human and chimpanzee genomes have established that the synonymous codons for a given amino acid appear with frequencies more biased than expected based on base composition differences (Chamary and Hurst 2005; Lu and Wu 2005; Parmley et al. 2006). This property has been seen in many species (Grantham et al. 1980; Bennetzen and Hall 1982; Blake and Hinds 1984; Eyre-Walker 1991; Akashi 2001; Mikkelsen et al. 2005).

Why certain codons are preferentially used is not completely understood, but reasons may include mutational bias, local GC content, and translational efficiency due to varying availability of tRNA for different codons. While mutational and compositional bias can explain some of the observed nonrandom synonymous codon usage, it is now abundantly clear that many synonymous mutations are eliminated by natural selection for, among other reasons, optimizing translational efficiency (e.g., Shields et al. 1988; Comeron et al. 1999; Carpen et al. 2006; Kimchi-Sarfaty et al. 2007).

Any existence of purifying selection on synonymous mutations compromises the assumption that the substitution rate at synonymous positions can be used directly as an estimate of mutation rate (Shields et al. 1988; Ochman et al. 1999; Tamura et al. 2004). A well-studied example of codon usage bias is from the genus *Drosophila*. In *D. melanogaster*, for example, the codon CTG for leucine is more than eight times as frequent as the codon TTA for the same amino acid. In *Drosophila*, synonymous distance (d_S) between orthologous genes correlates with the codon adaptation index (Tamura et al. 2004; Figure 11.1A). This is evidence that the codon usage bias constraint acts as negative selection. The relative effect of codon usage bias on the estimate of mutation rate from synonymous sequence divergence can be seen by comparing codon bias–corrected estimates of synonymous dis-

Figure 11.1 (A) Negative relationship of codon adaptation index and the number of ▶ substitutions per fourfold-degenerate site (d_S) between *Drosophila melanogaster* and *Drosophila simulans,* based on a study of 62 genes. ($R^2 = 0.148$). As codon constraint increases, synonymous rate goes down because more mutations become eliminated by purifying selection. (B) Effect of codon usage bias correction at the genomic level in *Drosophila*. Each plotted point represents a pair of species in the genus *Drosophila*. Distance between species was computed in two different ways: Tamura-Nei distance between orthologous genes from the species pair, averaged over all genes; and d_μ (codon usage bias–corrected Tamura-Nei distance between orthologous genes from the species pair, averaged over all genes). The dotted line shows identity and the solid line shows a linear best fit regression line. This is based on 9850 coding genes that have orthologs in all of the following twelve species: *D. melanogaster, D. simulans, D. sechellia, D. yakuba, D. erecta, D. ananassae, D. pseudoobscura, D. persimilis, D. willistoni, D. mojavensis, D. virilis,* and *D. grimshawi.* See http://rana.lbl.gov/drosophila/wiki/index.php/Datasets for more information about the gene set. (C) Histogram showing the distributions of mutation rates and nonsynonymous substitution rates in 9850 genes. Mutational distances were estimated by correcting the evolutionary distance at the fourfold-degenerate sites for the effect of codon usage bias and differences in base composition biases between sequences under a sophisticated model of nucleotide substitution (Tamura and Kumar 2002; Tamura et al. 2004). Nonsynonymous distances at the second codon positions were estimated using the Tamura and Kumar method for each gene (Tamura and Kumar 2002). For each gene, the mutational and nonsynonymous distances were regressed against the evolutionary time for each pair of species. Divergence times were estimated using the genomic mutation distance and a clock calibration of 0.011 mutations per site per million years (Tamura et al. 2004). Per-gene rates were computed as follows: Using the above interspecies times as the independent variable (x), we calculated regressions against evolutionary distances (either for fourfold-degenerate or second position) as the dependent variable obtained from individual genes. The best fit slope (evolutionary rate) for that gene was estimated via a regression forcing the y-intercept to zero. (D) Histogram showing the extent of linearity of rate with time for both second codon position and fourfold-degenerate differences, expressed as the linear coefficient of determination, R^2. This was done by taking, for each gene, all species pairs available for that gene, and computing a linear regression of the interspecies distance (second codon position or fourfold-degenerate) against interspecies times as the independent variable. Because of known problems in calculating and interpreting the R^2 coefficient of determination when the y-intercept is forced to zero, we calculated the R^2 for each regression without constraining the y-intercept to be zero (Eisenhauer 2003). (A, after Tamura et al. 2004.)

(A)

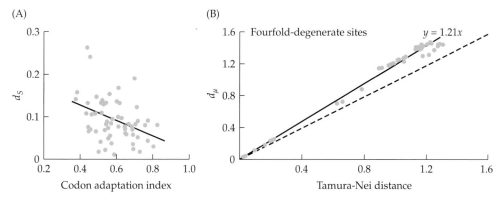

(B)

(C)

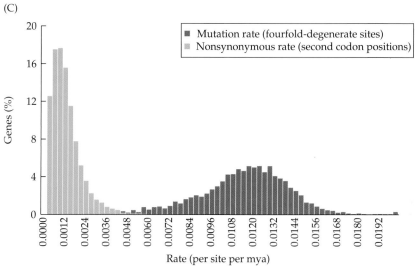

(D)

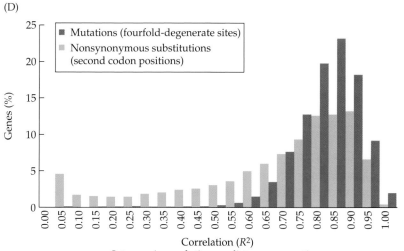

tances (corrected-d_S, which is referred to as d_μ as well) with those that only correct for multiple substitutions under sophisticated models of nucleotide substitutions (e.g., Keightley and Eyre-Walker 2000; Tamura et al. 2004).

A comparison of the estimates of selection intensities with and without correcting for codon usage bias for 9850 protein-coding genes from 12 *Drosophila* species provides a glimpse into the overall effect of codon usage bias on estimates of selection. Figure 11.1B shows that over all genes the bias correction increases the estimate of d_S by slightly more than 20 percent. Because d_S occurs in the denominator in the selection ratio, the failure to correct for codon usage bias will cause false positives when looking for genes and codons with positive selection. In the present data set, the 20 percent difference represents an average over all genes. However, much more dramatic effects are seen when we examine individual genes. For example, application of the correction to the *Adh* gene in the subgenus *Sophophora* (of genus *Drosophila*) approximately doubles the estimate of the mutation rate, which translates into much higher purifying selection on *Adh* than estimated based on synonymous divergences (Tamura et al. 2004).

Figure 11.1C shows the distribution of the rates of nonsynonymous and corrected-d_S for 9850 fruit fly genes. As expected, the distribution of nonsynonymous rates is highly skewed, with a large number of proteins evolving slowly, and only a few evolving with fast rates. In contrast, the distribution of mutation rates is symmetrical and shows a strong central tendency (average rate = 11.3 mutations per thousand base pairs [kbp] per million years [my]). Using the average nonsynonymous divergence (1.2 per kbp per my), it is clear that about 90 percent of all mutations have been eliminated by natural selection in *Drosophila*. Because different proteins may evolve with different relative evolutionary rates in the same set of species, we also present the correlation of nonsynonymous divergence with estimates of divergence times for the 12 *Drosophila* species (Figure 11.1D). Many proteins show rather low correlation between the nonsynonymous divergences with time, confirming a nonclocklike behavior.

In addition to the codon usage bias, synonymous mutations may be under selection for other reasons. For example, conserved RNA structures have been shown to overlap coding regions, and the functional relevance of both the structural RNA and the encoded protein has been determined (Konecny et al. 2000; Katz and Burge 2003; Chooniedass-Kothari et al. 2004; Meyer and Miklos 2005). The local structures on the mRNA in these cases serve as post-transcriptional regulatory signals for splicing, transport, and mRNA stabilization or destabilization, as well as translation efficiency (Cartegni et al. 2002; Chamary et al. 2006). Despite this evidence, genomic screens for structural RNAs suggest that, in general, purifying selection acts against secondary structures in coding regions (Babak et al. 2007). Also, a number of synonymous polymorphisms have been associated with phenotypic differences in humans for reasons that are not always clear (Oeffner et al. 2000; Carpen et al. 2006; Kimchi-Sarfaty et al. 2007). It is currently unclear how the effects of these (potentially minor) factors can be handled efficiently in correcting estimates of d_S.

Hypermutability of CpG Dinucleotides and the Estimates of Selection Ratio

In the above discussion, we assumed that the rate of synonymous mutations (μ_S) is identical to the rate of nonsynonymous mutations (μ_N). If this is not true, then synonymous divergence, even though estimated perfectly, will not be directly applicable for estimating μ_N. In vertebrates, and in nonanimal species, the existence of hypermutable CpG dinucleotides (a cytosine followed by a guanine on the same DNA strand; Figure 11.2) has been shown to cause a disparity between μ_S and μ_N (Subramanian and Kumar 2006).

Codon positions involved in CpG dinucleotides are expected to mutate at between five to twenty times the rate of other positions in the codon (Krawczak et al. 1998; Bird 1999; Subramanian and Kumar 2003), because CpG dinucleotides are often methylated in coding regions and are known to mutate rapidly to TpG and CpA. In this case, assuming that all CpG positions are likely to be methylated with equal probability, the rate of synonymous and nonsynonymous mutations in a given protein will only be equal,

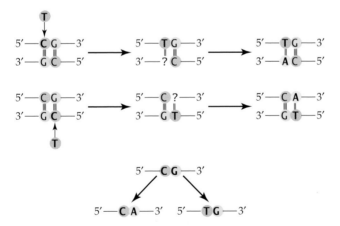

Figure 11.2 Hypermutability of CpG dinucleotides. The nucleotide sequence CpG is subject to mutation at a rate approximately ten times faster than other point mutations in mammals (Bird 1980; Krawczak et al. 1998; Anagnostopoulos et al. 1999). The methylation and subsequent mutation of the cytosine in the CpG configuration to thymine via spontaneous deamination can lead to either a C→T or G→A transitional mutation and thus cause the CpG to be transformed into either a CpA or a TpG. In the upper sequence, the cytosine of the CpG in the forward strand mutates to a thymine. The resulting mismatch with its complementary nucleotide is resolved by exchanging that complementary nucleotide for an adenine. In the lower sequence, the cytosine on the complementary DNA strand mutates to a thymine, and the resulting mismatch is again repaired by replacing its counterpart in the forward strand with an adenine.

Figure 11.3 Venn diagram showing the numbers of human genes under adaptive evolution as predicted by two different methods (silent divergence and estimated replacement mutation rate) to estimate the coefficient of selection. The numbers in the overlapping area indicate the genes that were identified by both the methods, and the numbers to the left and the right represent the genes that were predicted exclusively by one of the methods. (After Subramanian and Kumar 2006.)

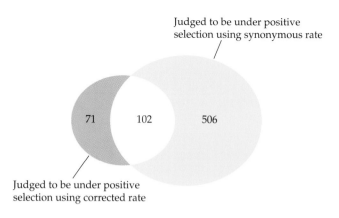

Judged to be under positive selection using synonymous rate

71   102   506

Judged to be under positive selection using corrected rate

on average, if the number of CpG sites per synonymous site is equal to the number of CpG sites per nonsynonymous site. This is unlikely to be true because the fraction of second codon positions involved in CpG dinucleotides will be greatly influenced by the amino acid frequencies in specific proteins. In contrast, the involvement of fourfold-degenerate positions in CpG dinucleotides is dependent on the mutational pattern of the genomic region, which dictates the G + C content of these sites, and the nucleotides in the flanking positions.

Incorporating the difference in the proportion of synonymous and nonsynonymous sites participating in CpG dinucleotides drastically changes the inferences about the proportion of genes predicted to be under positive selection (Subramanian and Kumar 2006). Figure 11.3 shows that, in a sample of human genes, the number judged to be under positive selection decreases to less than a third of its previous value when the CpG correction is applied. While the affects of CpG may be correctable when substitution rates are estimated over an entire polypeptide, the future looks unpromising when we begin to apply these concepts to the identification of individual codons under selection. We now need to know which CpG-involved codon positions are methylated in the germ line, and we need to ascertain whether the homologous codon positions in different species have the same methylation patterns. In the absence of any such information, we will not be able to reliably estimate nonsynonymous mutation rates for use in finding codons that have undergone positive selection.

Using Pseudogenes to Estimate Nonsynonymous Mutation Rates

Less than 25 percent of codon positions can experience synonymous mutations, and codons make up only about 2 percent of the genomes of higher organisms. Therefore, scientists have looked beyond the codons to generate estimates of nonsynonymous mutation rates, as mutations in the noncoding regions are considered to be not selected against (although this is now known to be an oversimplification; e.g., Andolfatto 2005). One possibility is to consider "dead" genes (pseudogenes), because these genes are expected to evolve without any selective pressures. For example, comparison of human and chimpanzee pseudogenes was used to estimate the mutation rate for these primates (Nachman and Crowell 2000b). Because pseudogenes are found to show higher divergence, they are sometimes considered to evolve with strict neutrality more readily than synonymous sites (e.g., Bustamante et al. 2002).

However, the hypermutable CpG positions also cause problems when using pseudogenes. In coding DNA, the CpG content is a function of the amino acid frequencies and codon usage; in noncoding DNA, this proportion is instead determined by GC content and the balance between mutations that establish CpG and those that disrupt it (Hwang and Green 2004; Fryxell and Moon 2005). This can vary considerably by organism and by location in the genome. For most vertebrates, between two and seven percent of the total cytosine nucleotides in the genome are methylated and subject to the acceleration of mutation rate caused by CpG hypermutability (Razin and Riggs 1980; Colot and Rossignol 1999). Furthermore, DNA methylation patterns are not distributed evenly over the genome. They tend to be targeted to mobile elements for host genome defense and to other regions for long-term silencing.

Moreover, pseudogenes have now been shown to evolve with very fast rates soon after they lose function, and with decreasing rates as further time passes (Figure 11.4). This nonlinear rate of evolution occurs because the CpG

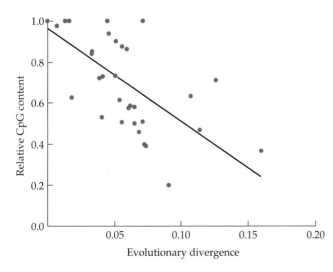

Figure 11.4 Relationship between evolutionary divergence and the ratio of CpG contents in pseudogenes and their functional counterparts, based on an analysis of 39 human pseudogenes. CpG content is seen to drop linearly as evolutionary divergence increases ($R^2 = 0.52$; $P < 0.01$). (After Subramanian and Kumar 2003.)

content at first and second codon positions of the pseudogene decays due to CpG→TpG and CpG→CpA mutations over time (Subramanian and Kumar 2003). Also, the rapid divergence rate makes the alignment of pseudogenes and functional genes difficult. Another drawback of using pseudogenes is that they may not be good indicators of mutation rate for a given protein, because they may reside in another part of the genome that has a different local mutation rate and pattern (Matassi et al. 1999).

Rate of Intron Divergence as a Proxy for Nonsynonymous Mutation Rates

In eukaryotes, comparison of homologous intron sequences is used as another candidate for estimating nonsynonymous mutation rates, because they are putatively neutral regions in proximity to the coding portion of a gene and not subject to cryptic selective pressures, especially if they are not located near exon boundaries (Castresana 2002). In fact, the mutation rate inferred from introns is less than the rate obtained from synonymous substitutions in mammals (Hoffman and Birney 2007), which is expected given that a smaller fraction of positions are involved in CpG dinucleotides in introns (Subramanian and Kumar 2003).

The mean observed (extrinsic) substitution rate within introns of a gene is a good estimate of the prevailing (intrinsic) mutation rate for the coding sequence of the gene if we assume that all intronic sites considered are strictly neutral. While restricting extrapolation of mutation rates from the introns to the exons of the same gene is more conservative than extrapolating over greater distances in the genome (Wolfe et al. 1989), it is clear that the GC content of third codon positions is significantly higher than that of introns (Bernardi 1986; Eyre-Walker 1991; Hughes and Yeager 1998). This would invalidate the assumption of equality of mutation rates between introns and exons.

Most recently, sequence divergence in introns and intergenic regions has been the preferred method to establish mutation rates for replacement positions in codons in human and chimpanzee lineages (Hellmann et al. 2003; Lu and Wu 2005; Mikkelsen et al. 2005; Parmley et al. 2006). However, the evolutionary divergence at introns (d_I) cannot always be directly equated with the rate at which mutations occur at codon positions that can experience nonsynonymous mutations. This is because nonsynonymous positions are found to be involved in CpG configurations twice as often as intronic sites. Furthermore, the per-gene frequency distribution of CpG content of introns is strikingly different from the distribution of CpG content of amino acid replacement positions, with the latter having a much higher CpG density (Figure 11.5A; Subramanian and Kumar 2006). Using the proportions of CpG contents in sites that experience nonsynonymous CpG mutations, rates can be corrected to account for the differences in CpG content. We previously have taken such an approach to compare two closely related species, human and chimpanzee, and found that the number of genes predicted to

(A)

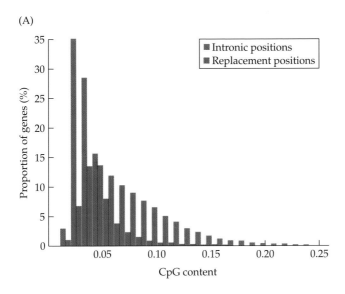

(B)

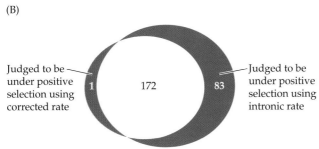

Figure 11.5 (A) The differential distribution of the fraction of intronic and replacement positions involved in CpG dinucleotide configurations from 10,196 functional human genes containing at least one intron. On average, the replacement positions involved in CpG configurations are two times higher than those of intronic positions (6.2% and 2.9%, respectively). The dispersion indices (the ratio of variance to mean) of the distributions of intronic and replacement positions are 0.015 and 0.024, respectively. (B) Venn diagram showing the numbers of common human genes inferred to be under adaptive evolution ($\omega > 1$), where ω estimated using intron distances is compared with the estimates derived using intrinsic mutational distances at nonsynonymous sites. The numbers in the overlapping area indicate the genes that were identified by both methods, and the numbers in the individual circles represent the genes that were predicted exclusively by one of the methods. (Data from Subramanian and Kumar 2006.)

be evolving with purifying selection changes considerably. This is shown in Figure 11.5B, which indicates that testing genes for selection based on CpG-adjusted versus unadjusted estimates of nonsynonymous mutation rates leads to both false positives and false negatives, primarily the former.

There are other potential problems with using introns as neutral regions. For one, introns may contain unknown conserved regulatory sites or RNA genes under purifying selection. For example, snoRNAs (small nucleolar RNAs) seem to be located exclusively in introns (Huang et al. 2005). Significant intron sequence conservation has also been detected in regions near exons (Hare and Palumbi 2003; Louie et al. 2003). Also, constraints from RNA structures (e.g., self-splicing introns), may also be sources of purifying selection. Similar to pseudogenes, the alignment of introns is more difficult than for exons, especially for more divergent species, and misalignment can have a significant effect on distance estimates. It has also been found that intronic rates differ more from species to species than synonymous rates; hence, the intron rates cannot be counted upon to be constant (Hoffman and Birney 2007).

How much difference is there between estimates of mutation rate derived from synonymous differences and those derived from presumably neutral intronic sites? This would depend on the extent of codon usage bias and the CpG density in different regions (among other factors). For 15,176 human–dog gene pairs, the median value of d_S (uncorrected for codon usage bias) was found to be 0.370, while median distance between intronic sites (d_I) was 0.305. For 16,183 mouse–rat orthologs, the distances are 0.212 and 0.158, respectively. Although correlations between the estimates of neutral distance are substantial (Spearman rank-order correlation coefficient of 0.57 for human–dog and 0.46 for mouse–rat), the two methods identify substantially different sets of genes as being under selection. In both the mouse–rat and human–dog cases, the sets of genes identified as being in the top 5 percent in terms of selection ratio do not overlap more than 65 percent between the two species (Hoffman and Birney 2007).

Direct Estimates of Mutation Rates in the Laboratory

Given the problems with comparative genomics and confounding factors, it is natural to ask why we don't just estimate mutation rates directly in the laboratory. Despite the inherent difficulties of large genomes and long generation times, some progress has been made in laboratory estimation of mutation rates for higher organisms. The first accurate experimental determination of a genome-wide mutation rate for a eukaryote (*Caenorhabditis elegans*) was only recently accomplished, which yielded a point mutation rate of 8.6×10^{-7} mutations per site per year. For reasons not yet completely clear, this estimate is considerably higher (by a factor of ten) than previous estimates using indirect methods (Denver et al. 2004). A similar method was used in 2007 for Drosophila, producing an annual rate of 5.8×10^{-8} (95% confidence interval 2.1×10^{-8} to 13.1×10^{-8}) under the assumption of 10 generations per year (Haag-Liautard et al. 2007). This rate is about 5 times higher than indirect estimates (Tamura et al. 2004).

Statistically significant variation was found among specific lineages within Drosophilids, however, with the slowest rate being 2.7×10^{-8}

(95% confidence interval 1.2×10^{-8} to 5.4×10^{-8}), and the fastest 11.7×10^{-8} (5.9×10^{-8} to 20.6×10^{-8}; Haag-Liautard et al. 2007). There are a number of additional difficulties with direct estimation: the cost and amount of time and data required for each species, the infeasibility of even maintaining mutation accumulation lines for many species, wide confidence intervals, uncertainty as to whether different generation times in the laboratory and in the wild affect the estimate, and the question of whether the sampled data are representative of the rest of the genome. Also, even if direct observation of mutation rate were easy to perform, it would give us only a snapshot of the rate for a species at the present time in the laboratory. The amount of error incurred by applying such rates to natural populations in the phylogenetic past is unknown.

Conclusion

Knowledge of mutation rates is the key not only to finding genes under selection, but to many other analyses, including detection of regulatory regions in genomic sequence data, identification of particular codons responsible for adaptive changes, and estimation of the number of deleterious mutations (Nei and Kumar 2000; Yang and Bielawski 2000; Nekrutenko et al. 2003; Yampolsky et al. 2005). For example, we may compute the number of deleterious mutations per diploid genome per generation (U) from the mutation rate, in order to examine whether the purging of synergistically interacting deleterious mutations can explain the maintenance of sex in a species. If U exceeds one, then the beneficial effect of sex in removing them can outweigh the cost to each organism of diluting its genetic material with that of a sex partner (Keightley and Eyre-Walker 2000). Understanding the strength of selection is also important, for example, in molecular clock analyses, since neutral substitution rates are expected to be less variable over lineages (over time) than nonsynonymous substitution rates (see Figure 11.1D).

How can we know the rate at which mutations occur in DNA? Mutation is a complex biochemical process that can be caused by inexact replication, chemical mutagens, or ionizing radiation (Brown 2002; Griffiths 2002). Mutation rates are further modulated by various repair mechanisms of differing efficacy (Brown 2002; Friedberg and Friedberg 2006). As a result, mutation rates may vary significantly among major taxa and within the same genome. They are not easy to determine experimentally in eukaryotes, because the rates are often so low that it takes a large amount of data and time to observe them in living systems.

Interspecific DNA comparison is a powerful tool for the estimation of the amount and type of selection on individual genes. Methods for doing this rely on accurate estimates of underlying mutation rates for the genes. This rate can vary from one lineage to another, from time to time, and over different regions of a single genome. Indirect methods for estimating mutation rate are based on the neutral theory and equate mutation rate with observed neutral substitution rate. The problem then becomes the identification of, and correction for, the effect of selective factors affecting putatively neutral substitutions. These factors include codon usage bias (in the case of synonymous changes in coding DNA) and CpG content (in the case of DNA in introns or pseudogenes). There are no doubt other factors of unknown magnitude to consider, but inroads have been made in correcting for these first two, and the resulting estimates of genes under selection are significantly affected.

Acknowledgments

We thank Kristi Garboushian for editorial support, Sven Findeiss for assistance with data preparation, and Wayne Parkhurst for drawing figures. This work was supported by a research grant from the National Institutes of Health (S. K.).

CHAPTER **12**

Gene Networks and Natural Selection

Andreas Wagner

Introduction

CAN QUALITATIVE INFORMATION ABOUT LARGE MOLECULAR NETWORKS inside cells teach us fundamentally new biology? In other words, is there a network biology (distinct from a network physics or network chemistry)? The answer to this question is important, because molecular networks are bridges between individual molecules—the lowest level of biological organization—and whole organisms. Both levels of organization, molecules and organisms, are intensively studied. Nonetheless, there is still an enormous gap in our understanding of how molecules collectively produce complex organismal phenotypes. Large molecular networks have the potential to fill this gap, because they contain most of the molecules that allow an organism to survive.

To find out whether large molecular networks can teach us new biology, we first need to answer a very basic question: Does natural selection influence the structure of biological networks, and if so, how? This question is key, because natural selection is the one central feature that distinguishes biological systems from all other, nonbiological systems—only biological systems have been shaped by natural selection, a process that acts on populations of organisms, and that requires heritable fitness differences among organisms.

In this chapter we use three examples to illustrate our progress in answering this question. First, we discuss recent work suggesting that the function of metabolic networks influences the rates at which its constituent enzymes evolve. Second, we show how multiple small transcriptional regulation networks may have arisen through convergent evolution. These examples demonstrate how natural selection can influence the small-scale, local structure of biological networks. Third, we discuss a number of candidate cases for the influence of natural selection on large-scale network structure—cases that illustrate the great challenges ahead.

Is There a Network Biology?

Efforts to understand the large-scale organization of living things are generating a wealth of information about how biologically important molecules interact to sustain life. One example concerns the interactions of proteins with DNA to promote transcriptional regulation. Such interactions have been elucidated for hundreds of regulators and thousands of target genes, using techniques such as chromatin immunoprecipitation (Lee et al. 2002). Another example concerns metabolic reactions, the interactions of enzymes with small molecules to convert food into energy and biosynthetic building blocks. A variety of information—from genome sequence to biochemical data—has been used to generate complete maps of metabolic reactions in different model organisms (Edwards and Palsson 1999, 2000a; Forster et al. 2003b). A third example concerns the physical interactions of proteins in protein complexes, which have been elucidated with techniques such as the yeast two-hybrid assay and affinity purification of tagged proteins (Uetz et al. 2000; Gavin et al. 2002).

Such information undoubtedly has many useful applications in both basic biology and medicine. For example, it could lead to the elucidation of new biochemical and regulatory pathways. It could speed the development of new and specific drugs targeted to key players of a regulatory system.

However, the potential of such information goes far beyond these extensions of business-as-usual in molecular biology. The reason is that we now have information about the interactions of thousands of molecules, which form large interaction networks akin to man-made information-processing networks and energy-distribution networks. The structural features of such networks may themselves contain biological information, information that would not be evident from studying individual molecules, or from studying whole organisms. Such networks could thus teach us fundamentally new biology. They might help build a bridge between our understanding of the molecular machinery of life and whole organisms.

The analysis of networks is not only popular in biology. It has seen an explosion of activity in fields ranging from physics to the social sciences (Albert and Barabasi 2002). This explosion has been greatly helped by graph theory, which provides students of networks with both a common language and a common set of tools. In addition, the common language of graph theory has greatly helped researchers cross disciplinary boundaries. Perhaps the most prevalent question in cross-disciplinary work has been whether networks in different disciplines share common features. The answer is a clear "yes," which is intriguing, because it means that the objects of disparate disciplines share common organizational laws. In contrast, there has been much less effort to study network architectures distinguishing, say, biological networks from physical networks. Put differently, little effort has been made to create a network biology that might be different from a network physics, network chemistry, or network sociology.

What distinguishes biological systems from all other systems, including physical, chemical, sociological, and man-made systems? The answer is natural selection: selective persistence of individuals with heritable features that coexist in populations. Natural selection is an absolutely essential ingredient to the creation and persistence of complex living systems. In contrast, it is required for neither the creation nor the persistence of any physical, chemical, or social system. This simple observation provides a unique perspective on the question of whether there is a network biology: If we want to find out whether biological networks have any features unique to them—features that they do not share with other kinds of networks—and thus features that may teach us new biology (as opposed to new physics or chemistry), we need to ask whether natural selection influences biological networks. This is easier said than done. Biology—and not only as practiced by physicists new to the field—is full of just-so stories, in which an organism's features are postulated to be shaped by natural selection without a shred of evidence other than plausibility. Nearly everyone can come up with such postulates, but relatively few try to provide supporting evidence.

In sum, information about large-scale biological networks might teach us qualitatively new biology. If this is the case—if there are laws that characterize and distinguish biological networks, and if there is a network biology— we'd better look for it by studying natural selection and its interaction with networks. In the following discussions, we illustrate three small steps in this direction, with different kinds of networks and at different levels of network organization. As will become obvious, natural selection influences the small-scale, local structure of networks. In contrast, on the larger, global scale of network organization, we have no lack of hypotheses but few solid answers. A new research program in evolutionary biology lies in wait here.

Natural Selection and Network Parts

In this section we discuss recent work that suggests how natural selection may influence the evolution of the smallest components of a metabolic network, namely its enzymes. Importantly, this work demonstrates that an understanding of network function may be essential to understand this influence.

Every living thing is sustained by a complex network of thousands of chemical reactions. In heterotrophic organisms, these reactions transform food into energy and new building blocks for growth and reproduction. Complete (or nearly so) maps of core metabolism, comprising hundreds of reactions and metabolites, are now available for several model organisms (Edwards and Palsson 1999, 2000a; Forster et al. 2003b). They can be used to study the structure, function, and evolution of large-scale metabolic networks. Some early work in this area focused only on network structure, for example, by characterizing one of a variety of graph representations of a metabolic network. Graphs (Figure 12.1) are mathematical objects that con-

Figure 12.1 (A) A simple graph comprising 7 nodes. Neighboring nodes are connected by edges. The number of edges emanating from a node is known as the node's degree. For example, the green node has degree 5. Nodes with especially high degree are also known as hubs. A simple way of characterizing a graph is through its degree distribution. (B) Three examples of small transcriptional regulation circuits that are overabundant in genome-scale transcriptional regulation networks. (After Milo et al. 2002; Shen-Orr et al. 2002.)

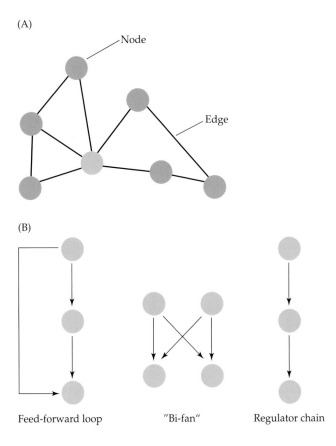

(A)

Node

Edge

(B)

Feed-forward loop   "Bi-fan"   Regulator chain

sist of nodes, and edges which connect neighboring nodes. (An example of a metabolic graph representation is a graph whose nodes are enzymes and metabolites, and where two nodes are connected if they participate in the same chemical reactions.) Such structural analysis, however, has one key limitation: it is poor at capturing the flow of matter through a metabolic network, which is at the heart of metabolic network function.

Metabolic network function can be computationally analyzed, even though information about enzymatic reaction rates in metabolic networks is very limited. Central to any such functional analysis are approaches such as flux balance analysis (Varma and Palsson 1993; Schilling et al. 1999). Flux balance analysis uses information about the stoichiometry and reversibility of chemical reactions to determine the possible rates (fluxes) at which individual chemical reactions can proceed if fundamental constraints such as that of mass conservation are taken into consideration. Within the limits of such constraints, flux balance analysis can then be used to determine the distribution of metabolic fluxes that will maximize some metabolic property of interest. The rate of biomass production is one of these properties. It is a proxy for cell growth rate, itself an important component of fitness in a single-celled organism. Flux balance analysis makes predictions that are often in good agreement with experimental evidence in *Escherichia coli* and the yeast *Saccharomyces cerevisiae* (Edwards and Palsson 2000c; Segre et al. 2002; Forster et al. 2003b). However, such predictions may fail if an organism has not been subject to natural selection to optimize growth in a particular environment.

Because natural selection clearly affects cell growth rates (the optimized codon usage bias of highly expressed genes in microbes is ample evidence), and because the distribution of metabolic fluxes in a network also bears a direct relation to cell growth rates, one can ask how natural selection may influence the rate of evolution of the enzymes responsible for these fluxes. Vitkup and collaborators (Vitkup et al. 2006) asked this question for the yeast metabolic network (Forster et al. 2003b), in which fluxes had been optimized to maximize growth. Specifically, they studied the relation between flux through individual enzymatic reactions and the ratio K_a/K_s, where K_s is the fraction of synonymous (silent) substitutions per silent nucleotide site in an enzyme-coding gene, and K_a is the fraction of amino acid replacement substitutions per replacement site. The ratio K_a/K_s is typically (much) smaller than one and a good indicator of the evolutionary constraint a protein is subject to: Proteins that can tolerate very few amino acid substitutions will have a smaller K_a/K_s than proteins that can tolerate more such substitutions. They also accumulate fewer amino acid substitutions per unit time and thus have a smaller K_a.

Is flux through any one enzymatic reaction associated with the evolutionary constraint (K_a/K_s) on the corresponding enzyme? The answer is "yes"— there is a significant and negative association between flux and evolutionary constraint. That is, enzymes associated with high metabolic flux under optimal growth conditions can tolerate fewer amino acid changes. This association persists if one takes into account that many enzymatic reactions are carried out by several isoenzymes. Furthermore, it is observed only for carbon sources, such as glucose and fructose, that are likely to be abundant in yeast's wild environment, but not for less-relevant carbon sources, such as acetate. A potential confounding factor in such an analysis is that many enzymes associated with high flux are expressed at high levels, and high expression is known to be associated with slow evolution. However, enzymes with high flux also evolve slowly if one corrects for differences in expression (Vitkup et al. 2006).

Why do enzymes associated with high metabolic flux evolve more slowly? An immediately obvious hypothesis is this: Most amino acid changes will lead to a reduction in the catalytic rate at which an enzymatic reaction proceeds. Enzymatic reactions associated with high metabolic flux occur preferentially in central metabolic pathways, whose products are distributed to many peripheral pathways. In contrast, enzymatic reactions in peripheral pathways often show lower (but no less essential) metabolic fluxes. Thus, a reduction in reaction rate caused by an amino acid substitution may have a greater effect in a central pathway, where it may affect the substrates available to multiple peripheral pathways. It may thus reduce the output of multiple pathways, as opposed to a mutation in a peripheral pathway, which may change only the output of the affected pathway.

Even though the distinction between central and peripheral pathways in a complex network may not be sharp, the hypothesis is clear: Reducing the flux through high-flux enzymes has a greater overall effect than reducing the flux through low-flux enzymes. This hypothesis produces a testable prediction: If it is correct, one would expect that mutations *increasing* flux through high-flux enzymes would also have a greater overall effect (except that these effects might be beneficial as opposed to deleterious). One such type of mutation is gene duplication, which potentially doubles the flux through an enzymatic reaction. If the hypothesis is correct, one would expect that such duplications can increase overall biomass production to a greater extent if they affect high-flux enzymes than if they affect low-flux enzymes. Because of the beneficial effects of increased biomass production, one would thus expect such mutations to become preferentially preserved in evolution. This is exactly what one observes, as we and others have shown (Papp et al. 2003b; Vitkup et al. 2006; Figure 12.2).

In sum, the rate at which enzymatic genes evolve varies. Some of this variation may be explicable through differences in protein structure or through differences in expression level. However, some fraction of this variation is explicable through differences in metabolic flux under conditions likely to have been important in the evolutionary history of yeast. Importantly, the distribution of metabolic fluxes optimal for cell growth cannot be explained by studying one enzyme. It cannot even be explained by

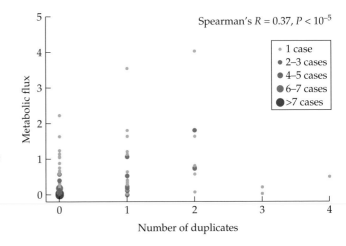

Figure 12.2 High-flux enzymes and gene duplication. The horizontal axis shows the number of duplicates of enzyme-coding genes in the yeast genome. The vertical axis shows the metabolic flux through the associated enzymes during aerobic growth on a minimal medium with glucose as the sole carbon source, as predicted by flux balance analysis. There is a highly significant positive association—that is, enzymes associated with high flux under these conditions have more isoenzymes. (After Wagner 2005c.)

Spearman's $R = 0.37$, $P < 10^{-5}$

- 1 case
- 2–3 cases
- 4–5 cases
- 6–7 cases
- >7 cases

Metabolic flux

Number of duplicates

studying an entire metabolic pathway. It emerges as a property of an entire complex metabolic reaction network comprising hundreds of reactions. In this sense, the network influences the evolution of its parts. Having information on a whole network thus clearly matters in explaining the evolution of its parts.

Natural Selection and Small-Scale, Local Network Features

In this section we focus not on the smallest parts of a network but on the local neighborhood of individual nodes. The network in question is that of transcriptional regulation in the yeast *S. cerevisiae* and in *E. coli*. The nodes are transcriptional regulators and their target genes. A transcriptional regulator is directly connected to a target gene (which may itself be a transcriptional regulator) if it regulates the target gene's transcription by binding to its regulatory region.

The work discussed here builds on recent studies that have identified small and highly abundant genetic circuit motifs in transcriptional regulation networks of the yeast *S. cerevisiae* (Ho et al. 2002) and the bacterium *E. coli* (Milo et al. 2002; Shen-Orr et al. 2002). These circuit motifs include regulatory chains, feed-forward circuits, and a "bi-fan" circuit (see Figure 12.1B). There are two extreme possibilities (and an entire spectrum of intermediates) for the evolutionary origin of these circuits. First, these circuits may have come about through the duplication and subsequent functional diversification of one or a few ancestral circuits (Figure 12.3)—that is, through the duplication of each of their constituent genes in a duplication event. This scenario is plausible, given the high frequency at which chromosome segments, large and small, and even whole genomes undergo duplication (Wolfe and Shields 1997; Lynch and Conery 2000; Dunham et

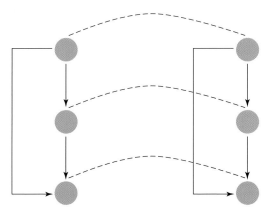

Duplication of a gene circuit

Figure 12.3 Hypothetical example of a duplicated gene circuit. Dashed lines connect paralogous genes: genes that arose in the duplication event that created the two feed-forward loops shown here.

al. 2002). Alternatively, most of these circuits may have arisen independently by recruitment of unrelated genes. In this case, abundant circuits would have arisen through *convergent evolution*.

Convergent evolution—the independent origin of similar organismal features—is a strong indicator of optimal "design" of a feature. It is ubiquitous at both the largest and the smallest levels of biological organization. For instance, eyes of similar basic design may have evolved multiple times independently; the wings of birds and bats have similar architectures; and both fish and whales—the latter descended from land mammals—have similarly streamlined body shapes (Futuyma 1998). On the smallest scale, lysozymes in foregut-fermenting herbivores have independently evolved digestive functions in bovids, colobine monkeys such as langurs, and a foregut-fermenting bird (Stewart et al. 1987; Kornegay et al. 1994). Another case in point is antifreeze glycoproteins, which have independently evolved similar amino acid sequences in Antarctic notothenioid fish and northern cods (Chen et al. 1997b). Evidence for convergent evolution is exciting to evolutionary biologists, because it can reveal the power of natural selection on an organismal feature.

A combination of genome and transcriptional regulation data allowed us (Conant and Wagner 2003) to test the hypothesis that abundant circuit types in transcriptional regulation networks have evolved through convergent evolution. Specifically, using available whole genome sequence information for yeast and *E. coli*, we asked, "How abundant are duplicated circuit pairs (see Figure 12.3) that may have shared a common ancestor?" Information on the abundance of such pairs allowed us to develop measures of common ancestry that quantify how many circuits in a genome may have shared a common ancestor (Figure 12.4).

Overall, in an analysis of 20 circuit types (18 in yeast and 2 in *E. coli*), preciously few circuit pairs shared any common ancestor. Specifically, for 17 circuit types not even one pair of circuits were duplicates of each other. Even for the remaining three circuit types, the vast majority of circuit pairs show independent ancestry. For example, 44 out of 48 feed-forward loops in yeast show independent ancestry, a number that is no greater than that expected by chance alone, due to the large number of single-gene duplications that have occurred in the yeast genome.

In sum, this analysis suggested that highly abundant regulatory circuit motifs in the transcriptional regulation network of two different organisms have arisen through convergent evolution (Conant and Wagner 2003). But what are the favorable functional properties of such networks—the properties that would drive such convergent evolution? Answers are beginning to emerge from a mix of computational and experimental work (Shen-Orr et al. 2002; Mangan and Alon 2003; Mangan et al. 2003). For example, a feed-forward loop may activate the regulated (downstream) genes only if the upstream-most regulator is persistently activated. It can thus act as a filter for ubiquitous intracellular gene expression noise. Conversely, this type of

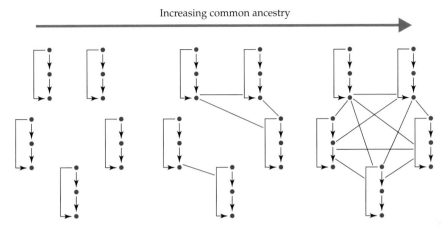

Figure 12.4 Quantifying common circuit ancestry as illustrated with a hypothetical example of $n = 5$ feed-forward loops. Each circuit is represented as a node in a graph. Nodes are connected if they are derived from a common ancestor, that is, if all k pairs of corresponding genes in the two circuits are duplicate genes. An index of common ancestry, A, can be defined that ranges from A = 0 if no circuits share a common ancestor (left group), to A ≈ 1 if all circuits share one common ancestor (right group). (After Conant and Wagner 2003.)

circuit can rapidly deactivate downstream genes when the upstream-most regulator is shut off.

Taken together, such functional information, plus the high abundance of some small regulatory circuit motifs and their independent evolutionary origins, indicates the importance of natural selection for the evolution of genetic networks. Importantly for the purpose of this chapter, it shows that natural selection can shape and maintain the small-scale features of a large biological network.

Natural Selection and Global Network Structure

Global network structure is an aspect of structure that cannot be reduced to characterizing individual nodes (proteins, genes) and their neighbors (see Figure 12.1A). Examples include a network's degree distribution; that is, the distribution of each node's number of immediate neighbors. Although each node's degree is a local property, the degree distribution is a property of the whole network. Similar examples include pairwise correlations among node degrees, or the distribution of the number of edges linking any two nodes in a network. We will discuss several candidate examples of natural selection acting on global network structure—all of them with important flaws—before outlining the lessons they provide.

Natural selection and the degree distribution

The connectivity or degree distribution of biological networks is often broad-tailed and sometimes consistent with a power law. In a power law degree distribution, the probability $P(d)$ that a node has d immediate neighbors is proportional to $P(d) \propto d^{-\gamma}$, γ being some constant. Among networks with a broad-tailed degree distribution are metabolic networks, in which nodes can be enzymes or metabolites, depending on the chosen representation; protein interaction networks (Figure 12.5), in which two nodes (proteins) are connected if they interact physically inside the cell and other cellular networks (Jeong et al. 2001; Rzhetsky and Gomez 2001; Wagner 2001; Wagner and Fell 2001; Wuchty 2001, 2002).

Because broad-tailed degree distributions are abundant among biological networks, the question arises whether they exist for some biological reason. Are networks with this degree distribution better suited than other networks for some biological function? In other words, has their degree distribution been shaped by natural selection? If so, their degree distribution may reveal profound organizational principles of biological networks.

Indeed, it has been recently suggested that the degree distribution of biological networks is optimal (Albert et al. 2000; Jeong et al. 2000; Jeong et al. 2001). At the base of this suggestion stands the following observation. The mean number of edges connecting pairs of nodes is very small and it increases only very little upon random node removal (Albert et al. 2000) in

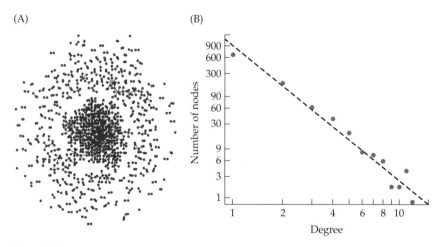

(A)　　　　　　　　　　(B)

Figure 12.5 (A) A graph representing the yeast protein interaction network as elucidated through high-throughput identification of interacting protein pairs. (B) The distribution of the number of neighbors each protein has. This distribution is broad-tailed and consistent with a power law. (Data from Uetz et al. 2000.)

networks with a broad-tailed degree distribution. This feature—a network's mean path length—is a measure of network compactness. In networks with narrow-tailed degree distributions (e.g., a Poisson distribution), random node removal leads to a more substantial increase in mean path length, and a ready fragmentation into disconnected components. The robust compactness of networks with broad-tailed degree distributions, so the hypothesis goes, might be advantageous for biological networks.

The problem with this hypothesis starts with the question of whether any such (unknown) advantage would be biologically sensible. For example, protein interaction networks are very heterogeneous mixtures of proteins, in which some interact to provide structural support to a cell, others interact to transmit signals, and yet others interact to catalyze chemical reactions. It is thus not clear that such networks have one clearly defined biological function that needs to be preserved by keeping the network compact and connected.

A second potential problem has been explored explicitly for the chemical reaction networks defined by metabolism. Is it possible that the broad-tailed degree distribution of metabolic networks is a feature of many or all large chemical reaction networks, whether or not they are formed by natural selection? If so, then metabolic networks would join a large group of other networks whose broad-tailed degree distribution is not due to a benefit they provide. In support of this possibility, Gleiss and collaborators (Gleiss et al. 2001) have studied the chemical reaction networks of planetary atmospheres, in which atmospheric photochemistry determines network structure. The available data stems from multiple planets, including Venus, Jupiter, and Earth. The respective chemical reaction networks, despite having very different structure, have a broad-tailed degree distribution (Gleiss et al. 2001). Such data suggests that broad-tailed degree distributions may be general features of chemical reaction networks, whether or not they exist in a living cell.

A third problem is an extension of the second problem. Broad-tailed degree distributions are generally highly abundant in systems ranging from physical to sociological networks. In contrast, networks that do not have this degree distribution are less frequent. This observation further weakens the argument that such distributions might be shaped by natural selection and are thus presumably specific to biological systems.

Finally, there are a growing number of mathematical models, grounded in empirical data, showing how the degree distribution of biological networks changes over evolutionary time (Sole et al. 2002; Wagner 2003; Berg et al. 2004; van Noort et al. 2004; Pagel et al. 2007). Although they differ in many details, these models demonstrate how power law degree distributions can emerge in evolution, without natural selection having shaped this distribution. In sum, several lines of evidence speak against the possibility that broad-tailed degree distributions in biological networks have been shaped by natural selection.

Global features of transcriptional regulation networks

The next example concerns the abundance of regulatory cycles in signal transduction networks. Regulatory cycles or feedback loops are important for biological networks (Savageau 1976; Fell 1997). They may endow a gene or network with robustness to environmental change or intracellular noise. Alternatively, they make multistability—the adoption of multiple stable states—in a biological network possible (Freeman 2000; Ferrell 2002).

Their biological importance makes cyclic structures good candidate features that might be under the influence of natural selection. Specifically, natural selection may influence the number and distribution of cycles in a biological network. One can find evidence for or against this hypothesis by asking if cycles are more abundant in biological networks than one would expect by chance alone (Wagner and Wright 2005). This question has been investigated for a variety of signal transduction networks, such as the c-Jun N-terminal MAPK network, and the B and T lymphocyte receptor signaling network, whose structure has been curated by experts and is publicly available (http://www.stke.org/cgi/cm/). We created from each of 15 biological networks 1000 randomized networks (Wagner and Wright 2005), and compared the number of cycles in the biological networks with the distribution of the number of cycles in the randomized networks. In general, exhaustive counting of cycles in large graphs may be computationally prohibitive, requiring alternative means to estimate the abundance of cycles (Gleiss et al. 2001). However, because of the moderate size of our networks, we were able to enumerate their cycles.

An example of the distribution of numbers for the B cell antigen receptor network is shown in Figure 12.6. Overall, 7 out of 15 signal transduction networks show a significantly smaller (!) number of cycles with length between two and ten than randomized networks, and 1 out of 15 networks showed more cycles than expected by chance alone.

As stated above, cycles may cause complex dynamical behavior in a network, in particular multistability (Thomas and D'Ari 1990; Ferrell 2002). On the one hand, multistability may allow a cell to adopt a stable response to extracellular information. On the other hand, complex dynamical behavior and multistability may render a network more fragile to perturbations. It is thus tempting to hypothesize that networks with an underabundance of cycles have experienced a selective purging of cycles for this reason. However, no evidence beyond the significant underabundance of cycles in biological networks currently speaks to this hypothesis.

Alternative pathways in transcriptional regulation networks

In a genome-scale transcriptional regulation network, a transcriptional regulator and its target molecule need not be immediate neighbors in the network. That is, the regulator may not regulate its target directly, but it may act through one or more intermediate regulators. If so, the path connecting

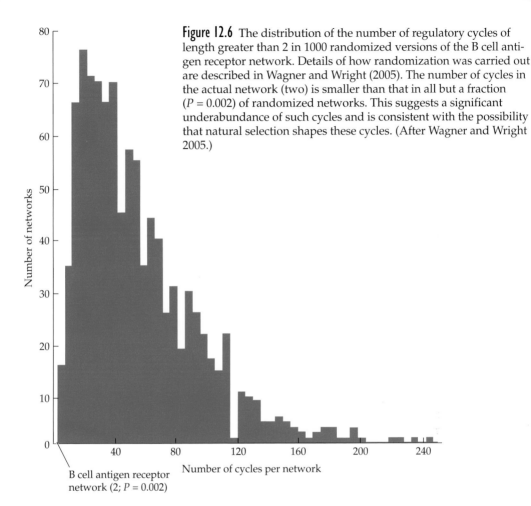

Figure 12.6 The distribution of the number of regulatory cycles of length greater than 2 in 1000 randomized versions of the B cell antigen receptor network. Details of how randomization was carried out are described in Wagner and Wright (2005). The number of cycles in the actual network (two) is smaller than that in all but a fraction ($P = 0.002$) of randomized networks. This suggests a significant underabundance of such cycles and is consistent with the possibility that natural selection shapes these cycles. (After Wagner and Wright 2005.)

the regulator and its target may be either unique or not (Figure 12.7). In the extreme case, regulator–target pairs may be connected by large bundles of alternative regulatory pathways. In the yeast transcriptional regulation network, such bundles of alternative pathways are not uncommon. For example, there are more than 100 regulator–target pairs connected by 10 or more alternative pathways (Wagner and Wright 2007). For such regulator–target pairs, perturbation of one of these pathways may be compensated through alternative regulatory routes.

To assess this possibility, we compared the rate at which amino acid substitutions accumulate in intermediate regulators of pathways in which regulators and targets are connected by a single pathway or by a pathway bundle. Using the ratio K_a/K_s as a measure of evolutionary rate, we found that

Figure 12.7 Regulatory molecules and their targets can either be linked directly (not shown) or indirectly through a single regulatory path involving intermediate regulators (A) or many such paths (B). The manner in which molecules are linked may have consequences for the rate at which intermediate regulators evolve.

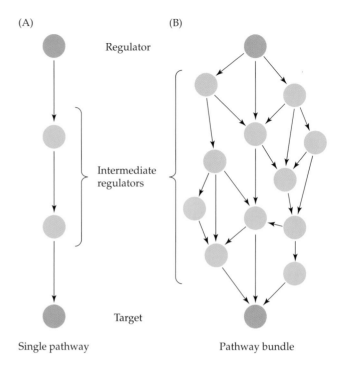

(A)

(B)

Regulator

Intermediate regulators

Target

Single pathway

Pathway bundle

the rate of evolution is significantly faster in intermediate regulators that are embedded in a bundle of pathways. (The same holds if one uses only K_a as a measure of amino acid divergence, or if one subdivides all regulator–target pairs according to the shortest distance of paths connecting them.) This means that in their evolutionary history intermediate regulators that are part of a pathway bundle have tolerated more amino acid changes (most of which have deleterious effects on protein function). We were able to exclude the possibilities that such regulators only evolve more rapidly because they are expressed in one of two ways: (1) at lower levels of either mRNA or protein; or (2) in a more limited spectrum of environmental conditions.

Taken together, this evidence suggests that alternative pathways between regulators and their targets can cause mutational robustness of the intermediate regulators. The distribution of the number of alternative pathways in a regulatory network is a global feature of the network. The key question for the present purpose is whether it has any adaptive significance. Has it been shaped by natural selection? And this is where an otherwise intriguing story collapses like a soufflé.

We tested in two ways the hypothesis that the distribution of alternative pathways in the yeast transcriptional regulation network has adaptive significance. First, if so, the number of alternative pathways should be significantly different from that in randomized networks with the same degree distribution. This is not the case. For example, more than 50 percent of randomized networks have more regulator–target pairs with greater than 10, 20, or 30 connecting paths than the actual transcriptional regulation network. Second, if the hypothesis is correct, then many alternative pathways between regulator–target gene pairs should exist for target genes that are important to the cell (as indicated by their strong deletion effects or slow rate of evolutionary change). Again, this is not the case. Specifically, there is no statistical association between the number of paths terminating at a target gene and its deletion effect or its rate of evolution K_a/K_s. We thus have no evidence to support the notion that the alternative pathway structure of the yeast transcriptional regulation network is shaped by natural selection.

Conclusion

The first of the preceding sections showed how natural selection can influence the rate at which enzyme-coding genes, the smallest parts of large-scale metabolic networks, evolve. Incidentally, it also illustrated how information about a biological network's function can help explain evolutionary rates of network parts. (In other words, it showed that networks matter.) The second section of this chapter showed how natural selection can shape small, local parts of a network's larger-scale structure.

These are intriguing results, but they leave the biggest challenge unmet. This challenge is to demonstrate that natural selection shapes the large-scale structure of biological networks. Only if this challenge is met, arguably, can we speak of a network biology created by new and large-scale information about molecular interactions, a field of investigation distinct from a physics or chemistry of networks. (Small-scale features of networks have been accessible for a long time through the tools of conventional molecular biology.) Notice that this influence of natural selection on large-scale network features need not exist. In other words, natural selection may well shape the small, local structure of biological networks, but not global network features. The earlier example of a postulated advantage—robust compactness— of the broad-tailed degree distribution of biological networks provides a case in point. As noted, there are several lines of evidence speaking against the notion that it might be important specifically for biological systems, one being that such distributions are ubiquitous in all natural and man-made systems. In addition, an increasing number of mathematical models can explain this and other large-scale features of biological networks without recourse to selection on any one feature.

The above example of the degree distribution is symptomatic of many others in the literature, where the influence of natural selection is postulated without proof. Just like the examples that followed (cycles in signal transduction networks, and alternative pathways in transcriptional regulation networks), it illustrates both the seductive appeal of such hypotheses—anybody could come up with one—and the difficulty in providing proof. In the signal transduction networks, where many networks we examined contain few cycles, the hypothesis is that cycles can cause complex dynamics and multistability, which may not always be desirable. However, without further evidence, this is mere speculation. We cannot examine the regulatory dynamics of such networks, because we have incomplete information about their structure and no good, quantitative models. In addition, not enough is known about their biology to *prove* that multistability is a bad thing. And not even our efforts to ask whether such cycles are significantly less abundant than expected "by chance alone" help much. The reason for this is that in order to do so we need to randomize network structure, and there we make an implicit assumption about the structure of networks (the randomized networks) that are not under the influence of natural selection. If this assumption were correct, we would be done. But there are many ways of generating randomized networks, and they may yield substantially different results (Artzy-Randrup et al. 2004). We have no way of knowing which one yields the correct "null" network that we would expect if selection is absent.

In the final example, the hypothesis is that the alternative pathways structure of transcriptional regulation networks has been shaped by natural selection. The availability of molecular evolution data in this example allows different routes of attack, which lead to the failure of the hypothesis, at least with current data. The evidence is thus consistent with the notion that the robustness (as indicated by evolutionary constraint) of intermediate regulators is a mere consequence, not a cause, of the structure of this network. In sum, for the examples discussed here, there is either evidence against, or lack of evidence for, the action of natural selection on global network features. We know of no example where there is convincing positive evidence.

What information would we need to make the case for natural selection shaping large biological networks, and thus the case for a network biology? A look at the rich history of studies in both organismal and molecular evolution (Li 1997; Futuyma 1998) provides some answers. First, one needs a reference standard of how a network would evolve without the influence of natural selection, in order to compare it to existing networks. Such a standard has long been available for molecules, through the neutral theory of molecular evolution (Kimura 1983). It is absent for networks. Second, by far the most successful approach in evolutionary biology is the comparative approach, which consists of comparing organismal features that are subject to different selective pressures. Comparative data is conspicuously absent for networks. Chances are that we won't get far without it. And finally, the truly compelling cases for natural selection always use a variety of evidence, including comparative data and a functional characterization of the feature of interest. On this count also, we have a long way to go for genetic networks. The generation of this kind of information encompasses a research program that could take evolutionary biology to a completely new level, and that could build the long-sought bridge between molecules and living organisms.

CHAPTER **13**

Human Evolutionary Genomics

Ines Hellmann and Rasmus Nielsen

Introduction

THE SEQUENCING OF THE HUMAN GENOME NECESSARILY CONTRIBUTED a great deal
to our understanding of genome structure, but it was only with the availability of genomes from related species that we began to gain a deeper
understanding of the evolutionary processes that have shaped the human
genome. The comparative data gave researchers a chance to identify what
was unique to humans and to gain an increased understanding of the
processes that create differences among species. Now we are beginning to
understand how chromosomal rearrangements, insertions/deletions, and
point mutations impact the evolution of genomes. These data have also
given us an opportunity to determine where in the genome Darwinian selection may have been operating during human evolution, thus informing us
about functional changes in the evolution of modern man.

The most important tool used in comparative genomics is the estimation
of rates of substitutions, that is, the accumulation of differences between
species. For example, if we encounter noncoding regions in the genome that
are highly conserved among otherwise divergent species, this would
strongly suggest that these regions have a functional role. Identification of
conserved regions is one of the primary tools used to identify transcription factor binding sites, protein-coding regions, and other functional sites.
However, there is an increasing awareness that in identifying the causes of
evolutionary change and species-specific phenotypes, it is not sufficient to
concentrate on conserved genomic elements. Instead, we must also look at
the regions with high variability and large differences among species. In
particular, we should identify the regions of the genome that have been subject to positive selection.

In addition to the comparative genomic data, genome-wide population genetic data from humans have helped to identify regions affected by selection. There are numerous inferential tools used in population genetics to detect selection (known as neutrality tests; reviewed in Nielsen 2005). Recent positive selection may be inferred from the existence of a single very long haplotype (e.g., Sabeti et al. 2002; Voight et al. 2006), unusual allele frequency distributions (e.g., Tajima 1989; Fu and Li 1993; Fay and Wu 2000; Williamson et al. 2005), increased levels of population subdivision (e.g., Akey et al. 2004), and in various other ways. These methods can be used for identifying selection both in coding and in noncoding regions. They have been used to compare the effect of selection among different types of mutations, and to detect the effect of single strongly selected mutations that have recently increased in frequency in the population—so-called *selective sweeps*.

In this chapter we review some of the important observations regarding human evolutionary genomics from recent comparative and population genetic studies. Much of this review focuses on the effect of natural selection, but we start by outlining what is known regarding mutational processes and recombination affecting the human genome.

Mutations—the Sources of Genome Variability

Point mutations

The ultimate source of variability in our genome is mutation. An understanding of the basic processes generating variability is necessary to grasp genome structure and the effect of natural selection on the genome. Mutation rates are hard to measure directly; hence, they are usually inferred from the divergence between species. For simplicity, the focus so far has been on single nucleotide substitutions and thus point mutations. From these studies, it is clear that not all mutations occur at the same rate: they vary across the genome and over time.

Although mutations accrue at a roughly constant rate (Zuckerkandl and Pauling 1965), there is considerable evidence for variation in the pattern and rate of mutation within and between species. For example, the mutation rate per year seems to be considerably higher in rodents than in primates, due to the difference in generation time (e.g., Li et al. 1987; Yi et al. 2002) and/or higher metabolic rates (e.g., Martin and Palumbi 1993). Likewise, mutation rates and patterns of mutation seem to vary at many scales, and for many different reasons, in the human genome (Eyre-Walker and Hurst 2001; Ebersberger et al. 2002; Reich et al. 2002; Hwang and Green 2004).

Mutation rates may vary among different chromosomes because point mutations are thought to occur more frequently in the male than in the female germ line, probably because male germ cells undergo more cell divisions (reviewed in Li et al. 2002). This may explain why the Y chromosome has the highest divergence between humans and chimpanzees (1.9%) and

the X chromosome the lowest (0.94%), with the autosomes being intermediate. However, most variation in mutation rate seems to occur at a subchro-
mosomal level (Mikkelsen et al. 2005). Reich and colleagues (2002) found
that a positive spatial autocorrelation of substitution rates between humans
and chimpanzees persists up to a distance of 70–100 kb.

Variation in mutation rates along the genome is not random. For instance,
it is commonly observed that mutation rates seem to increase with GC content. This effect is relatively weak at small scales (Reich et al. 2002), but when
investigating variation in substitution rates at a scale of 3 Mb, the combination of various sequence motifs (including GC content), the positioning relative to telomeres and centromeres, and the recombination rate explain a larger
fraction of the variance (53%; Hellmann et al. 2005). One of the sequence motifs
associated with increased mutation rates is the dinucleotide CG (CpG). The
mutation rate of CpGs in humans is about 18 times higher than that of any
other sequence motif (Mikkelsen et al. 2005), representing the strongest context dependence of mutation rate in our genome (Hwang and Green 2004).
The primary cause for the increased rate of CpG mutations is deamination of
methylated Cs (Shen et al. 1994); cytosines within CpGs are frequently methylated in mammals (Ehrlich and Wang 1981). However, context dependen-
cies stretching over only a couple of nucleotides are not likely to fully explain
variation in mutation rate at the 3Mb scale (Hellmann et al. 2005). Other factors such as chromatin structure may also be of importance.

Chromatin structures, as measured by chromosome banding patterns,
are tightly bound to GC content, and there is considerable heterogeneity in
GC content in mammalian genomes (Filipski et al. 1973). The causes for this
heterogeneity have been the subject of much debate. There are three competing theories (reviewed in Eyre-Walker and Hurst 2001) to explain these
so-called isochores (regions of similar high or low GC content):

1. Isochores could be the result of a mutational bias.

2. Isochores could be the result of weak selection on genome structure.

3. Isochores could be due to biased gene conversion.

Biases in DNA repair and mutation timing are well documented and are
consistent with a mutational origin of isochores (Eyre-Walker and Hurst
2001). The selection theory is supported by the observation that GC→AT
SNPs (single nucleotide polymorphisms) have a lower frequency than
AT→GC SNPs (Webster et al. 2003). Biased gene conversion can lead to patterns very similar to selection (Nagylaki 1983), because the GC allele is preferentially retained when a heteroduplex is formed during recombination.
Whatever mechanism led to the isochore structures, it seems to be no longer
a strong force, because isochores apparently are vanishing (Duret et al. 2002;
Webster et al. 2003).

Recently, it was reported that the mutation process in humans is not in
equilibrium. Comparisons of human and chimpanzee sequences, using

Figure 13.1 The base composition of the human genome is not in equilibrium. There are more mutations from GC→AT, predicting that the genome will move from a GC content of 35–55 percent to a lower one of 33–42 percent. (After Meunier and Duret 2004.)

baboon as an outgroup, demonstrate that GC→AT substitutions occur more frequently than AT→GC substitutions in most regions of the genome (Figure 13.1; Duret et al. 2002; Smith et al. 2002). Thus the genome is evolving towards a lower GC content. Only a small part of this bias seems to be due to CpG transitions. The remaining nonequilibrium might be explained by selection due to a change in the optimal GC content or a change in methylation of CpGs. Alternatively, the effect may be due to biased gene conversion, which predicts that the equilibrium GC content changes with changing recombination rates (Meunier and Duret 2004). There is evidence that recombination rates are indeed subject to fast changes (see later sections in this chapter). However, the causes of the nonequilibrium condition in humans and other mammals are still controversial. Lastly, we want to note here that a nonequilibrium in base composition, as observed by Webster and colleagues (2003), may in itself explain the observation that substitution rates covary with GC content (Hardison et al. 2003; Hellmann et al. 2005).

Insertions, deletions, and inversions

Humans and chimpanzees differ not only through point substitution, but also through chromosomal rearrangements, insertions, and deletions. Similar to point mutations, small-scale insertions up to 10 bp (microindels) accrue at an approximately constant rate, but they are 20 times less common than point substitutions (Cooper et al. 2004). Furthermore, comparisons of the human, mouse, and rat genomes show no correlation between the rate of point substitutions and microindels, suggesting that different forces may be responsible for variation in the rate of these two processes.

Cheng and colleagues (2005) created a list of duplicated regions in humans and chimpanzees, where duplicates were required to be more than 95 percent similar over 20 kb of sequence. The analysis of this list shows that humans and chimpanzees have similar rates of segmental duplication, and about one percent of the human genome has been duplicated since the divergence from chimpanzees. Among the 515 human-specific duplications identified, many have remained unstable. This is probably because the repeat structures, which might have mediated the rearrangement in the first place, are still present. Also, duplications themselves make a region more prone to nonhomologous recombination, resulting in further copy

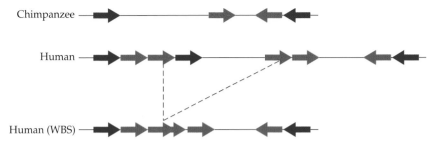

Figure 13.2 Williams-Beuren syndrome (WBS) occurs with a frequency of about 1:20,000, and most cases are caused by a 1.55 Mb deletion on chromosome 7q11 (Bayes et al. 2003). The large blocks of segmental duplications in this region are depicted as thick arrows, indicating their relative orientation. The complicated arrangement of duplications, which predisposes humans for this deletion, has occurred recently in the human lineage and is not present in the chimpanzee. (After Antonell et al. 2005.)

number changes. For instance, the Williams-Beuren syndrome (WBS) is caused by a common 1.55 Mb deletion on chromosome 7q (Pérez Jurado et al. 1996; Robinson et al. 1996); part of the complicated structure of the duplication that leads to the deletion is specific to humans (Antonell et al. 2005; Figure 13.2).

Further evidence for more recent large-scale insertions or deletions in the human genome comes from copy number polymorphisms (CNPs). Recent studies have identified a total of about 1500 CNPs ranging in size from 70 bp to 1900 kb in healthy individuals (reviewed in Eichler 2006). Large CNPs (>50 kb) have been identified by comparative genome hybridization techniques (Iafrate et al. 2004; Sebat et al. 2004; Sharp et al. 2005). Small- to medium-sized CNPs (70 bp–10.5 kb) have been identified from fosmid paired-end sequences (Tuzun et al. 2005), non-Mendelian inheritance in pedigree studies based on large-scale SNP data sets (Conrad et al. 2006; McCarroll et al. 2006), and genome tiling array analyses (Hinds et al. 2006). All of these studies have most likely underestimated the true number of CNPs. For example, the three studies examining small- to medium-sized CNPs cannot identify CNPs involving repeats and duplicated regions (Conrad et al. 2006; Hinds et al. 2006; McCarroll et al. 2006).

There is evidence from the distribution of alleles within populations that CNPs are affected by selection (Conrad et al. 2006); CNPs are linked to an excess of low-frequency polymorphisms compared to other putatively neutral SNP markers. This observation suggests that many CNP mutations are deleterious and might, therefore, be of particular interest in disease-mapping studies. There is also direct evidence that a large, 900 kb inversion on chromosome 17q has been under positive selection in Iceland (Stefansson et al. 2005). From pedigree studies it was found that females carrying the inversion have more children than females not carrying the inversion. The authors hypothesize that the fitness advantage is due to an increase in the recombination rate of the females carrying the mutation (Kong et al. 2004).

Recombination

Knowledge about recombination in the genome is crucial for our understanding of genome evolution and selection (see discussion later in this chapter). The human recombination landscape has been studied extensively, by comparing recombination maps from pedigrees with physical maps of the genome (Broman et al. 1998; Yu et al. 2001; Kong et al. 2002), through sperm genotyping, and indirectly through the analysis of patterns of linkage disequilibrium (LD). These methods differ in their spatial and temporal resolution, and in their presumptions. For example, recombination rate estimates from sperm typing and from the analysis of pedigrees reflect recombination within one individual and in the last two or three generations, respectively. Meanwhile, recombination rates inferred from LD reflect past recombination in the population during thousands of generations.

Estimates based on LD rely on explicit models describing the history and demography of the populations examined, and the absence of natural selection. Violations of these presumptions can lead to incorrect inferences of the recombination landscape (e.g., Reed and Tishkoff 2006). For this reason, more direct estimates based on pedigrees or sperm typing may be preferable. Unfortunately, sperm typing is expensive and methods based on pedigrees are limited in terms of their spatial resolution. With the current data from pedigrees, recombination can be confidently estimated for windows not much smaller than 3 Mb (Kong et al. 2002), while the resolution from sperm typing (Jeffreys et al. 2001) and LD data (Crawford et al. 2004; McVean et al. 2004) is on a scale of 1–2 kb. The recombination rate estimates from the pedigree data vary among genomic regions by roughly two orders of magnitude, while the LD-based estimates vary over four orders of magnitude (McVean et al. 2004). Of the recombination events, 80 percent occur in only 10 to 20 percent of the sequence (Myers et al. 2005). Such short regions, in which the recombination rate is elevated 10–1000 times above the background level, are termed *recombination hotspots*.

Sperm-typing studies have shown that a single SNP can determine whether there is a hotspot or not (Jeffreys and Neumann 2002). Furthermore, Jeffreys and Neumann suggest that when gene conversion occurs between hotspot and nonhotspot alleles, the nonhotspot allele is retained more frequently. This mechanism will induce a type of meiotic drive which will eventually eliminate the hotspot allele from the population. So far this mechanism leaves unresolved how hotspot alleles can reach sufficiently high frequencies in the population to be detected by LD studies at all.

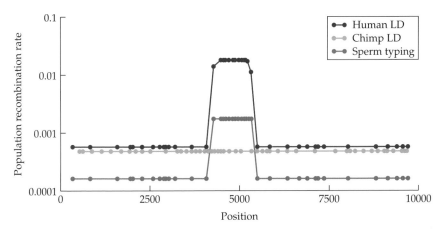

Figure 13.3 The recombination landscape of about 9 kb within the *TAP2* locus, as inferred from sperm genotyping in humans (Jeffreys et al. 2000) and LD analysis in humans and chimpanzees (Ptak et al. 2004). The recombination hotspot must have been present in humans for a long enough time to leave its mark on the patterns of LD, while there is no indication for a hotspot at this position in chimpanzees. (Courtesy of Susan Ptak.)

In any case, hotspots appear to have a high turnover rate that leads to a quickly changing fine-scale recombination landscape. Evidence for this comes from the observation that recombination hotspots are not conserved between humans and chimpanzees (Ptak et al. 2004; Ptak et al. 2005; Winckler et al. 2005; Figure 13.3), and some of the hotspots identified by sperm typing cannot be detected in LD studies, which may be expected if hotspot locations change very quickly (Jeffreys et al. 2005).

It is still unknown what determines the location of recombination hotspots. There are some suggestions that hotspots could be affected by rather complex interactions between repeats or short sequence motifs (Myers et al. 2005). The fact that the hotspot distribution changes quickly suggests that many mutable sites affect hotspot locations. More work is needed to reconcile observations from LD and sperm-typing studies to fully understand the forces determining the frequency and distribution of recombination hotspots in humans.

This said, there is good concordance between the LD and pedigree data on a large scale (Myers et al. 2005), suggesting that while the location of recombination hotspots may change rapidly through time, the overall density of hotspots remains relatively constant. Population differences in large-scale recombination rates are mainly attributable to rearrangements, most prominently an inversion on chromosome 8p23 (Jorgenson et al. 2005; Serre et al. 2005). Furthermore, a major proportion of the large-scale variance in recombination rate can be explained by simple sequence motifs such as GC- and CpG- content as well as gene content (Kong et al. 2002; Myers et al. 2005). These sequence features change little over evolutionary time and large-scale recombination rates should, therefore, also be moderately stable.

Transposable Elements

At least 45 percent of our genome is derived from transposable elements (TE) (International Human Genome Sequencing Consortium 2001). Most of the TEs within the genome are so degraded that they have lost the ability to self-propagate. However, some TEs are still active within the human genome. These are the LINE element L1, the SINE element AluY, and retroviral elements (long terminal repeat–containing elements or LTRs) of the type HERV-K. Functional L1 elements are about 6 kb long and carry their own reverse transcription and reintegration machinery. Most newly inserted L1 elements are less than half as long as the one from which they originated, and hence cannot propagate autonomously. SINEs depend on the replication machinery from fully functional L1 elements and are not able to propagate autonomously. HERVs are autonomous and propagate through the formation of actual viral particles.

Transposable elements contribute to genome dynamics in many ways, most notably through new insertions—which could be viewed as small structural mutations—and ectopic recombination. Since the split from the chimpanzee, the human genome experienced insertions of about 2000 L1, about 7000 AluY, and about 80 HERV-K elements (Mikkelsen et al. 2005). Given the draft nature of the chimpanzee genome, these are likely underestimates. Nevertheless, it seems that the human genome experienced roughly three times as many new Alu insertions as the chimpanzee genome, mainly due to the proliferation of two new subfamilies (AluYa5 and AluYb8). In humans many new Alu insertions are polymorphic (reviewed in Batzer and Deininger 2002). There are also several instances in which an Alu insertion into a gene was identified as disease-causing. One example is the insertion of an Alu element into the gene *NF1*, which caused neurofibrimatosis (Wallace et al. 1991). For a comprehensive review on Alu insertions as disease mutations see Deininger and Batzer (1999).

New insertions of transposable elements can also promote the evolution of new genes or new gene functions. For example, the primate placental gene syncytin was derived from the insertion of a HERV-W element (Mi et al. 2000). Furthermore, about four percent of all human coding regions contain transposable elements (Nekrutenko and Li 2001). The majority of these elements are located within introns and are spliced into the mature mRNA as an additional exon (Nekrutenko and Li 2001). Exonized Alus are often spliced into alternative transcripts (Sorek et al. 2002), thus they can contribute to protein diversity without deleterious effects. HERVs also have the potential to act as promoters. Most known transcription factor binding sites are overrepresented within HERVs as compared to the genome average (Thornburg et al. 2006). Indeed, the human placental gene *MID1* derives its promoter from an HERV-E element (Landry et al. 2002). SVA elements also have the potential to serve as promoters and are still active in the human genome. Three such elements have been inserted in the promoter region of genes since the human–chimpanzee split; however, it is unknown to what degree they affect transcription (Mikkelsen et al. 2005).

In addition, TEs affect genome plasticity, as evidenced by their being found in the proximity of many large-scale rearrangements. Due to their high sequence similarity and frequency in the genome, TEs may frequently be involved in nonhomologous recombination (reviewed in Stankiewicz and Lupski 2002; Eichler and Sankoff 2003) and gene conversion (e.g., Hayakawa et al. 2001). In particular, the young Alus might have caused many of the large segmental duplications in the human genome (Bailey et al. 2003). Furthermore, the LINE and Alu elements have the potential to delete a portion of the site where they integrate (Gilbert et al. 2002; Symer et al. 2002; Callinan et al. 2005); two deletions of exons in the human elastin gene are likely to be Alu mediated (Szabo et al. 1999). Likewise, an Alu-mediated gene conversion event between an AluSq and an AluY element led to disruption of the open reading frame of the CMP-*N*-acetylneuraminic acid gene (Figure 13.4; Hayakawa et al. 2001). However, most of the known 612 human-specific Alu-mediated deletions do not affect exons (Mikkelson et al. 2005).

Genes can also duplicate through retrotransposition using the L1 machinery, and the human genome contains numerous retrogenes (a retrotranscribed mRNA) of various ages (reviewed in Zhang and Gerstein 2004). The majority of these are pseudogenes that were either dead-on-arrival or have degenerated after retrotransposition. However, some of the pseudogenes have remained functional or have even acquired new functions, as will be discussed in the next part of this chapter.

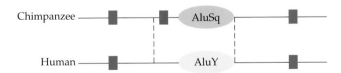

Figure 13.4 A schematic drawing of the CMP-Neu5Ac hydroxylase. Boxes represent exons and the ellipses show the AluSq and the AluY elements, which probably initiated a gene conversion event that also led to the deletion of a 92 bp exon (red) in humans (Hayakawa et al. 2001). This deletion rendered the human gene nonfunctional.

Evolution through Gene Duplication

Gene duplication is thought to be pivotal in the evolution of new gene functions. Indeed, both duplicated copies of a gene remain functional in an unexpectedly large number of cases (reviewed in Prince and Pickett 2002; Zhang 2003). Three mechanisms are invoked to explain this observation. First, duplications might immediately be beneficial because of the higher dosage of the gene. This scenario would lead to the maintenance of highly similar gene copies. This might explain the case for rRNA repeats. Second, duplicates might be kept due to subfunctionalization—a process in which the gene copy takes on one of the functions previously carried out by the ancestral gene copy (see Lynch and Conery 2000; Figure 13.5). Third, one of the gene copies may evolve a new function (neofunctionalization; see Figure 13.5).

As reviewed by Li and colleagues (2005), conservation of function is unlikely on theoretical grounds and there seem to be more reported instances of neofunctionalization than subfunctionalization. However, the latter two models are hard to distinguish. Both predict accelerated protein evolution following duplication, as is frequently observed (e.g., Kondrashov et al. 2002), and both can account for asymmetry in the rate of protein evolution between duplicates. Several studies have used expression divergence between duplicates as a measure of functional divergence between copies to determine the relative role of sub- and neofunctionalization in human evolution (Huminiecki and Wolfe 2004; He and Zhang 2005). Based on such data, He and Zhang (2005) argued for a composite model, in which duplicates are first maintained because of subfunctionalization and then eventually develop a new function.

One of the best-described instances of the development of a new function through gene duplication is trichromatic vision in old world monkeys (the group humans belong to). Old world monkeys have three opsin genes coding for proteins that detect different ranges of wavelength (Nathans et al. 1986). The green and the red genes originate from a head-to-tail duplica-

Figure 13.5 In subfunctionalization, a gene is duplicated and each new copy takes on a specialized function (symbolized by the different colors) from the original gene, possibly involving different motifs, exons, or regulatory elements of the gene. In neofunctionalization, the gene is duplicated and one copy retains the old function while the other copy evolves to have a new function.

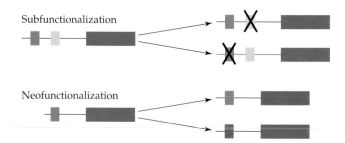

tion event after the split of old world and new world monkeys (Nei et al. 1997). Another interesting example is the retrogene *GLUD2*. This gene occurred on the lineage leading to humans and great apes and shows clear signs of adaptive evolution (Burki and Kaessmann, 2004). Two out of the six observed amino acid changes make the protein more suitable for the biochemical environment of the brain, which is the only tissue where this gene is found to be expressed.

There are other examples of functional retrogenes (Marques et al. 2005), many of which originated from a burst in retrotransposition on the primate lineage about 40–50 mya (Zhang et al. 2004). Interestingly, functional retrogenes are preferentially expressed in testes (Marques et al. 2005; Vinckenbosch et al. 2006). This bias is stronger in younger than older genes, and for genes originating on the X chromosome as compared to genes of autosomal origin. The excess of testes-specific genes originating from the X chromosome may be caused by selection favoring avoidance of X inactivation during male meiosis (Vinckenbosch et al. 2006).

Approximately five percent of the human genome resides within recent segmental duplications and an unexpectedly high fraction of these duplications contain genes or gene fragments (Bailey et al. 2002). Furthermore, the genes contained within duplications exhibit a functional bias, which would argue that the enrichment of genes in duplications is not due to the mechanisms that promote segmental duplications, but rather is due to selection. Among other protein classes, there is an enrichment in cytochrome P450 genes, which are involved in drug detoxification. Genes that have been copied through segmental duplications, rather than retrotranspositions, can carry some of the original regulatory elements and hence could lead to dosage problems. However, genes that have been duplicated in the human lineage rarely appear to be more highly expressed in humans than in chimpanzees. Only 56 percent of the human duplicates show a significant expression difference to the corresponding chimpanzee gene. Of these, 83 percent have—as would be expected—a higher combined expression (Cheng et al. 2005).

Analyses of Protein-Coding Regions

Evidence for selection in human protein-coding genes has been accumulated over the past few decades (Table 13.1). Famous examples include the human leukocyte antigen (HLA) locus, where selection seems to have increased diversity, possibly to increase the range of antigenic responses; and HBB (α-globin), where a mutant causing partial resistance to malaria in the heterozygous state may be favored by overdominant selection in areas with high prevalence of malaria (e.g., Webster et al. 2002). With the availability of genomic data, the range of candidate loci possibly affected by selection has been increased dramatically. The primary tool for identifying positive selection is a comparison of the rates of substitution. In coding regions,

Table 13.1 Examples of suggested cases of recent positive selection in the human population

Gene	Gene function	Selection hypothesis	Reference[a]
ADH	Alcohol dehydrogenase	Selection for more efficient breakdown of alchohol	Osier et al. 2002
ASPM	Mitotic spindle function	Selection for increased brain size[b]	Mekel-Bobrov et al. 2005
CAPN10	Cysteine protease	Selection for more efficient use of supplied nutrients	Fullerton et al. 2002
CASP12P1	Apoptosis-related cysteine protease	Less prone to sepsis	Xue et al. 2006
CD40LG	Stimulates B cells in the immune response	Malaria resistance	Sabeti et al. 2002
CDK5RAP2	Regulator of a cell cycle kinase	Selection for increased brain size[b]	Voight et al. 2006
CENPJ	Involved in microtubule nucleation	Selection for increased brain size[b]	Voight et al. 2006
CYP3A5	Most common cytochrome P450 oxidase, metabolizes >50% of clinically used drugs	Adaptation to differential sodium availability	Thompson et al. 2004
DMD	Anchor protein in the membrane of muscle cells	—	Nachman and Crowell 2000a
DRD4	Neurotransmitter (dopamine) receptor	Selection for personality traits such as novelty seeking and perseverance	Ding et al. 2002
FGFR2	Fibroblast growth factor receptor	Increased male fertility	Goriely et al. 2005

in particular, the rate of substitutions that lead to an amino acid change (nonsynonymous substitutions) is compared to the rate of substitutions that do not change the amino acid sequence (synonymous substitutions). Using synonymous mutations as a neutral proxy to give the background level of substitution in the absence of selection, an increased ratio of nonsynonymous to synonymous substitution can be interpreted as evidence for positive selection at the amino acid level (Table 13.2). Given that nonsynonymous and

Table 13.2 The relationship between selection and d_N/d_S

	Selection
$d_N/d_S > 1$	Positive selection
$d_N/d_S = 1$	Neutrality, no selection
$d_N/d_S < 1$	Negative (purifying) selection

Table 13.1 *Continued*

Gene	Gene function	Selection hypothesis	Reference[a]
F IX	Blood-clotting factor	Pathogen resistance	Harris and Hey 2001
FOXP2	Transcription factor	Selection because of its involvement in the development of speech	Enard et al. 2002b
G6PD	NADPH-producing glycolysis enzyme	Malaria resistance	Tishkoff et al. 2001; Sabeti et al. 2002; Saunders et al. 2002
HBB	Part of the adult hemoglobin	Malaria resistance	Webster et al. 2002
HFE	Iron uptake	Increased survival of children because of lower risk of iron deficiency	Toomajian and Kreitman 2002
LCT	Digestion of milk sugar	Adults in dairy-farming populations can drink fresh milk	Bersaglieri et al. 2004
MAOA	Oxidative deamination of catecholamine neurotransmitters	Selection on behavioral traits	Gilad et al. 2002
MC1R	Regulates pigmentation and adrenocortical function	Adaptation of skin and hair pigmentation to different climatic zones	Harding et al. 2000; Makova et al. 2001
MCPH1	Cell cycle checkpoint	Selection for increased brain size[b]	Evans et al. 2005
TTL.6	Testis-specific apoptosis gene	Increased male fertility	Chen et al. 2006

[a]References are for the studies providing evidence for a selective sweep.

[b]Variants of these loci have been connected to microcephaly. Because all these are testis-expressed cell cycle proteins, sperm competition or meiotic drive could also have been the reason for selection.

synonymous sites are interspersed among each other, differences in the rate at which these two types of mutations occur can be attributed to selection and not just differences in the mutation rate. However, it should be noted that the hypermutable dinucleotide CpG tends to overlap more frequently with synonymous than nonsynonymous sites (Hellmann et al. 2003). Estimates of synonymous substitution rates may, therefore, be inflated, leading to more conservative tests of positive selection.

Comparative data

Clark and colleagues (2003) analyzed data from more than 7000 human, chimpanzee, and mouse protein-coding genes to find evidence for selection on the lineage leading to humans compared to the lineages leading to chimpanzees and mice. They compared the rates of nonsynonymous and synonymous substitution and found significant evidence for an accelerated rate of nonsynonymous substitution in humans in genes related to olfaction (especially olfactory receptors) and other forms of sensory perception, such as hearing; in signaling and transport molecules; in genes relating to developmental processes; and in several other categories of genes. A number of genes relating to neurogenesis and skeletal development also showed some evidence of positive selection. Some of the genes suggested to be under positive selection in this study had previously been identified as possible targets of selection (see section on human-specific evolution).

Nielsen and colleagues (2005a) compared humans and chimps to identify genes that may be under selection in either or both of the evolutionary lineages leading to humans and chimpanzees. Again, the primary inferential tool in this comparison is the rate ratio of nonsynonymous to synonymous mutations (d_N/d_S). As already mentioned, when $d_N/d_S > 1$, this is taken as evidence of positive selection. In general, the genes associated with positive selection tend to be related to functions in immune defense, olfaction, and other types of sensory perception, as in the analysis by Clark and colleagues (2003). In addition, positive selection was associated with genes involved in the production and function of germ cells, apoptosis and its inhibition, and in KRAB-box transcription factors. Interestingly, if the same test is done in a comparison of rat and mouse, the same functional categories as in the human–chimpanzee comparison often exhibit an unusually high or low d_N/d_S ratio (Mikkelsen et al. 2005).

The evidence for positive selection on immune defense–related processes is expected. It has long been hypothesized that the immune system is engaged in an evolutionary arms race with various pathogens. The immune defense system is constantly adapting to an ever-changing pathogenic environment, leading to a constant turnover of alleles and a high rate of substitution. The factors causing positive selection on the other systems are more controversial. Genes involved in germ cell development and function may be under positive selection due to sexual selection (Darwin 1871; Swanson et al. 2001; Swanson et al. 2003), sperm competition (reviewed in Wigby and Chapman 2004), or mutations causing segregation distortion (reviewed in Hurst et al. 1996). The gene showing the highest d_N/d_S in comparisons of humans and chimpanzees is *PRTM1* (Wyckoff et al. 2000; Nielsen et al. 2005a), which encodes a protein that substitutes for histones in sperm. Wyckoff and colleagues (2000) suggested that sperm competition drives the positive selection acting on this gene.

It also appears that genes on the X chromosome are experiencing more positive selection than genes on autosomes, presumably because selection acts more efficiently on recessive advantageous mutations with hemizygous expression (Torgerson and Singh 2003; Nielsen et al. 2005a; Torgerson and Singh 2006).

Reassuringly, many of the categories of genes identified as targets of positive selection by Clark and colleagues (2003) and Nielsen and coworkers (2005a) were verified after the sequencing of the chimpanzee genome (Mikkelsen et al. 2005). However, Mikkelsen and colleagues (2005) did not

	Polymorphisms within species	Divergence between species
Nonsynonymous mutations	A	B
Synonymous mutations	C	D

Figure 13.6 Tests based on McDonald–Kreitman tables are among the most well-established tests of neutrality. If the divergence-to-polymorphisms ratio differs significantly between nonsynonymous and synonymous mutations, this is interpreted as evidence for selection.

Neutrality: $\dfrac{A}{B} = \dfrac{C}{D}$

employ any statistical tests to detect positive selection and did not attempt to distinguish between a relaxation of constraints and positive selection. Nonetheless, the categories of genes showing increased rates of amino acid substitution between human and chimpanzee compared to mouse and rat were very similar to the categories found to show evidence for lineage-specific positive selection in chimpanzees or humans (Clark et al. 2003). Similarly, the categories showing the highest overall rates of substitution were similar to the categories with evidence of positive selection identified by Nielsen and colleagues (2005a), despite differences in the methods of analysis.

Statistical tests for selection based on the d_N/d_S ratio are quite conservative when applied to one or a few species, because many amino acid sites of functional proteins tend to be highly constrained. However, power can be gained by comparing polymorphism data (within species) with comparative data (between species). The McDonald–Kreitman test (McDonald and Kreitman 1991) compares the d_N/d_S ratio in comparative and polymorphism data (Figure 13.6). Under a strictly neutral model (i.e., a model with no selection except for that against strongly deleterious mutations), these two ratios should be identical. If there are more nonsynonymous substitutions between species than expected, given the polymorphism data, this provides evidence for directional positive selection under most demographic models. Likewise, a relative deficiency of nonsynonymous substitutions provides evidence for negative selection. In contrast to analyses of d_N/d_S ratios applied solely to comparative data, the McDonald–Kreitman inference framework, therefore, also allows inference of negative selection on segregating mutations.

In the first large-scale survey of variation in human protein-coding genes, Bustamante and colleagues (2005) used this type of method. The categories of genes they found to be affected by positive selection were similar to the categories identified in the previous studies based on comparative data. One of the categories showing the most evidence of positive selection was KRAB-box transcription factors, which are involved in numerous processes, including developmental regulation. Confirming results by Gilad and coworkers

(2003), Bustamante and colleagues (2005) also found that olfactory receptors show strong evidence of positive selection.

However, most human genes showed a deficiency of nonsynonymous substitutions, indicating that many slightly deleterious mutations are segregating within populations. The functional categories of genes with the most evidence of negative selection include genes involved in cell structure and motility, vesicle transport, intracellular protein traffic, and ectoderm development. Negative selection on segregating mutations may be of particular medical interest because negatively selected segregating mutations are likely to be disease associated. In fact, Bustamante and colleagues (2005) showed that there was a clear correlation between disease status and the strength of selection affecting genes.

One of the advantages of the McDonald–Kreitman test is that it will only reject a neutral model in the presence of selection, independent of demographic assumptions (McDonald and Kreitman 1991; Akashi 1999; Nielsen 2001; Williamson et al. 2004). Most other tests of neutrality based on population genetic data are highly sensitive to assumptions regarding human demography. However, there is still some controversy over how robustly the direction and magnitude of selection are estimated using the McDonald–Kreitman inference framework.

Detecting selective sweeps from SNP data

Studies based on comparative data can detect selection occurring in a particular evolutionary lineage, for example, the lineage leading from the ancestor of humans and chimpanzees to modern humans. However, they cannot determine more precisely when selection took place. In contrast, population genetic data can be used to detect selection in humans within the last 100,000–500,000 years, to identify recent episodes of selection. Furthermore, using population genetic data it may be possible to detect selection acting on just a single mutation, while comparative methods require selection to act on multiple mutations to have any appreciable power. This is because an advantageous mutation increasing in frequency in the population will affect the variation at linked sites. A mutation linked in the coupling phase to an advantageous mutation (i.e., on the same background as the advantageous mutation) will increase in frequency with the advantageous mutation, while mutations linked in the repulsion phase (i.e., on a different background from the selected mutation) will decrease in frequency—a phenomenon known as the *hitchhiking* effect (Maynard Smith and Haigh 1974; Kaplan et al. 1989). The effect diminishes with genomic distance, or more precisely with the probability of recombination. At the time of fixation, there will be very little variability left in the region around the advantageous mutation (Figure 13.7). Subsequently, most of the observed mutations will be new and thus at low frequency (Braverman et al. 1995). A selective sweep will, therefore, leave a pattern of reduced variability and an

Before sweep   After sweep

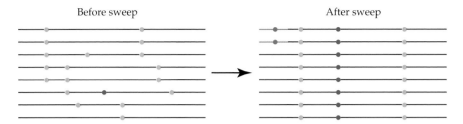

Figure 13.7 DNA sequences before and after a selective sweep. The new advantageous mutation (blue) arises on a haplotype. As this haplotype increases in frequency in the population, linked mutations (orange) are either being driven to fixation with the selected mutation or lost from the population. However, variability in positions linked to the advantageous mutation may be preserved due to recombination (green).

excess of low-frequency mutations. However, it will also increase LD and affect haplotype structure. There are statistical methods aimed at detecting selective sweeps based on all of these effects (reviewed in Nielsen 2005).

A number of projects are aimed at obtaining large-scale human population genetic data using direct sequencing, such as the ENCODE project (ENCODE project consortium 2004), and there are several large-scale SNP typing projects, in particular the HapMap project (The International HapMap Consortium 2005). These data provide a unique opportunity for identifying the regions of the human genome that have been targeted by selection. Mostly, this involves the search for traces of selective sweeps, such as a reduced variability, as well as unexpected patterns of allele frequencies and haplotype structure. Unfortunately, most SNP data have not been obtained using direct sequencing. Consequently, the data suffers from an ascertainment bias that enriches the data with high-frequency polymorphisms and also affects levels of LD, haplotype structure, and levels of population differentiation (Nielsen and Signorovitch 2003; Nielsen, 2004). These biases render standard population genetic methods inapplicable, although the problem can be rectified by properly accounting for the ascertainment bias when full information regarding the ascertainment protocol is available (e.g., Nielsen and Signorovitch 2003; Nielsen et al. 2004). Unfortunately, for much of the data, including data in dbSNP (The International SNP MAP Working Group 2001) and data from the HapMap project (The International HapMap Consortium 2005), it has so far been impossible to fully reconstruct the complicated ascertainment schemes employed (The International HapMap Consortium 2005; Clark et al. 2005). Nonetheless, subsets of the data have been obtained using ascertainment schemes that allow an unbiased population genetic analysis of the data (Hinds et al. 2005).

The first genomic survey of polymorphism data in humans analyzed 26,530 SNPs (Akey et al. 2002). These researchers identified regions of the genome with extreme levels of population differentiation (high F_{ST} values) among the major human ethnic groups; recent selection may in some cases increase the level of population differentiation. Akey and colleagues (2002) found a relatively high level of population subdivision on the X chromosome and identified a number of genes with elevated levels of population differentiation, including several disease-associated genes, such as the *CFTR* gene. They also found elevated levels of population differentiation in immune- and defense-related genes. Weir and colleagues (2005) made a similar survey based on a much larger data set and found significant heterogeneity among genomic regions.

Mikkelsen and coworkers (2005) compared the divergence among humans, estimated from genome-wide, low-coverage, shotgun sequencing of DNA from multiple individuals, to the divergence between humans and chimpanzees. Reduced diversity within humans as compared to the human–chimp divergence may indicate that the region in question has been subject to a selective sweep. Using this approach, Mikkelsen and coworkers (2005) identified six large genomic regions (> 1 Mb) that may have been targeted by a selective sweep.

Altschuler and colleagues (2005) used a test based on haplotype structure to scan the HapMap data for evidence of a recent selective sweep. They found a number of candidate genes, including the lactase gene on chromosome 2, which previously had been found to show signs of a selective sweep (Bersaglieri et al. 2004). Using a new test that accounts for ascertainment biases applied to the HapMap data set, Nielsen et al. (2005b) found the strongest evidence for a selective sweep on chromosome 2, again, in the lactase locus.

Voight et al. (2006) used the HapMap data to screen the human genome for ongoing selective sweeps, using another statistic based on haplotype structure. The regions that were identified in this study ranged in size from 100–3000 kb and often contained multiple genes, but some contained no genes. In this study, one of the strongest signals again came from the lactase gene locus. Other signals of selection could be recovered for the inversion on chromosome 17 (Stefansson et al. 2005) and some olfactory gene clusters (Gilad et al. 2003). However, other well-documented cases of selection such as selection on G6PD (Tishkoff et al. 2001; Sabeti et al. 2002; Saunders et al. 2002) could not be detected.

A caveat in all these studies is that demographic forces may also affect the results and potentially cause spurious evidence for selection (Nielsen 2001; Przeworski 2002; Nielsen 2005). In addition, results of studies based on publicly available SNP data sets that do not explicitly account for ascertainment biases caused by the SNP selection process may be difficult to interpret.

Human-Specific Genetic Adaptations

Naturally, there is a lot of interest in identifying the genes underlying the important adaptations unique to humans. Recently, good candidates for such genes have emerged, for example, a number of genes that may be related to brain size (reviewed in Woods et al. 2005), of which *ASPM* and *MCPH1* are the most well-known examples. Mutations in these genes cause microcephaly (reduced brain size) and there is an increasing amount of population genetic evidence suggesting that these genes have been under selection in the recent history of mankind (Evans et al. 2005; Mekel-Bobrov et al. 2005). Recently, *CENPJ* has also been shown to control brain size together with another gene, *CDK5RAP2*. Both genes show clear signs of positive selection in a recent genome-wide screen for ongoing selective sweeps in the human genome (Voight et al. 2006). Further, *CENPJ* was shown to have an elevated d_N/d_S ratio (albeit not significantly so), making this another good candidate gene for brain size differences between humans and chimpanzees (Nielsen et al. 2005a). *ASPM* and *MCPH1* do not show evidence for $d_N/d_S > 1$ on the human lineage and some controversy has recently emerged regarding the degree to which these genes truly have been targeted by positive selection (Currat et al. 2006).

The *FOXP2* gene has been associated with an unusual speech disorder (Lai et al. 2001). Mutations in this gene cause problems with speech articulation in the affected individuals (Developmental Verbal Dyspraxia). *FOXP2* shows a clear acceleration of amino acid substitution ($d_N/d_S > 1$) on the human lineage (Enard et al. 2002b; Zhang et al. 2002a). Two amino acid substitutions occurred on the human lineage after the split from the common ancestor with chimpanzees, while only one further substitution distinguishes us from mice. Although a speedup of amino acid substitution in itself does not provide evidence for positive selection, additional analyses of population genetic data further support the case for selection acting on this gene, and suggest that the last fixation in the human lineage must have occurred within the last 200,000 years (Enard et al. 2002b). It has been hypothesized that selection on this gene has been associated with the evolution of speech. Fisher and Marcus (2006) provide a more detailed discussion of this and other genes that may explain cognitive and behavioral differences between humans and chimpanzees.

There are also a number of genes that may help to explain morphological differences between humans and our closest relatives; for example, the myosin heavy chain (*MHY16*) gene that has been inactivated by a frameshift mutation in the lineage leading to humans (Stedman et al. 2004). The loss of this protein is associated with marked size reductions in individual muscle fibers and entire masticatory muscles (Stedman et al. 2004). Another loss-of-function mutation, which is currently sweeping through the human population, is a nonsense mutation in the caspase 12 gene (Xue et al. 2006). The truncated variant apparently reduces the risk of sepsis (Saleh et al. 2004) and hence improves the chances of survival of bacterial infections. Several other candidate genes for explaining phenotypic differences between humans and apes have been discussed by other investigators (Clark et al. 2003; Bustamante et al. 2005; Nielsen et al. 2005a; Varki and Altheide 2005).

Although the list of candidate genes that may explain phenotypic differences between humans and apes is growing very fast, most evidence for causative effects is weak. The most convincing stories are based on evidence for Darwinian selection acting on the gene, combined with some knowledge of its functional role, usually inferred from the effects of disease-causing mutations. However, the disease-causing mutations are in general not reversals to an ancestral state. Out of 12,164 known human disease variants, only 15 appear to reflect the healthy, normal state in the chimpanzee (Mikkelsen et al. 2005). For example, the mutations in the gene *FOXP2* that lead to language deficiency probably cause a loss of function of the affected allele, while the functional effects of the two human-specific amino acid substitutions are unknown. In most known cases of Darwinian selection there is very little evidence that the human-specific mutations have a functional effect that can explain phenotypic differences between species. Most of the adaptive stories remain just-so stories because the functional effects of individual mutations on complex systems are hard to determine. This does not imply that the stories are wrong—but it is entirely possible that many or most of the phenotypic differences between humans and apes, such as the important cognitive differences, cannot be ascribed to specific genes, but are caused by multiple changes in complex networks of interacting genes affecting both expression and amino acid sequence.

Selection on Noncoding Sequences

So far we have mainly assumed that selection has acted on changes in the amino acid sequence of proteins. However, there is considerable evidence for selection acting on noncoding regions (e.g., Lewinsky et al. 2005; Siepel et al. 2005). King and Wilson (1975) were startled by the high similarity of proteins between humans and chimpanzees, and suggested that many functional differences might be mediated by changes in gene regulation. With the availability of genome sequences from many species, this notion seemed to have found support in the observation that organism complexity is not necessarily reflected in gene numbers. Many of the adaptive changes between humans and chimpanzees may be located in noncoding regions affecting gene regulation. Additionally, selection may affect non-protein-coding regions because of the presence of RNA genes, or because of structural requirements, for example, those relating to GC content. Finally, selection might be expected to act on nonfunctional regions to avoid a number of motifs that otherwise would serve as binding sites for proteins (Hahn et al. 2003).

The comparative approach is ideally suited to detect selection in noncoding regions, because selection should act to increase or decrease the rate of substitution among species. Several studies have made use of this property to screen the human genome for conserved noncoding sequences. Initially, conserved sequence elements were identified as regions with a relatively low divergence between humans and rodents (e.g., Dermitzakis et al. 2002;

Waterston et al. 2002; Bejerano et al. 2004; Cooper et al. 2004). Many of these elements were shown to be conserved also in more distantly related species (Dermitzakis et al. 2003). This and more recent work, showing that the allele frequency distribution in the conserved regions is skewed towards rare alleles, suggest that mutations in these regions are indeed deleterious (Keightley et al. 2005a; Drake et al. 2006). Hence, the reduced divergence is neither spurious nor due to variations in mutation rates.

With the availability of more genomes, the methodologies to identify conserved regions have become more sophisticated. Siepel and colleagues (2005) developed a phylogenetic hidden Markov model to identify conserved regions from multiple genome alignments (Figure 13.8). This method makes use of the information contained in a phylogenetic tree and eliminates the need for arbitrary cutoffs for sequence similarity and window sizes. The conclusion from this study is that 4.3 percent of our genome is conserved and 70 percent of the conserved regions fall outside of exons.

Although coding sequences represent a high fraction (58%) of the 5000 most conserved elements, they constitute only about half of the most conserved parts of the human genome. Curiously, some of the most conserved elements were located in gene deserts (regions of low gene density), while

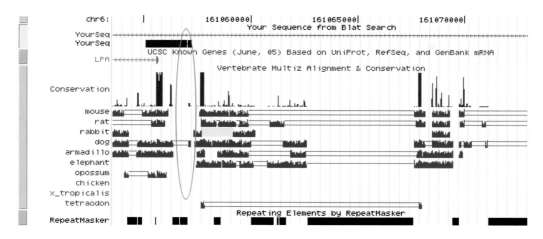

Figure 13.8 Screenshot from the UCSC genome browser (http://genome.ucsc.edu). This screenshot shows the upstream region of the human gene *LPA*. The conservation track shows the conservation scores as calculated from up to 17 vertebrate species (Siepel et al. 2005). Underneath the overall conservation score one can see the pairwise comparisons of the various species to the human genome. The red circle highlights a region with important regulatory elements in humans, which is completely deleted in rodents and partly so even in chimpanzees. (From Boffelli et al. 2003.)

the closest known genes to such deserts were frequently involved in transcription and development (Bejerano et al. 2004; Siepel et al. 2005; Woolfe et al. 2005). Minimal promoter constructs to test the conserved elements for enhancer activity showed that most of them indeed were active and many even mimic the expression pattern of the closest known gene when transfected into mice or fish (Nobrega et al. 2003; Woolfe et al. 2005).

Some highly conserved elements overlap with known microRNAs (miRNAs) and other structural RNAs, which is also reflected in an overall higher potential of these sequences to form RNA secondary structures (Siepel et al. 2005). Another set of the conserved noncoding regions overlaps with matrix-scaffold attachment regions (Glazko et al. 2003).

So far, most approaches to identify conserved noncoding regions have been based on models of nucleotide substitutions. Lunter and colleagues (2006) found yet another approach to identify conserved regions in the genome, reasoning that functional sequences should be relatively devoid of insertions or deletions. Comparing the human, mouse, and dog genomes, Lunter and colleagues (2006) found that about 2.6–3.3 percent of the human genome is conserved. Ultimately, it will be necessary to unite all information from sequence similarity, the underlying phylogeny, and indel frequencies to sift through the vast amounts of noncoding sequences in the human genome in search of conserved and putatively functional elements.

Regulatory sequences

Not all functional elements are conserved across mammals or vertebrates. Of human transcription factor binding sites (TFBSs), 32–40 percent do not appear to be functional in the mouse (Dermitzakis and Clark 2002). Such primate-specific, or even human-specific, functional sequences might be of primary interest to those studying human adaptation.

Boffelli and colleagues (2003) compared multiple primate species, and identified primate-specific exons and regulatory elements of the gene *Apo(a)*, which has regulatory functions in liver cells. Primate-specific evolution may also take the form of loss of regulatory elements. For example, some regulatory elements of the human gene *SIM2* are conserved between humans, mouse, and dog, but are deleted in chimpanzees and the rhesus macaque (Frazer et al. 2004).

More general models of TFBS evolution suggest that they also evolve rather quickly. These short sequences are likely to appear de novo within reasonable evolutionary times (Rockman and Wray 2002). Hence, the order of TFBSs within a promoter or enhancer does not necessarily need to be conserved, suggesting that methods are needed which incorporate models of TFBS evolution into the alignment procedure.

Eddy (2005) provides a theoretical framework for calculating the tree size needed to identify elements of varying size and conservation, assuming that constraints remain approximately constant. For example, if the rate of evolution of all bases of a regulatory element is 20 percent of the neutral rate, a total tree length of about 4 substitutions per site would be needed to identify a TFBS of 8 bp. Based on these calculations, Margulies and colleagues (2005) suggested sequencing 25 mammalian genomes—16 more than currently available.

The importance of a good scheme to identify putative TFBSs is illustrated by two recent studies that used different approaches to identify functional elements. The two studies arrived at contradictory conclusions regarding constraints on human *cis*-regulatory sequences, through comparisons with the chimpanzee genome (Bush and Lahn 2005; Keightley et al. 2005b). Keightley and colleagues (2005b) analyzed windows of 500 bp around putative transcription start sites (TSS) and came to the conclusion that human *cis*-regulatory sites evolve under little constraint as compared to rodents. Bush and Lahn (2005) analyzed only sequences flanking highly conserved 16-mers within 10 kb upstream of the TSS. The latter study concluded that there are significant amounts of negative selection acting on human regulatory regions. Clearly, more studies will be required to resolve this apparent contradiction.

Wong and Nielsen (2004) implemented a method for identifying negative or positive selection in noncoding sequences. Conceptually, the method is similar to other statistical phylogenetic footprinting methods, but differs in contrasting the rate of evolution in the noncoding regions to rates of synonymous evolution in coding regions. If synonymous mutations do in fact evolve approximately neutrally, this might be a promising approach for detecting selection in upstream regulatory sequences.

Despite all the hardships involved with the analysis of *cis*-regulatory elements, at least four cases of recent positive selection on *cis*-regulatory elements have been documented through a combination of phylogenetic and population genetic analysis, together with good hunches regarding the selection of the candidate genes (Rockman et al. 2003; Hahn et al. 2004b; Rockman et al. 2004; Rockman et al. 2005). For a more comprehensive review on the evolution of *cis*-regulatory sequences see Wray and colleagues (2003).

Expression Data

So far, most of the noncoding functional sequences that have been analyzed—enhancer elements, miRNAs, and matrix-scaffold adhesion sequences—act to regulate transcript levels. Recently, several studies have been aimed at detecting human-specific changes in gene expression (e.g., Enard et al. 2002a). Although these studies are still in their infancy, several interesting results have already emerged. Alongside protein-coding and regulatory sequence evolution, the level of selective constraints on gene expression is highly dependent on the tissue in which a gene is expressed (Duret and Mouchiroud 1999; Khaitovich et al. 2005). This parallel between protein sequence evolution and expression evolution suggests that both are

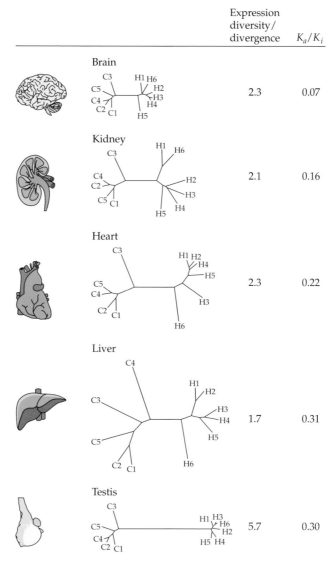

Figure 13.9 These trees illustrate the variation in gene expression among and between humans and chimpanzees in five different tissues. The distances of these neighbor-joining trees represent the mean of the squared difference of expression intensities of all detected genes (Khaitovich et al. 2005). The trees are drawn in proportion to each other. Differences in the diversity-to-divergence ratio can be taken as an indication of variations in the amount of positive selection. The K_a/K_i is, similar to d_N/d_S, a measure of selection on protein sequence evolution, but using intron divergence as neutral reference. Brain tissue evolves under the most constraint (it has the smallest tree and K_a/K_i), but expression diversity/divergence is not significantly different from kidney, heart, and liver. In contrast, the ratio of divergence to diversity in testis is 5.7, which is significantly different from the ratio in all other tissues.

shaped by similar forces (Castillo-Davis et al. 2004; Khaitovich et al. 2005). Adjustments in protein activity levels can be achieved either by changing the protein sequence to work more efficiently or by simply producing more protein. Whether it is the protein or the expression level that changes will most likely depend on which mutation occurs first.

An open question is the degree to which changes in expression evolve neutrally. Kimura (1983) argued that most observed amino acid differences are functionally neutral, using the apparent constancy of the molecular clock as an argument for his theory (although Gillespie [1984] pointed out that many selection models also predict a constant molecular clock). Khaitovich and colleagues (2004) assessed the neutrality of expression changes by testing for constancy in the rate of accumulation of expression differences in expression data from humans, chimpanzees, orangutans, and rhesus macaques (Figure 13.9). Similar to Kimura's arguments for amino acid data, Khaitovich and colleagues (2004) argued that most expression differences among species are selectively neutral because they were found to accumulate approximately linearly with time in their study. By analogy to the d_N/d_S test Khaitovich and colleagues (2004) also compared diversity and divergence. As a neutral category, akin to synonymous sites, they used pseudogene expression and found no significant difference in the divergence in expression levels of pseudogenes and functional genes. This was again interpreted as evidence against selection acting on the evolution of expression level. However, there should be a note of caution in the use of pseudogenes as a proxy for neutral evolution of gene expression. Expression of pseudogenes may occur because they are coupled to the regulatory elements of functional genes and are therefore under similar constraints. Nevertheless, there is some evidence for positive selection on expression evolution. Khaitovich and colleagues (2005) found that testis-specific genes evolve with a higher divergence-to-diversity ratio in expression levels than genes in other tissues, possibly reflecting more positive selection acting on testis-specific genes.

Gilad and colleagues (2006) analyzed expression levels among primates and found expression levels to be significantly conserved among species and not the subject of clocklike changes. These researchers used species-specific probes, while previous studies had relied on commercial human microarrays based on 25-mer probes. Although probes that were not identical between the human and the chimpanzee genome sequences were excluded from most of the analyses, a bias due to array design cannot be excluded. However, whether such a bias can explain the apparent discrepancy between the results of these studies needs to be determined.

Amount of Selection Affecting the Human Genome and the Genetic Load of Humans

Several recent studies have attempted to assess the total amount of selection affecting the human genome. Eyre-Walker and Keightley (1999) compared the number of nonsynonymous and synonymous mutations in protein-coding genes on the human lineage and found that at least 38 percent of all new nonsynonymous mutations are being eliminated by selection, resulting in 1.6 new deleterious mutations per diploid genome per generation in coding regions. It should be noted that this estimate might be slightly inflated, because it is based on an older estimate of the number of protein-coding genes in the human genome. Comparing data in McDonald–Kreitman tables, Fay and colleagues (2001) estimated that only 20 percent of all new nonsynonymous mutations are neutral and that roughly another 20 percent of them are slightly deleterious and often attain frequencies of 1–10 percent. Williamson and colleagues (2005) compared the allele frequency distributions in nonsynonymous sites to the allele frequency distribution in noncoding regions and found the average selection coefficient (scaled by the population size) acting on segregating nonsynonymous mutations in the human genome to be approximately –9. These results suggest that a large amount of selection is acting on the standing variation in the human population.

Overall, the evidence suggests that negative selection is very common in the human genome and that, at any point in time, there are many deleterious mutations segregating in human populations. This raises the important question of whether this much selection is even theoretically possible. Population genetic theory tells us that if the effects on fitness of different mutations are independent of each other (multiplicative), the proportion of a population dying due to selection acting on new deleterious mutations should equal approximately $1 - e^{U}$, where U is the rate of deleterious mutations per genome per generation. In addition, there should be a considerable number of deaths from other types of selection, including positively selected mutations going to fixation, mutations under balancing selection, and so forth. Under simplifying assumptions, estimates of the number of selective deaths required to maintain the number of segregating mutations found in humans far exceeds the population size (Williamson et al. 2005).

With this much selection, and a relatively low number of births per individual, how are we surviving as a species? The main reason is probably that the effects of mutations on fitness are not independent (e.g., Ewens 2004, pp. 78–86). It is entirely possible that most systems have so much inherent redundancy that most deleterious mutations will not have any immediate effect on fitness, except if they occur in conjunction with multiple other mutations affecting the same system. This may explain the observation that highly conserved elements can be deleted in the mouse without apparent effects on viability (Nobrega et al. 2004). Likewise, it is highly likely that positively selected mutations need to occur in combination to have a positive effect on fitness. If that is the case, there is no theoretical limit to how much selection can affect the human genome. The human species is not only alive and doing well—it is also continuously evolving.

Conclusion

The human genome has been formed through a complex interaction between mutation, genetic drift, and selection. Many of the main issues regarding the role of selection in human evolution are still controversial, but there is increasing evidence that selection has played an important role in shaping the human genome and is still acting. As we are learning more about functional genomics, and the mutational processes acting on the human genome, we are also developing a better understanding of the evolutionary forces affecting the human genome. Likewise, evolutionary inferences are becoming more and more important in the identification of functional sequences, such as RNA or protein-coding genes, or in the identification of upstream regulatory elements. With the availability of comparative genome sequencing, and large-scale SNP data sets, evolutionary genetics has found an important new role as one of the basic analytical tools in the genomic sciences. Evolution is no longer a topic reserved for dedicated evolutionary biologists, but is of core importance for any person interested in developing a full understanding of genomics.

References

Note: The numbers in brackets indicate the chapters in which the references are cited.

Adorján, P. and 25 others. 2002. Tumour class prediction and discovery by microarray-based DNA methylation analysis. *Nucleic Acids Res.* 30: e21. [5]

Agrawal, A. F. 2001. Sexual selection and the maintenance of sexual reproduction. *Nature* 411: 692–695. [1]

Akashi, H. 1994. Synonymous codon usage in *Drosophila melanogaster*: natural selection and translational accuracy. *Genetics* 136: 927–935. [2]

Akashi, H. 1999. Within- and between-species DNA sequence variation and the 'footprint' of natural selection. *Gene* 238: 39–51. [13]

Akashi, H. 2001. Gene expression and molecular evolution. *Curr. Opin. Genet. Dev.* 11: 660–666. [11]

Akashi, H. 2003. Translational selection and yeast proteome evolution. *Genetics* 164: 1291–1303. [2]

Akerley, B. J., E. J. Rubin, V. L. Novick, K. Amaya, N. Judson and J. J. Mekalanos. 2002. A genome-scale analysis for identification of genes required for growth or survival of *Haemophilus influenzae*. *Proc. Natl. Acad. Sci. USA.* 99: 966–971. [7]

Akey, J. M., G. Zhang, K. Zhang, L. Jin and M. D. Shriver. 2002. Interrogating a high-density SNP map for signatures of natural selection. *Genome Res.* 12: 1805–1814. [13]

Akey, J. M., M. A. Eberle, M. J. Rieder, C. S. Carlson, M. D. Shriver, D. A. Nickerson and L. Kruglyak. 2004. Population history and natural selection shape patterns of genetic variation in 132 genes. *PLoS Biol.* 2: e286. [13]

Alba, M. M. and J. Castresana. 2005. Inverse relationship between evolutionary rate and age of mammalian genes. *Mol. Biol. Evol.* 22: 598–606. [7]

Albert, R. and A. L. Barabasi. 2002. Statistical mechanics of complex networks. *Rev. Modern Physics* 74: 47–97. [12]

Albert, R., H. Jeong and A. L. Barabasi. 2000. Error and attack tolerance of complex networks. *Nature* 406: 378–382. [7, 12]

Alberts, B., A. Johnson, J. Lewis, M. Raff, K. Roberts and P. Walter. 2002 *Molecular Biology of the Cell.* 4th ed., Garland Science: New York, NY. [5]

Alon, U., M. G. Surette, N. Barkai and S. Leibler. 1999. Robustness in bacterial chemotaxis. *Nature* 397: 168–171. [7]

Alonso, J. M. and 38 others. 2003. Genome-wide insertional mutagenesis of *Arabidopsis thaliana*. *Science* 301: 653–657. [1, 7]

Alvarez, N., B. Benrey, M. Hossaert-McKey, A. Grill, D. McKey and N. Galtier, N. 2006. Phylogeographic support for horizontal gene transfer involving sympatric bruchid species. *Biol. Direct* 1: 21. [4]

Amsterdam, A., R. M. Nissen, Z. X. Sun, E. C. Swindell, S. Farrington and N. Hopkins. 2004. Identification of 315 genes essential for early zebrafish development. *Proc. Natl. Acad. Sci. USA* 101: 12792–12797. [7]

Anagnostopoulos, T., P. M. Green, G. Rowley, C. M. Lewis and F. Giannelli. 1999. DNA variation in a 5-Mb region of the X chromosome and estimates of sex-specific/type-specific mutation rates. *Am. J. Hum. Genet.* 64: 508–517. [11]

Ancel, L. W. and W. Fontana. 2000. Plasticity, evolvability, and modularity in RNA. *J. Exp. Zool.* 288: 242–283. [7]

Andersson, J. O. 2005. Lateral gene transfer in eukaryotes. *Cell. Mol. Life Sci.* 62: 1182–1197. [4]

Andersson, J. O. and A. J. Roger. 2002. Evolutionary analyses of the small subunit of glutamate synthase: Gene order conservation, gene fusions, and prokaryote-to-eukaryote lateral gene transfers. *Eukaryot. Cell* 1: 304–310. [4]

Andersson, J. O. and A. J. Roger. 2003. Evolution of glutamate dehydrogenase genes: evidence for lateral gene transfer within and between prokaryotes and eukaryotes. *BMC Evol. Biol.* 3: 14. [4]

Andersson, J. O., W. F. Doolittle and C. L. Nesbo. 2001. Genomics: Are there bugs in our genome? *Science* 292: 1848–1850. [4]

Andersson, J. O., A. M. Sjogren, L. A. Davis, T. M. Embley and A. J. Roger. 2003. Phylogenetic analyses of diplomonad genes reveal frequent lateral gene transfers affecting eukaryotes. *Curr. Biol.* 13: 94–104. [4]

Andersson, J. O., S. W. Sarchfield and A. J. Roger. 2005. Gene transfers from nanoarchaeota to an ancestor of diplomonads and parabasalids. *Mol. Biol. Evol.* 22: 85–90. [4]

Andersson, J. O., R. P. Hirt, P. G. Foster and A. J. Roger. 2006. Evolution of four gene families with patchy phylogenetic distributions: Influx of genes into protist genomes. *BMC Evol. Biol.* 6: 27. [4]

Andolfatto, P. 2005. Adaptive evolution of non-coding DNA in *Drosophila*. *Nature* 437: 1149–1152. [5, 11]

Annoura, T., T. Nara, T. Makiuchi, T. Hashimoto and T. Aoki. 2005. The origin of dihydroorotate dehydrogenase genes of kinetoplastids, with special reference to their biological significance and adaptation to anaerobic, parasitic conditions. *J. Mol. Evol.* 60: 113–127. [4]

Antonell, A., O. de Luis, X. Domingo-Roura and L. A Pérez Jurado. 2005. Evolutionary mechanisms shaping the genomic structure of the Williams-Beuren syndrome chromosomal region at human 7q11.23. *Genome Res.* 15: 1179–1188. [13]

Aphasizhev, R. 2007. RNA editing. *Mol. Biol.* 41: 227–239. [5]

Aravin, A. A. and 8 others. 2003. The small RNA profile during *Drosophila melanogaster* development. *Dev. Cell* 5: 337–350. [8]

Aravin, A. and 16 others. 2006. A novel class of small RNAs bind to MILI protein in mouse testes. *Nature* 442: 203–207. [8]

Aravin, A. A., R. Sachidanandam, A. Girard, K. Fejes-Toth and G. J. Hannon. 2007. Developmentally regulated piRNA clusters implicate MILI in transposon control. *Science* 316: 744–747. [8]

Aravind, L. and G. Subramanian. 1999. Origin of multicellular eukaryotes—insights from proteome comparisons. *Curr. Opin. Gen. Devel.* 9: 688–694. [6]

Arber, W. and S. Linn. 1969. DNA modification and restriction. *Annu. Rev. Biochem.* 38: 467–500. [8]

Archibald, J. M., M. B. Rogers, M. Toop, K. Ishida and P. J. Keeling, P. J. 2003. Lateral gene transfer and the evolution of plastid-targeted proteins in the secondary plastid-containing alga *Bigelowiella natans*. *Proc. Natl. Acad. Sci. USA* 100: 7678–7683. [4]

Arguello, J. R., Y. Chen, S. Yang, W. Wang and M. Long. 2006. The origin of young testes-expressed chimerical genes by illegitimate recombination in *Drosophila*. *PLoS Genet.* 2: e77. [3]

Arigo, J. T., K. L. Carroll, J. M. Ames and J. L. Corden. 2006. Regulation of yeast NRD1 expression by premature transcription termination. *Mol. Cell* 21: 641–651. [5]

Aris-Brosou, S. 2005. Determinants of adaptive evolution at the molecular level: the extended complexity hypothesis. *Mol. Biol. Evol.* 22: 200–209. [2]

Arita, M. 2004. The metabolic world of *Escherichia coli* is not small. *Proc. Natl. Acad. Sci. USA.* 101: 1543–1547. [7]

Artzy-Randrup, Y., S. J. Fleishman, N. Ben-Tal and L. Stone. 2004. Comment on "Network motifs: Simple building blocks of complex networks" and "Superfamilies of evolved and designed networks". *Science* 305: 1107. [12]

Atanassov, I., C. Delichère, D. A. Filatov, D. Charlesworth, I. Negrutiu and F. Monéger. 2001. A putative monofunctional fructose-2,6-bisphosphatase gene has functional copies located on the X and Y sex chromosomes in white campion (*Silene latifolia*). *Mol. Biol. Evol.* 18: 2162–2168. [10]

Athanasiadis, A., A. Rich and S. Maas. 2004. Widespread A-to-I RNA editing of Alu-containing mRNAs in the human transcriptome. *PLoS Biol* 2: e391. [8]

Aubin-Horth N., C. R. Landry, B. H. Letcher and H. A. Hofmann. 2005. Alternative life histories shape brain gene expression profiles in males of the same population. *Proc. Royal Soc. B Biol. Sci.* 272: 1655–1662. [5]

Aufsatz, W., M. F. Mette, J. van der Winden, A. J. Matzke and M. Matzke. 2002. RNA-directed DNA methylation in Arabidopsis. *Proc. Natl. Acad. Sci. USA* 99: 16499–16506. [8]

Baba, T. and 9 others. 2006. Construction of *Escherichia coli* K-12 in-frame, single-gene knockout mutants: the Keio collection. *Mol. Syst. Biol.* 2: 2006–2008. [7]

Babak, T., B. J. Blencowe and T. R. Hughes. 2007. Considerations in the identification of functional RNA structural elements in genomic alignments. *BMC Bioinformatics* 8: 33. [11]

Bachtrog, D. 2003. Accumulation of Spock and Worf, two novel non-LTR retrotransposons, on the neo-Y chromosome of *Drosophila miranda*. *Mol. Biol. Evol.* 20: 173–181. [10]

Bachtrog, D. 2004. Evidence that positive selection drives Y-chromosome degeneration in *Drosophila miranda*. *Nature Genet.* 36: 518–522. [10]

Bachtrog, D. 2005. Sex chromosome evolution: Molecular aspects of Y-chromosome degeneration in Drosophila. *Genome Res.* 15: 1393–1401. [10]

Bachtrog, D. 2006. Expression profile of a degenerating neo-Y chromosome in Drosophila. *Current Biology* 16: 1694–1699. [10]

Bailey, J. A. and 9 others. 2002. Recent segmental duplications in the human genome. *Science* 297: 1003–1007. [13]

Bailey, J. A., G. Liu and E. E. Eichler. 2003. An Alu transposition model for the origin and expansion of human segmental duplications. *Am. J. Hum. Genet.* 73: 823–834. [13]

Bányai, L., A. Váradi and L. Patthy. 1983. Common evolutionary origin of the fibrin-binding structures of fibronectin and tissue-type plasminogen activator. *FEBS Lett.* 163: 37–41. [6]

Bapteste, E., Y. Boucher, J. Leigh and W. F. Doolittle. 2004. Phylogenetic reconstruction and lateral gene transfer. *Trends Microbiol.* 12: 406–411. [4]

Bapteste, E., E. Susko, J. Leigh, D. MacLeod, R. L. Charlebois and W. F. Doolittle. 2005. Do orthologous gene phylogenies really support tree-thinking? *BMC Evol. Biol.* 5: 33. [4]

Barabasi, A. L. 2002. *Linked: The New Science of Networks.* Perseus Press, New York. [2]

Barabasi, A. L. and R. Albert. 1999. Emergence of scaling in random networks. *Science* 286: 509–512. [2]

Barabasi, A. L. and Z. N. Oltvai. 2004. Network biology: understanding the cell's functional organization. *Nat. Rev. Genet.* 5: 101–113. [2, 5, 6]

Barbon, A., I. Vallini, L. La Via, E. Marchina and S. Barlati. 2003. Glutamate receptor RNA editing: A molecular analysis of GluR2, GluR5 and GluR6 in human brain tissues and in NT2 cells following in vitro neural differentiation. *Brain Res. Mol. Brain Res.* 117: 168–178. [5]

Barkai, N. and S. Leibler. 1997. Robustness in simple biochemical networks. *Nature* 387: 913–917. [7]

Barrett, M. and 12 others. 2004. Comparative genomic hybridization using oligonucleotide microarrays and total genomic DNA. *Proc. Natl. Acad. Sci. USA* 101: 17765–17770. [3]

Barski, A. and 8 others. 2007. High-resolution profiling of histone methylations in the human genome. *Cell* 129: 823–837. [5]

Bartel, D. P. 2004. MicroRNAs: Genomics, biogenesis, mechanism, and function. *Cell* 116: 281–297. [5]

Bartolomé, C. and B. Charlesworth. 2006. Evolution of amino acid sequences and codon usage on the *Drosophila miranda* neo-sex chromosomes. *Genetics* 174: 2033–2044. [10]

Batada, N. N., L. D. Hurst and M. Tyers. 2006a. Evolutionary and physiological importance of hub proteins. *PloS Comp. Biol.* 2: 748–756. [7]

Batada, N. N., T. Reguly, A. Breitkreutz, L. Boucher, B. J. Breitkreutz, L. D. Hurst and M. Tyers. 2006b. Stratus not altocumulus: A new view of the yeast protein interaction network. *PLoS Biology* 4: 1720–1731. [7]

Bateson, W. and C. Pellew. 1915. On the genetics of 'rogues' among culinary peas (*Pisum sativum*). *J. Genet.* 5: 15–36. [5]

Bathe, S., M. Lebuhn, J. W. Ellwart, S. Wuertz and M. Hausner. 2004. High phylogenetic diversity of transconjugants carrying plasmid pJP4 in an activated sludge-derived microbial community. *FEMS Microbiol. Lett.* 235: 215–219. [4]

Batzer, M. A. and P. L. Deininger. 2002. Alu repeats and human genomic diversity. *Nat. Rev. Genet.* 3: 370–379. [13]

Bayes, M., L. F. Magano, N. Rivera, R. Flores and L. A. Pérez Jurado. 2003. Mutational mechanisms of Williams-Beuren syndrome deletions. *Am. J. Hum. Genet.* 73: 131–151. [13]

Bazykin, G. A. and A. S. Kondrashov. 2006. Rate of promoter class turn-over in yeast evolution. *BMC Evol. Biol.* 6: 14. [5]

Bazykin, G. A., F. A. Kondrashov, A. Y. Ogurtsov, S. Sunyaev and A. S. Kondrashov. 2004. Positive selection at sites of multiple amino acid replacements since rat–mouse divergence. *Nature* 429: 558–562. [1]

Bean, C. J., C. E. Schaner and W. G. Kelly. 2004. Meiotic pairing and imprinted X chromatin assembly in *Caenorhabditis elegans*. *Nat. Genet.* 36: 100–105. [9]

Begun, D. J. 1997. Origin and evolution of a new gene descended from alcohol dehydrogenase in Drosophila. *Genetics* 145: 375–382. [3]

Begun, D. J., H. A. Lindfors, A. D. Kern and C. D. Jones. 2007. Evidence for de novo evolution of testis-expressed genes in the *Drosophila yakuba/Drosophila erecta* clade. Genetics176:1131–1137

Beiko, R. G., T. J. Harlow and M. A. Ragan. 2005. Highways of gene sharing in prokaryotes. *Proc. Natl. Acad. Sci. USA* 102: 14332–14337. [4]

Bejerano, G., M. Pheasant, I. Makunin, S. Stephen, W. J. Kent, J. S. Mattick and D. Haussler. 2004. Ultraconserved elements in the human genome. *Science* 304: 1321–1325. [5, 13]

Beldade, P. and P. M. Brakefield. 2002. The genetics and evo-devo of butterfly wing patterns. *Nat. Rev. Genet.* 3: 442–452. [5]

Bennetzen, J. 2005. Transposable elements, gene creation and genome rearrangement in flowering plants. *Curr. Op. Gene. Devel.* 15: 621–627. [3]

Bennetzen, J. L. and B. D. Hall. 1982. Codon selection in yeast. *J. Biol. Chem.* 257: 3026–3031. [11]

Berg, J., M. Lassig and A. Wagner. 2004. Structure and evolution of protein interaction networks: A statistical model for link dynamics and gene duplications. *BMC Evol. Biol.* 4:51. [12]

Bergero, R., A. Forrest, E. Kamau and D. Charlesworth. 2007. Evolutionary strata on the X chromosomes of the dioecious plant Silene latifolia: evidence from new sex-linked genes. *Genetics* 175:1945–1954. [10]

Bergmann, S., J. Ihmels and N. Barkai. 2004. Similarities and differences in genome-wide expression data of six organisms. *PLoS Biol.* 2: e9. [5]

Bergström, T. F., R. Erlandsson, H. Engkvist, A. Josefsson, H. A. Erlich and U. Gyllensten. 1999. Phylogenetic history of hominoid DRB loci and alleles inferred from intron sequences. *Immunol. Rev.* 167: 351–365. [11]

Bergthorsson, U., K. L. Adams, B. Thomason and J. D. Palmer. 2003. Widespread horizontal transfer of mitochondrial genes in flowering plants. *Nature* 424: 197–201. [3, 4]

Bergthorsson, U., A. O. Richardson, G. J. Young, L. R. Goertzen and J. D. Palmer. 2004. Massive horizontal transfer of mitochondrial genes from diverse land plant donors to the basal angiosperm *Amborella*. *Proc. Natl. Acad. Sci. USA* 101: 17747–17752. [4]

Berlin, S. and H. Ellegren. 2004. Chicken W: A genetically uniform chromosome in a highly variable genome. *Proc. Natl. Acad. Sci. USA* 101: 15967–15969. [10]

Bernardi, G. 1986. Compositional constraints and genome evolution. *J. Mol. Evol.* 24: 1–11. [11]

Bernatchez, L. and C. Landry. 2003. MHC studies in nonmodel vertebrates: What have we learned about natural selection in 15 years? *J. Evol. Biol.* 16: 363–377. [5]

Bernstein, B. E. and 11 others. 2005. Genomic maps and comparative analysis of histone modifications in human and mouse. *Cell* 120: 169–181. [5]

Bernstein, E., A. A. Caudy, S. M. Hammond and G. J. Hannon. 2001. Role for a bidentate ribonuclease in the initiation step of RNA interference. *Nature* 409: 363–366. [8]

Berriman, M. and 101 others. 2005. The genome of the African trypanosome *Trypanosoma brucei*. *Science* 309: 416–422. [4]

Bersaglieri, T., P. C. Sabeti, N. Patterson, T. Vanderploeg, S. F. Schaffner, J. A. Drake, M. Rhodes, D. E. Reich and J. N. Hirschhorn. 2004. Genetic signatures of strong recent positive selection at the lactase gene. *Am. J. Hum. Genet.* 74: 1111–1120. [13]

Bertran, E., K. Thorton and M. Long. 2002. Retroposed new genes out of the X in Drosophila. *Genome Res.* 12: 1854–1859. [9]

Bestor, T. H. 2003. Cytosine methylation mediates sexual conflict. *Trends Genet.* 19: 185–190. [8]

Bestor, T. H. 2005. Transposons reanimated in mice. *Cell* 122: 322–325. [8]

Betran, E. and M. Long. 2003. *Dntf-2r*, a young *Drosophila* retroposed gene with specific male expression under positive Darwinian selection. *Genetics* 164: 977–988. [3]

Betran, E., K. Thornton and M. Long. 2002. Retroposed new genes out of the X in *Drosophila*. *Genome Res.* 12: 1854–1859. [3]

Betran, E, J. J. Emerson, H. Kaessmann and M. Long. 2004. Sex chromosomes and male functions: where do new genes go? *Cell Cycle* 3: 873–875. [3]

Betran, E., Y. Bai and M. Motiwale. 2006. Fast protein evolution and germ line expression of a *Drosophila* parental gene and its young retroposed paralog. *Mol. Biol. Evol.* 23: 2191–2202. [3]

Bhowmick, B., Y. Satta and N. Takahata. 2007. The origin and evolution of human ampliconic gene families and ampliconic structure. *Genome Res.* 17: 441–450. [10]

Billy, E., V. Brondani, H. Zhang, U. Muller and W. Filipowicz, W. 2001. Specific interference with gene expression induced by long, double-stranded RNA in mouse embryonal teratocarcinoma cell lines. *Proc. Natl. Acad. Sci. USA* 98: 14428–14433. [8]

Bird, A. 1999. DNA methylation de novo. *Science* 286: 2287–2288. [11]

Bird, A. P. 1980. DNA methylation and the frequency of CpG in animal DNA. *Nucleic Acids Res.* 8: 1499–1504. [11]

Bishop, K. N., R. K. Holmes, A. M. Sheehy, N. O. Davidson, S. J. Cho and M. H. Malim. 2004. Cytidine deamination of retroviral DNA by diverse APOBEC proteins. *Curr. Biol.* 14: 1392–1396. [8]

Blake, R. D. and P. W. Hinds. 1984. Analysis of the codon bias in E. coli sequences. *J. Biomol. Struct. Dyn.* 2: 593–606. [11]

Blank, L. M., L. Kuepfer and U. Sauer. 2005. Large-scale ^{13}C-flux analysis reveals mechanistic princi-

ples of metabolic network robustness to null mutations in yeast. *Genome Biol.* 6: R49. [7]

Bloom, J. D. and C. Adami. 2003. Apparent dependence of protein evolutionary rate on number of interactions is linked to biases in protein-protein interactions data sets. *BMC Evol. Biol.* 3: 21. [2]

Bloom, J. D. and C. Adami. 2004. Evolutionary rate depends on number of protein-protein interactions independently of gene expression level: Response. *BMC Evol. Biol.* 4: 14. [2]

Bode, J., S. Goetze, H. Heng, S. A. Krawetz and C. Benham. 2003. From DNA structure to gene expression: Mediators of nuclear compartmentalization and dynamics. *Chromosome Res.* 11: 435–445. [5]

Boffelli, D., J. McAuliffe, D. Ovcharenko, K. D. Lewis, I. Ovcharenko, L. Pachter and E. M. Rubin. 2003. Phylogenetic shadowing of primate sequences to find functional regions of the human genome. *Science* 299: 1391–1394. [5, 13]

Bolzer, A. and 10 others. 2005. Three-dimensional maps of all chromosomes in human male fibroblast nuclei and prometaphase rosettes. *PLoS Biol.* 3: e157. [5]

Bond, D. R., T. Mester, C. L. Nesbo, A. V. Izquierdo-Lopez, F. L. Collart and D. R. Lovley. 2005. Characterization of citrate synthase from *Geobacter sulfurreducens* and evidence for a family of citrate synthases similar to those of eukaryotes throughout the Geobacteraceae. *Appl. Environ. Microbiol.* 71: 3858–3865. [4]

Borneman, A. R. and 8 others. 2007. Divergence of transcription factor binding sites across related yeast species. *Science* 317: 815–819. [1, 5]

Boucher, Y. and W. F. Doolittle. 2000. The role of lateral gene transfer in the evolution of isoprenoid biosynthesis pathways. *Mol. Microbiol.* 37: 703–716. [4]

Bourc'his, D. and T. H. Bestor. 2004. Meiotic catastrophe and retrotransposon reactivation in male germ cells lacking Dnmt3L. *Nature* 431: 96–99. [8]

Bourguet, D. 1999. The evolution of dominance. *Heredity* 83: 1–4. [1, 7]

Bourque, G., P. A. Pevzner and G. Tesler. 2004. Reconstructing the genomic architecture of ancestral mammals: lessons from human, mouse, and rat genomes. *Genome Res.* 14: 507–516. [10]

Brakhage, A. A., Q. Al-Abdallah, A. Tuncher and P. Sprote. 2005. Evolution of beta-lactam biosynthesis genes and recruitment of trans-acting factors. *Phytochemistry* 66: 1200–1210. [4]

Braverman, J. M., R. R. Hudson, N. L. Kaplan, C. H. Langley and W. Stephan. 1995. The hitchhiking effect on the site frequency spectrum of DNA polymorphisms. *Genetics* 140: 783–796. [13]

Brem, R. B., G. Yvert, R. Clinton and L. Kruglyak. 2002. Genetic dissection of transcriptional regulation in budding yeast. *Science* 296: 752–755. [5]

Brem, R. B., J. D. Storey, J. Whittle and L. Kruglyak. 2005. Genetic interactions between polymorphisms that affect gene expression in yeast. *Nature* 436: 701–713. [5]

Brenner, S. E., C. Chothia and T. J. P. Hubbard. 1997. Population statistics of protein structures: lessons from structural classifications. *Curr. Opin. Struct. Biol.* 7: 369–376. [6]

Brenowitz, M., D. F. Senear, M. A. Shea and G. K. Ackers. 1986. Quantitative DNase footprint titration: A method for studying protein-DNA interactions. *Methods Enzymol.* 130: 132–181. [5]

Brink, R. A. 1959. Paramutation at the R locus in maize plants trisomic for chromosome 10. *Proc. Natl. Acad. Sci. USA* 45: 819–827. [5]

Brink, R. A. 1973. Paramutation. *Annu. Rev. Genet.* 7: 129–152. [5]

Brock, T. D. 1990. *The Emergence of Bacterial Genetics.* Cold Spring Harbor Laboratory Press, Cold Spring Harbor, New York. [4]

Broman, K. W., J. C. Murray, V. C. Sheffield, R. L. White and J. L. Weber. 1998. Comprehensive human genetic maps: individual and sex-specific variation in recombination. *Am. J. Hum. Genet.* 63: 861–869. [13]

Brookfield, J. F. Y. 1997. Genetic redundancy. *Adv. Genet.* 36:137–155. [7]

Brown, T. A. 2002. *Genomes.* New York, Wiley-Liss. [11]

Bruggeman, F. J. and H. V. Westerhoff. 2007. The nature of systems biology. *Trends Microbiol.* 15: 45–50. [2]

Brunetti, C. R., J. E. Selegue, A. Monteiro, V. French, P. M. Brakefield and S. B. Carroll. 2001. The generation and diversification of butterfly eyespot color patterns. *Curr Biol* 11: 1578–1585. [5]

Brüssow, H., C. Canchaya and W. D. Hardt. 2004. Phages and the evolution of bacterial pathogens: From genomic rearrangements to lysogenic conversion. *Microbiol. Mol. Biol. Rev.* 68: 560–602. [4]

Bull, J. J. 1983. *Evolution of Sex Determining Mechanisms.* Benjamin/Cummings, Menlo Park, CA. [10]

Burch, C. L. and L. Chao. 2004. Epistasis and its relationship to canalization in the RNA virus phi 6. *Genetics* 167: 559–567. [7]

Burki, F. and H. Kaessmann. 2004. Birth and adaptive evolution of a hominoid gene that supports high neurotransmitter flux. *Nat. Genet.* 36: 1061–1063. [13]

Burton, M., T. M. Rose, N. J. Faergeman and J. Knudsen. 2005. Evolution of the acyl-CoA binding protein (ACBP). *Biochem. J.* 392: 299–307. [4]

Bush, E. C. and B. T. Lahn. 2005. Selective constraint on noncoding regions of hominid genomes. *PLoS Comput. Biol.* 1: e73. [13]

Bustamante, C. D., R. Nielsen and D. L. Hartl. 2002. A maximum likelihood method for analyzing pseudogene evolution: implications for silent site evolution in humans and rodents. *Mol. Biol. Evol.* 19: 110–117. [11]

Bustamante, C. D. and 13 others. 2005. Natural selection on protein-coding genes in the human genome. *Nature* 437: 1153–1157. [1, 11, 13]

Callinan, P. A., J. Wang, S. W. Herke, R. K. Garber, P. Liang and M. A. Batzer. 2005. Alu retrotransposition-mediated deletion. *J. Mol. Biol.* 348: 791–800. [13]

Calteau, A., M. Guoy and G. Perrière. 2005. Horizontal transfer of two operons coding for hydrogenases between bacteria and archaea. *J. Mol. Evol.* 60: 557–565. [4]

Cambareri, E. B., B. C. Jensen, E. Schabtach and E. U. Selker. 1989. Repeat-induced G-C to A-T mutations in Neurospora. *Science* 244: 1571–1575. [8]

Cao, X. and S. E. Jacobsen. 2002. Role of the arabidopsis DRM methyltransferases in de novo DNA methylation and gene silencing. *Curr. Biol.* 12: 1138–1144. [8]

Carleton, K. L. and T. D. Kocher. 2001. Cone opsin genes of african cichlid fishes: Tuning spectral sensitivity by differential gene expression. *Mol. Biol. Evol.* 18: 1540–1550. [5]

Carlson, R. P. 2007. Metabolic systems cost-benefit analysis for interpreting network structure and regulation. *Bioinformatics* 23: 1258–1264. [7]

Carlton, J. M. and 64 others. 2007. Draft genome sequence of the sexually transmitted pathogen *Trichomonas vaginalis*. *Science* 315: 207–212. [4]

Carmell, M. A., A. Girard, H. J. van de Kant, D. Bourc'his, T. H. Bestor, D. G. de Rooij and G. Hannon. 2007. MIWI2 is essential for spermatogenesis and repression of transposons in the mouse male germline. *Dev. Cell.* 12: 503–514. [8]

Carmichael, A. N., A.-K. Fridolfsson, J. Halverson and H. Ellegren. 2000. Male-biased mutation rates revealed from Z- and W-chromosome-linked ATP synthase α subunit (ATP5A1) sequences in birds. *J. Mol. Evol.* 50: 443–447. [10]

Carninci, P. and 40 others. 2006. Genome-wide analysis of mammalian promoter architecture and evolution. *Nat. Genet.* 38: 626–635. [5]

Carpen, J. D., M. von Schantz, M. Smits, D. J. Skene and S. N. Archer. 2006. A silent polymorphism in the PER1 gene associates with extreme diurnal preference in humans. *J. Hum. Genet.* 51: 1122–1125. [11]

Carrel, L. and H. F. Willard. 2005. X-inactivation profile reveals extensive variability in X-linked gene expression in females. *Nature* 434: 400–404.

Carrel, L., A. Cottle, K. C. Goglin and H. F. Willard. 1999. A first-generation X-inactivation profile of the human X chromosome. *Proc. Natl. Acad. Sci. USA* 96: 14440–14444. [10]

Cartegni, L., S. L. Chew and A. R. Krainer. 2002. Listening to silence and understanding nonsense: exonic mutations that affect splicing. *Nat. Rev. Genet.* 3: 285–298. [11]

Carvalho, A. B. 2002. Origin and evolution of the Drosophila Y chromosome. *Curr. Opin. Genet. Dev.* 12: 664–668. [10]

Castillo-Davis, C. I. and D. L. Hartl. 2003. Conservation, relocation and duplication in genome evolution. *Trends Genet.* 19: 593–597. [7]

Castillo-Davis, C. I., D. L. Hartl and G. Achaz. 2004. *cis*-Regulatory and protein evolution in orthologous and duplicate genes. *Genome Res.* 14: 1530–1536. [5, 13]

Castresana, J. 2002. Estimation of genetic distances from human and mouse introns. *Genome Biol.* 3: R28. [11]

Castro, J. P. and C. M. Carareto. 2004. *Drosophila melanogaster* P transposable elements: Mechanisms of transposition and regulation. *Genetica* 121: 107–118. [8]

Catalanotto, C., M. Pallotta, P. ReFalo, M. S. Sachs, L. Vayssie, G. Macino and C. Cogoni. 2004. Redundancy of the two dicer genes in transgene-induced posttranscriptional gene silencing in *Neurospora crassa*. *Mol. Cell Biol.* 24: 2536–2545. [8]

Cavalier-Smith, T. 1985. *The Evolution of Genome Size.* John Wiley & Sons, New York. [1]

Cazalet, C. and 13 others. 2004. Evidence in the Legionella pneumophila genome for exploitation of host cell functions and high genome plasticity. *Nat. Genet.* 36: 1165–1173. [4]

Chamary, J. V. and L. D. Hurst 2005. Evidence for selection on synonymous mutations affecting stability of mRNA secondary structure in mammals. *Genome Biol.* 6: R75. [11]

Chamary, J. V., J. L. Parmley and L. D. Hurst. 2006. Hearing silence: non-neutral evolution at synonymous sites in mammals. *Nat. Rev. Genet.* 7: 98–108. [11]

Chandler, V. L. 2007. Paramutation: From maize to mice. *Cell* 128: 641–645. [5]

Chandler, V. L. and M. Stam. 2004. Chromatin conversations: Mechanisms and implications of paramutation. *Nat. Rev. Genet.* 5: 532–544. [5]

Chandler, V. L., W. B. Eggleston and J. E. Dorweiler. 2000. Paramutation in maize. *Plant Mol. Biol.* 43: 121–145. [5]

Charlebois, R. L. and W. F. Doolittle. 2004. Computing prokaryotic gene ubiquity: rescuing the core from extinction. *Genome Res.* 14: 2469–2477. [4]

Charlebois, R. L., R. G. Beiko and M. A. Ragan. 2003. Microbial phylogenomics: Branching out. *Nature* 421: 217. [4]

Charlesworth, B. 1978. A model for the evolution of Y chromosomes and dosage compensation. *Proc. Natl. Acad. Sci. USA* 75: 5618–5622. [10]

Charlesworth, B. 1979. Evidence against Fisher's theory of dominance. *Nature* 278: 848–849. [7]

Charlesworth, B. 1992. Evolutionary rates in partially self-fertilizing species. *Amer. Nat.* 140: 126–148. [10]

Charlesworth, B. 1996. The evolution of chromosomal sex determination and dosage compensation. *Current Biology* 6: 149–162. [10]

Charlesworth, B. and D. Charlesworth. 1978. A model for the evolution of dioecy and gynodioecy. *Amer. Nat.* 112: 975–997. [10]

Charlesworth, B. and D. Charlesworth. 1997. Rapid fixation of deleterious alleles can be caused by Muller's ratchet. *Genet. Res.* 70: 63–73. [10]

Charlesworth, B. and D. Charlesworth. 2000. The degeneration of Y chromosomes. *Phil. Trans. Roy. Soc. Lond. B* 355: 1563–1572. [10]

Charlesworth, D. and B. Charlesworth. 2005. Sex chromosomes: Evolution of the weird and wonderful. *Curr. Biol.* 15: R129–R131. [9]

Charlesworth, D., B. Charlesworth and G. Marais. 2005. Steps in the evolution of heteromorphic sex chromosomes. *Heredity* 95: 118–128. [9]

Cheah, M. T., A. Wachter, N. Sudarsan and R. R. Breaker. 2007. Control of alternative RNA splicing and gene expression by eukaryotic riboswitches. *Nature* 447: 497–500. [5]

Chen, F. C., E. J. Vallender, H. Wang, C. S. Tzeng and W.-H. Li. 2001. Genomic divergence between human and chimpanzee estimated from large-scale alignments of genomic sequences. *J. Hered.* 92: 481–489. [11]

Chen, I., P. J. Christie and D. Dubnau. 2005. The ins and outs of DNA transfer in bacteria. *Science* 310: 1456–1460. [4]

Chen, L., A. L. DeVries and C. H. Cheng. 1997a. Convergent evolution of antifreeze glycoproteins in Antarctic notothenioid fish and Arctic cod. *Proc. Natl. Acad. Sci. USA* 94: 3817–3822. [3, 6]

Chen, L., A. L. DeVries and C. H. Cheng. 1997b. Evolution of antifreeze glycoprotein gene from a trypsinogen gene in Antarctic notothenioid fish. *Proc. Natl. Acad. Sci. USA* 94: 3811–3816. [3, 12]

Chen, P. Y. and G. Meister. 2005. microRNA-guided posttranscriptional gene regulation. *Biol. Chem.* 386: 1205–1218. [8]

Chen, W. Q. J. and 8 others. 2005. Contribution of transcriptional regulation to natural variations in *Arabidopsis*. *Gen. Biol.* 6: R32. [5]

Chen, X., H. Shi, X. Liu and B. Su. 2006. The testis-specific apoptosis related gene *TTL.6* underwent adaptive evolution in the lineage leading to humans. *Gene* 370: 58–63. [13]

Chen, Y. and D. Xu. 2005. Understanding protein dispensability through machine-learning analysis of high-throughput data. *Bioinformatics* 21: 575–581. [7]

Cheng, Z. and 10 others. 2005. A genome-wide comparison of recent chimpanzee and human segmental duplications. *Nature* 437: 88–93. [13]

Cherry, L. M., S. M. Case and A. C. Wilson. 1978. Frog perspective on morphological differentiation between humans and chimpanzees. *Science* 200: 209–211. [5]

Cheung, V. G., L. K. Conlin, T. M. Weber, M. Arcaro, K. Y. Jen, M. Morley and R. S. Spielman. 2003. Natural variation in human gene expression assessed in lymphoblastoid cells. *Nat. Gen.* 33: 422–425. [5]

Chin, C. S., J. H. Chuang and H. Li. 2005. Genome-wide regulatory complexity in yeast promoters: Separation of functionally conserved and neutral sequence. *Genome Res.* 15: 205–213. [5]

Chooniedass-Kothari, S. and 9 others. 2004. The steroid receptor RNA activator is the first functional RNA encoding a protein. *FEBS Lett.* 566: 43–47. [11]

Ciccarelli, F. D., T. Doerks, C. von Mering, C. J. Creevey, B. Snel and P. Bork. 2006. Toward automatic reconstruction of a highly resolve tree of life. *Science.* 311: 1283–1287. [4]

Clark, A. G. 1994. Invasion and maintenance of a gene duplication. *Proc. Natl. Acad. Sci. USA* 91: 2950–2954. [7]

Clark, A. G. and 16 others. 2003. Positive selection in the human genome inferred from human-chimp-mouse orthologous gene alignments. *Cold Spring Harb. Symp. Quant. Biol.* 68: 471–477. [13]

Clark, A. G., M. J. Hubisz, C. D. Bustamante, S. H. Williamson and R. Nielsen. 2005. Ascertainment bias in studies of human genome-wide polymorphism. *Genome Res.* 15: 1496–1502. [13]

Clarke, G. D., R. G. Beiko, M. A. Ragan and R. L. Charlebois. 2002. Inferring genome trees by using a filter to eliminate phylogenetically discordant sequences. *J. Bacteriol.* 184: 2072–2080. [4]

Claverie, J. M. 2000. Gene number. What if there are only 30,000 human genes? *Science* 291: 1255–1257. [6]

Cohan, F. M. 2002. What are bacterial species? *Annu. Rev. Microbiol.* 56: 457–487. [4]

Colot, V. and J. L. Rossignol. 1999. Eukaryotic DNA methylation as an evolutionary device. *Bioessays* 21: 402–411. [11]

Comeron, J. M., M. Kreitman and M. Aguadé. 1999. Natural selection on synonymous sites is correlated with gene length and recombination in Drosophila. *Genetics* 151: 239–249. [11]

Conant, G. C. and A. Wagner. 2003. Convergent evolution in gene circuits. *Nat. Genet.* 34: 264–266. [12]

Conant, G. C. and A. Wagner. 2004. Duplicate genes and robustness to transient gene knock-downs in *Caenorhabditis elegans. Proc. Roy. Soc. Lond. Ser. B. Bio. Sci.* 271: 89–96. [7]

Conrad, D. F., T. D. Andrews, N. P. Carter, M. E. Hurles and J. K. Pritchard. 2006. A high-resolution survey of deletion polymorphism in the human genome. *Nat. Genet.* 38: 75–81. [13]

Cooper, G. M., M. Brudno, E. A. Stone, I. Dubchak, S. Batzoglou and A. Sidow. 2004. Characterization of evolutionary rates and constraints in three mammalian genomes. *Genome Res.* 14: 539–548. [13]

Copley, R. R., J. Schultz, C. P. Ponting and P. Bork. 1999. Protein families in multicellular organisms. *Curr. Opin. Struct. Biol.* 9: 408–415. [6]

Coppins, R. L., K. B. Hall and E. A. Groisman. 2007. The intricate world of riboswitches. *Curr. Opin. Microbiol.* 10: 176–181. [5]

Coulomb, S., M. Bauer, D. Bernard and M. C. Marsolier-Kergoat. 2005. Gene essentiality and the topology of protein interaction networks. *Proc. Biol. Sci.* 272: 1721–1725. [7]

Covert, M. W., E. M. Knight, J. L. Reed, M. J. Herrgard and B. O. Palsson. 2004. Integrating high-throughput and computational data elucidates bacterial networks. *Nature* 429: 92–96. [7]

Crawford, D. C., T. Bhangale, N. Li, G. Hellenthal, M. J. Rieder, D. A. Nickerson and M. Stephens. 2004. Evidence for substantial fine-scale variation in recombination rates across the human genome. *Nat. Genet.* 36: 700–706. [13]

Crayton, M. E., 3rd, B. C. Powell, T. J. Vision and M. C. Giddings. 2006. Tracking the evolution of alternatively spliced exons within the Dscam family. *BMC Evol. Biol.* 6: 16. [5]

Creevey, C. J. and 8 others. 2004. Does a tree-like phylogeny only exist at the tips in the prokaryotes? *Proc. Biol. Sci.* 271: 2551–2558. [4]

Crow, J. 1999. The odds of losing and genetic roulette. *Nature* 397: 293–294. [1]

Currat, M., L. Excoffier, W. Maddison, S. P. Otto, N. Ray, M. C. Whitlock and S. Yeaman. 2006. Comment on "Ongoing adaptive evolution of ASPM, a brain size determinant in *Homo sapiens*" and "Microcephalin, a gene regulating brain size, continues to evolve adaptively in humans". *Science* 313: 172. [13]

Dagan, T. and W. Martin. 2006. The tree of one percent. *Genome Biol.* 7: O118.

Dagan, T. and W. Martin. 2007. Ancestral genome sizes specify the minimum rate of lateral gene transfer during prokaryote evolution. *Proc. Natl. Acad. Sci. USA.* 104: 870–875. [4]

Dai, H., T. F. Yoshimatsu and M. Long. 2006. Retrogene movement within- and between-chromosomes in the evolution of *Drosophila* genomes. *Gene* 385: 96–102. [3]

Daimon, T., S. Katsuma, M. Iwanaga, W. Kang and T. Shimada. 2005. The BmChi-h gene, a bacterial-type chitinase gene of *Bombyx mori*, encodes a functional exochitinase that plays a role in the chitin degradation during the molting process. *Insect. Biochem. Mol. Biol.* 35: 1112–1123. [4]

Darwin, C. 1871. *The Descent of Man and Selection in Relation to Sex.* J. Murray, Publishers, London. [9, 13]

Davidson, A. R. and R. T. Sauer. 1994. Folded proteins occur frequently in libraries of random amino acid sequences. *Proc. Natl. Acad. Sci. USA* 91: 2146–2150. [6]

Davis, C. C. and K. J. Wurdack. 2004. Host-to-parasite gene transfer in flowering plants: Phylogenetic evidence from Malpighiales. *Science* 305: 676–678. [4]

Davis, C. C., W. R. Anderson and K. J. Wurdack. 2005. Gene transfer from a parasitic flowering plant to a fern. *Proc. Biol. Sci.* 272: 2237–2242. [4]

de Felipe, K. S, S. Pampou, O. S. Jovanovic, C. D. Pericone, S. F. Ye, S. Kalachikov and H. A. Shuman. 2005. Evidence for acquisition of *Legionella* type IV secretion substrates via interdomain horizontal gene transfer. *J. Bacteriol.* 187: 7716–7726. [4]

de Koning, A. P., F. S. Brinkman, S. J. Jones and P. J. Keeling. 2000. Lateral gene transfer and metabolic adaptation in the human parasite *Trichomonas vaginalis*. *Mol. Biol. Evol.* 17: 1769–1773. [4]

de Koning, D. J. and C. S. Haley. 2005. Genetical genomics in humans and model organisms. *Trends Genet.* 21: 377–381. [5]

De Preter, K. and 13 others. 2006. Human fetal neuroblast and neuroblastoma transcriptome analysis confirms neuroblast origin and highlights neuroblastoma candidate genes. *Genome Biol.* 7: R84. [2]

Decottignies, A., I. Sanchez-Perez and P. Nurse. 2003. *Schizosaccharomyces pombe* essential genes: A pilot study. *Genome Res.* 13: 399–406. [7]

Deeds, E. J., O. Ashenberg and E. I. Shakhnovich. 2006. A simple physical model for scaling in protein-protein interaction networks. *Proc. Natl. Acad. Sci. USA* 103: 311–316. [2]

Deininger, P. L. and M. A. Batzer. 1999. Alu repeats and human disease. *Mol. Genet. Metab.* 67: 183–193. [13]

Deitsch, K., C. Driskill and T. Wellems. 2001. Transformation of malaria parasites by the spontaneous uptake and expression of DNA from human erythrocytes. *Nucleic Acids Res.* 29: 850–853. [4]

Delaval, K. and R. Feil. 2004. Epigenetic regulation of mammalian genomic imprinting. *Curr. Opin. Genet. Dev.* 14: 188–195. [5]

Delsuc, F., H. Brinkmann and H. Philippe. 2005. Phylogenomics and the reconstruction of the tree of life. *Nature Reviews Genetics* 6: 361–375. [4]

Denver, D. R., K. Morris, M. Lynch and W. K. Thomas. 2004. High mutation rate and predominance of insertions in the *Caenorhabditis elegans* nuclear genome. *Nature* 430: 679–682. [11]

Denver, D. R., K. Morris, J. T. Streelman, S. K. Kim, M. Lynch and W. K. Thomas. 2005. The transcriptional consequences of mutation and natural selection in *Caenorhabditis elegans*. *Nat. Genet.* 37: 544–548. [5]

Deppenmeier, U. and 21 others. 2002. The genome *of Methanosarcina mazei*: Evidence for lateral gene transfer between bacteria and archaea. *J. Mol. Microbiol. Biotechnol.* 4: 453–461. [4]

Deragon, J. M., D. Sinnette and D. Labuda. 1990. Reverse transcriptase activity from human

embryonal carcinoma cells NTera2D1, *EMBO J.* 9: 3363–3368. [8]

Dermitzakis, E. T. and A. G. Clark. 2002. Evolution of transcription factor binding sites in Mammalian gene regulatory regions: conservation and turnover. *Mol. Biol. Evol.* 19: 1114–1121. [13]

Dermitzakis, E. T. and 10 others. 2002. Numerous potentially functional but non-genic conserved sequences on human chromosome 21. *Nature* 420: 578–582. [13]

Dermitzakis, E. T., A. Reymond, N. Scamuffa, C. Ucla, E. Kirkness, C. Rossier and S. E. Antonarakis. 2003. Evolutionary discrimination of mammalian conserved non-genic sequences (CNGs). *Science* 302: 1033–1035. [5, 13]

Dernburg, A. F., J. Zalevsky, M. P. Colaiacovo and A. M. Villeneuve. 2000. Transgene-mediated cosuppression in the *C. elegans* germ line. *Genes Dev.* 14: 1578–1583. [8]

Deutschbauer, A. M., D. F. Jaramillo, M. Proctor, J. Kumm, M. E. Hillenmeyer, R. W. Davis, C. Nislow and G. Giaever. 2005. Mechanisms of haploinsufficiency revealed by genome-wide profiling in yeast. *Genetics* 169: 1915–1925. [7]

Ding, Y. C. and 11 others. 2002. Evidence of positive selection acting at the human dopamine receptor D4 gene locus. *Proc. Natl. Acad. Sci. USA* 99: 309–314. [13]

Dion, M. F., S. J. Altschuler, L. F. Wu and O. J. Rando. 2005. Genomic characterization reveals a simple histone H4 acetylation code. *Proc. Natl. Acad. Sci. USA* 102: 5501–5506. [5]

Djikeng, A., H. Shi, C. Tschudi and E. Ullu. 2001. RNA interference in *Trypanosoma brucei*: cloning of small interfering RNAs provides evidence for retroposon-derived 24-26-nucleotide RNAs. *RNA* 7: 1522–1530. [8]

Dobzhansky, T. 1973. Nothing in biology makes sense except in the light of evolution. *The American Biology Teacher* 35: 125–129. [2]

Doench, J. G. and P. A. Sharp. 2004. Specificity of microRNA target selection in translational repression. *Genes Dev.* 18: 504–511. [8]

Doolittle, W. F. 1998. You are what you eat: A gene transfer ratchet could account for bacterial genes in eukaryotic nuclear genomes. *Trends Genet.* 14: 307–311. [4]

Doolittle, W. F. 1999. Phylogenetic classification and the universal tree. *Science* 284: 2124–2129. [4]

Doolittle, W. F. 2000. The nature of the universal ancestor and the evolution of the proteome. *Curr. Opin. Struct. Biol.* 10: 355–358. [4]

Doolittle, W. F. and E. Bapteste. 2007. Pattern pluralism and the Tree of Life hypothesis. *Proc. Natl. Acad. Sci. USA.* 104: 2043–2049. [4]

Doolittle, W. F. and R. T. Papke. 2006. Genomics and the bacterial species problem. *Genome Biology* 7: 116. [4]

Doolittle, W. F. and C. Sapienza. 1980. Selfish genes, the phenotype paradigm and genome evolution. *Nature* 284: 601–603. [8]

Drake, J. A. and 10 others. 2006. Conserved noncoding sequences are selectively constrained and not mutation cold spots. *Nat. Genet.* 38: 223–227. [13]

Drummond, D. A., J. D. Bloom, C. Adami, C. O. Wilke and F. H. Arnold. 2005. Why highly expressed proteins evolve slowly. *Proc. Natl. Acad. Sci. USA* 102: 14338–14343. [2]

Drummond, D. A., A. Raval and C. O. Wilke. 2006. A single determinant dominates the rate of yeast protein evolution. *Mol. Biol. Evol.* 23: 327–337. [2, 7]

Dudley, A. M., D. M. Janse, A. Tanay, R. Shamir and G. M. Church. 2005. A global view of pleiotropy and phenotypically derived gene function in yeast. *Mol. Syst. Biol.* 1: 1. [7]

Duncan, I. W. 2002. Transvection effects in Drosophila. *Annu. Rev. Genet.* 36: 521–56. [5]

Dunham, M. J., H. Badrane, T. Ferea, J. Adams, P. O. Brown, F. Rosenzweig and D. Botstein. 2002. Characteristic genome rearrangements in experimental evolution of *Saccharomyces cerevisiae*. *Proc. Natl. Acad. Sci. USA* 99: 16144–16149. [12]

Duret, L. and D. Mouchiroud. 1999. Expression pattern and, surprisingly, gene length shape codon usage in *Caenorhabditis*, *Drosophila*, and *Arabidopsis*. *Proc. Natl. Acad. Sci. USA* 96: 4482–4487. [13]

Duret, L. and D. Mouchiroud. 2000. Determinants of substitution rates in mammalian genes: Expression pattern affects selection intensity but not mutation rate. *Mol. Biol. Evol.* 17: 68–74. [7]

Duret, L., M. Semon, G. Piganeau, D. Mouchiroud and N. Galtier. 2002. Vanishing GC-rich isochores in mammalian genomes. *Genetics* 162: 1837–1847. [13]

Dutko, J. A., A. Schafer, A. E. Kenny, B. R. Cullen and M. J. Curcio. 2005. Inhibition of a yeast LTR retrotransposon by human APOBEC3 cytidine deaminases. *Curr. Biol.* 15: 661–666. [8]

Dykhuizen, D. E. and A. M. Dean. 1990. Enzyme-activity and fitness-evolution in solution. *Trends Ecol. Evol.* 5: 257–262. [7]

Dykhuizen, D. E. and L. Green. 1991. Recombination in *Escherichia coli* and the definition of biological species. *J. Bacteriol.* 173: 7257–7268. [4]

Dykhuizen, D. E., A. M. Dean and D. L. Hartl. 1987. Metabolic flux and fitness. *Genetics* 115: 25–31. [7]

Ebersberger, I., D. Metzler, C. Schwarz and S. Paabo. 2002. Genomewide comparison of DNA sequences between humans and chimpanzees. *Am. J. Hum. Genet.* 70: 1490–1497. [13]

Eddy, S. R. 2005. A model of the statistical power of comparative genome sequence analysis. *PLoS Biol.* 3: e10. [13]

Edwards, J. S. and B. O. Palsson. 1999. Systems properties of the *Haemophilus influenzae* Rd metabolic genotype. *J. Biol. Chem.* 274: 17410–17416. [7, 12]

Edwards, J. S. and B. O. Palsson. 2000a. The *Escherichia coli* MG1655 in silico metabolic genotype: Its definition, characteristics, and capabilities. *Proc. Natl. Acad. Sci. USA* 97: 5528–5533. [7, 12]

Edwards, J. S. and B. O. Palsson. 2000b. Metabolic flux balance analysis and the in silico analysis of *Escherichia coli* K-12 gene deletions. *BMC Bioinformatics* 1:1. [7]

Edwards, J. S. and B. O. Palsson. 2000c. Robustness analysis of the *Escherichia coli* metabolic network. *Biotechnol. Prog.* 16: 927–939. [12]

Edwards, J. S., R. U. Ibarra and B. O. Palsson. 2001. In silico predictions of *Escherichia coli* metabolic capabilities are consistent with experimental data. *Nat. Biotechnol.* 19:125–130. [7]

Edwards, R. A. and F. Rohwer. 2005. Viral metagenomics. *Nat. Rev. Microbiol.* 3: 504–510. [4]

Ehrlich, M. and R. Y. Wang. 1981. 5-Methylcytosine in eukaryotic DNA. *Science* 212: 1350–1357. [13]

Eichinger, L. and A. A. Noegel. 2005. Comparative genomics of *Dictyostelium discoideum* and *Entamoeba histolytica*. *Curr. Opin. Microbiol.* 8: 606–611. [4]

Eichinger, L. and 94 others. 2005. The genome of the social amoeba *Dictyostelium discoideum*. *Nature* 435: 43–57. [4]

Eichler, E. E. 2006. Widening the spectrum of human genetic variation. *Nat. Genet.* 38: 9–11. [13]

Eichler, E. E. and D. Sankoff. 2003. Structural dynamics of eukaryotic chromosome evolution. *Science* 301: 793–797. [13]

Eisenhauer, J. 2003. Regression through the origin. *Teaching Statistics* 25: 76–80. [11]

Eldar, A., B. Z. Shilo and N. Barkai. 2004. Elucidating mechanisms underlying robustness of morphogen gradients. *Curr. Opin. Genet. Dev.* 14: 435–439. [7]

Elena, S. F. and R. E. Lenski. 2001. Epistasis between new mutations and genetic background and a test of genetic canalization. *Evolution* 55: 1746–1752. [7]

Elena, S. F., P. Carrasco, J. A. Daros and R. Sanjuan. 2006. Mechanisms of genetic robustness in RNA viruses. *EMBO Rep.* 7: 168–173. [7]

Elena, S. F., C. O. Wilke, C. Ofria and R. E. Lenski. 2007. Effects of population size and mutation rate on the evolution of mutational robustness. *Evolution* 61: 666–674. [7]

Elhaik, E., N. Sabath and D. Graur. 2006. The "inverse relationship between evolutionary rate and age of mammalian genes" is an artifact of increased genetic distance with rate of evolution and time of divergence. *Mol. Biol. Evol.* 23: 1–3. [7]

Emerson, J. J., H. Kaessmann, E. Betran and M. Long. 2004. Extensive gene traffic on the mammalian X chromosome. *Science* 303: 537–540. [3]

Emmons, S. W. and L. Yesner. 1984. High-frequency excision of transposable element Tc 1 in the nematode *Caenorhabditis elegans* is limited to somatic cells. *Cell* 36: 599–605. [8]

Enard, W. and 12 others. 2002a. Intra- and interspecific variation in primate gene expression patterns. *Science* 296: 340–343. [2, 13]

Enard, W., M. Przeworski, S. E. Fisher, C. S. Lai, V. Wiebe, T. Kitano, A. P. Monaco and S. Paabo. 2002b. Molecular evolution of *FOXP2*, a gene involved in speech and language. *Nature* 418: 869–872. [13]

ENCODE Project Consortium. 2004. The ENCODE (ENCyclopedia Of DNA Elements) Project. *Science* 306: 636–640. [13]

ENCODE Project Consortium. 2007. Identification and analysis of functional elements in 1% of the human genome by the ENCODE pilot project. *Nature* 447: 799–815. [1]

Esser, C. and 15 others. 2004. A genome phylogeny for mitochondria among alpha-proteobacteria and a predominantly eubacterial ancestry of yeast nuclear genes. *Mol. Biol. Evol.* 21: 1643–1660. [4]

Evangelisti, A. M. and A. Wagner. 2004. Molecular evolution in the yeast transcriptional regulation network. *J. Exp. Zool. B Mol. Dev. Evol.* 302B: 392–411. [7]

Evans K, S. Ott, A. Hansen, G. Koentges and L. Wernisch. 2007. A comparative study of S/MAR prediction tools. *BMC Bioinformatics* 8: 71. [5]

Evans, P. D. and 8 others. 2005. Microcephalin, a gene regulating brain size, continues to evolve adaptively in humans. *Science* 309: 1717–1720. [13]

Ewens, W. J. 2004. *Mathematical Population Genetics*. Springer, New York. [13]

Eyre-Walker, A. C. 1991. An analysis of codon usage in mammals: selection or mutation bias? *J. Mol. Evol.* 33: 442–449. [11]

Eyre-Walker, A. and L. D. Hurst. 2001. The evolution of isochores. *Nat. Rev. Genet.* 2: 549–555. [13]

Eyre-Walker, A. and P. D. Keightley. 1999. High genomic deleterious mutation rates in hominids. *Nature* 397: 344–347. [1, 13]

Fan, C. and M. Long. 2007. A new retroposed gene in *Drosophila* heterochromatin detected by microarray-based comparative genomic hybridization. *J. Mol. Evol.* 64: 272–283. [3]

Fang, G., E. Rocha and A. Danchin. 2005. How essential are nonessential genes? *Mol. Biol .Evol.* 22: 2147–2156. [7]

Fares, M. A., M. X. Ruiz-Gonzalez, A. Moya, S. F. Elena and E. Barrio. 2002. Endosymbiotic bacteria: GroEL buffers against deleterious mutations. *Nature* 417: 398. [7]

Fast, N. M., J. S. Law, B. A. Williams and P. J. Keeling. 2003. Bacterial catalase in the microsporidian *Nosema locustae*: Implications for microsporidian metabolism and genome evolution. *Eukaryot. Cell* 2: 1069–1075. [4]

Fay, J. C. and C. I. Wu. 2000. Hitchhiking under positive Darwinian selection. *Genetics* 155: 1405–1413. [13]

Fay, J. C., G. J. Wyckoff and C. I. Wu. 2001. Positive and negative selection on the human genome. *Genetics* 158: 1227–1234. [13]

Fay, J. C., G. J. Wyckoff and C. I. Wu. 2002. Testing the neutral theory of molecular evolution with genomic data from Drosophila. *Nature* 415: 1024–1026. [1]

Fay, J. C., H. L. McCullough, P. D. Sniegowski and M. B. Eisen. 2004. Population genetic variation in gene expression is associated with phenotypic variation in *Saccharomyces cerevisiae*. *Gen. Biol.* 5: R26. [5]

Feil, R. 2006. Environmental and nutritional effects on the epigenetic regulation of genes. *Mutat. Res.* 600: 46–57. [5]

Feinberg, E. H. and C. P. Hunter. 2003. Transport of dsRNA into cells by the transmembrane protein SID-1. *Science* 301: 1545–1547. [8]

Fell, D. 1997. *Understanding the Control of Metabolism*. Portland Press, Miami, FL. [12]

Felsenfeld, G. and M. Groudine. 2003. Controlling the double helix. *Nature* 421: 448–453. [5]

Ferea, T. L., D. Botstein, P. O. Brown and R. F. Rosenzweig. 1999. Systematic changes in gene expression patterns following adaptive evolution in yeast. *Proc. Natl. Acad. Sci. USA* 96: 9721–9726. [5]

Ferrell, J. E. 2002. Self-perpetuating states in signal transduction: Positive feedback, double-negative feedback and bistability. *Curr. Opin. Cell Biol.* 14: 140–148. [12]

Ferris, P. J., E. V. Armbrust and U. Goodenough. 2002. Genetic structure of the mating-type locus of *Chlamydomonas reinhardtii*. *Genetics* 160: 181–200. [10]

Fickett, J. W. and A. G. Hatzigeorgiou. 1997. Eukaryotic promoter recognition. *Genome Res.* 7: 861–878. [5]

Field, J., B. Rosenthal and J. Samuelson. 2000. Early lateral transfer of genes encoding malic enzyme, acetyl-CoA synthetase and alcohol dehydrogenases from anaerobic prokaryotes to *Entamoeba histolytica*. *Mol. Microbiol.* 38: 446–455. [4]

Figge, R. M. and R. Cerff. 2001. GAPDH gene diversity in spirochetes: A paradigm for genetic promiscuity. *Mol. Biol. Evol.* 18: 2240–2249. [4]

Figge, R. M., M. Schubert, H. Brinkmann and R. Cerff. 1999. Glyceraldehyde-3-phosphate dehydrogenase gene diversity in eubacteria and eukaryotes: Evidence for intra- and inter-kingdom gene transfer. *Mol. Biol. Evol.* 16: 429–440. [4]

Filatov, D. A. 2005. Evolutionary history of *Silene latifolia* sex chromosomes revealed by genetic mapping of four genes. *Genetics* 170: 975–979. [10]

Filatov, D. A. and D. Charlesworth. 2002. Substitution rates in the X- and Y-linked genes of the plants

Silene latifolia and *S. dioica*. *Mol. Biol. Evol.* 19: 898–907. [10]

Filatov, D. A., F. Monéger, I. Negrutiu and D. Charlesworth. 2000. Evolution of a plant Y-chromosome: variability in a Y-linked gene of *Silene latifolia*. *Nature* 404: 388–390. [10]

Filatov, D. A., V. Laporte, C. Vitte and D. Charlesworth. 2001. DNA diversity in sex linked and autosomal genes of the plant species *Silene latifolia* and *S. dioica*. *Mol. Biol. Evol.* 18: 1442–1454. [10]

Filipski, J., J. P. Thiery and G. Bernardi. 1973. An analysis of the bovine genome by Cs_2SO_4-Ag density gradient centrifugation. *J. Mol. Biol.* 80: 177–197. [13]

Fire, A., S. Xu, J. S. Montgomery, S. A. Kostas, S. E. Driver and C. C. Mello. 1998. Potent and specific genetic interference by double-stranded RNA in *Caenorhabditis elegans*. *Nature* 391: 806–811. [5, 8]

Fischle, W., Y. Wang and C. D. Allis. 2003. Histone and chromatin cross-talk. *Curr. Opin. Cell. Biol.* 15: 172–183. [5]

Fisher, R. A. 1930. *The Genetical Theory of Natural Selection*. The Clarendon Press, Oxford University Press, Oxford. [1, 11]

Fisher, S. E. and G. F. Marcus. 2006. The eloquent ape: genes, brains and the evolution of language. *Nat. Rev. Genet.* 7: 9–20. [13]

FitzGerald, P. C., D. Sturgill, A. Shyakhtenko, B. Oliver and C. Vinson. 2006. Comparative genomics of Drosophila and human core promoters. *Genome Biol.* 7: R53. [5]

Fleishmann, R. D. and 39 others. 1995. Whole-genome random sequencing and assembly of *Haemophilus influenzae* Rd. *Science* 269: 496–512. [1]

Foat, B. C., S. S. Houshmandi, W. M. Olivas and H. J. Bussemaker. 2005. Profiling condition-specific, genome-wide regulation of mRNA stability in yeast. *Proc. Natl. Acad. Sci. USA* 102: 17675–17680. [5]

Formstecher, E. and 31 others. 2005. Protein interaction mapping: a Drosophila case study. *Genome Res.* 15: 376–384. [6]

Forne, T., J. Oswald, W. Dean, J. R. Saam, B. Bailleul, L. Dandolo, S. M. Tilghman, J. Walter and W. Reik. 1997. Loss of the maternal H19 gene induces changes in *Igf2* methylation in both cis and trans. *Proc. Natl. Acad. Sci. USA* 94: 10243–10248. [7]

Forrest, E. C., C. Cogoni and G. Macino. 2004. The RNA-dependent RNA polymerase, QDE-1, is a rate-limiting factor in post-transcriptional gene silencing in *Neurospora crassa*. *Nucleic Acids Res.* 32: 2123–2128. [8]

Forster, J., I. Famili, B. O. Palsson and J. Nielsen. 2003a. Large-scale evaluation of in silico gene deletions in *Saccharomyces cerevisiae*. *Omics* 7: 193–202. [7]

Forster, J., I. Famili, P. Fu, B. Palsson and J. Nielsen. 2003b. Genome-scale reconstruction of the *Saccharomyces cerevisiae* metabolic network. *Genome Res.* 13: 244–253. [12]

Forsyth, R. A. and 21 others. 2002. A genome-wide strategy for the identification of essential genes in *Staphylococcus aureus*. *Mol. Microbiol.* 43: 1387–1400. [7]

Fraser, C., W. P. Hanage and B. G. Spratt. 2007. Recombination and the nature of bacterial species. *Science* 315: 476–480. [4]

Fraser, H. B. 2005. Modularity and evolutionary constraint on proteins. *Nat. Genet.* 37: 351–352. [2]

Fraser, H. B., A. E. Hirsh, L. M. Steinmetz, C. Scharfe and M. W. Feldman. 2002. Evolutionary rate in the protein interaction network. *Science* 296: 750–752. [2]

Fraser, H. B., D. P. Wall and A. E. Hirsh. 2003. A simple dependence between protein evolution rate and the number of protein-protein interactions. *BMC Evol. Biol.* 3: 11. [2]

Fraser, J. A. and J. Heitman. 2004. Evolution of fungal sex chromosomes. *Mol. Microbiol.* 51: 299. [10]

Frazer, K. A., H. Tao, K. Osoegawa, P. J. de Jong, X. Chen, M. F. Doherty and D. R. Cox. 2004. Noncoding sequences conserved in a limited number of mammals in the *SIM2* interval are frequently functional. *Genome Res.* 14: 367–372. [13]

Freeman, M. 2000. Feedback control of intercellular signalling in development. *Nature* 408: 313–319. [12]

Freeze, H. H. 2006. Genetic defects in the human glycome. *Nat. Rev. Genet.* 7: 537–551. [2]

Freilich, S., T. Massingham, S. Bhattacharyya, H. Ponsting, P. A. Lyons, T. C. Freeman and J. M. Thornton. 2005. Relationship between the tissue-specificity of mouse gene expression and the evolutionary origin and function of the proteins. *Genome Biol.* 6: R56. [7]

Freitag, M., R. L. Williams, G. O. Kothe and E. U. Selker. 2002. A cytosine methyltransferase homologue is essential for repeat-induced point mutation in *Neurospora crassa*. *Proc. Natl. Acad. Sci. USA* 99: 8802–8807. [8]

Friedberg, E. C. and E. C. Friedberg. 2006. *DNA Repair and Mutagenesis*. ASM Press, Washington, D.C. [11]

Frisch, M., K. Frech, A. Klingenhoff, K. Cartharius, I. Liebich and T. Werner. 2002. In silico prediction of scaffold/matrix attachment regions in large genomic sequences. *Genome Res.* 12: 349–354. [5]

Frost, L. S., R. Leplae, A. O. Summers and A. Toussaint. 2005. Mobile genetic elements: The agents of open source evolution. *Nat. Rev. Microbiol.* 3: 722–732. [2, 4]

Fryxell, K. J. and W. J. Moon. 2005. CpG mutation rates in the human genome are highly dependent on local GC content. *Mol. Biol. Evol.* 22: 650–658. [11]

Fu, Y. X. and W.-H. Li. 1993. Statistical tests of neutrality of mutations. *Genetics* 133: 693–709. [13]

Fullerton, S. M. and 8 others. 2002. Geographic and haplotype structure of candidate type 2 diabetes susceptibility variants at the calpain-10 locus. *Am. J. Hum. Genet.* 70: 1096–1106. [13]

Futuyma, D. J. 1998. *Evolutionary Biology.* Sinauer Associates, Sunderland, MA. [12]

Galagan, J. E. and E. U. Selker. 2004. RIP: The evolutionary cost of genome defense. *Trends Genet.* 20: 417–423. [8]

Gallagher, L. A., E. Ramage, M. A. Jacobst, R. Kaul, M. Brittnacher and C. Manoil. 2007. A comprehensive transposon mutant library of *Francisella novicida*, a bioweapon surrogate. *Proc. Natl. Acad. Sci. USA* 104:1009–1014. [7]

Gandhi, T. K. and 20 others. 2006. Analysis of the human protein interactome and comparison with yeast, worm and fly interaction datasets. *Nat. Genet.* 38: 285–93. [6]

Garcia-Vallve, S., A. Romeu and J. Palau, J. 2000. Horizontal gene transfer of glycosyl hydrolases of the rumen fungi. *Mol. Biol. Evol.* 17: 352–361. [4]

Gavin, A. C. and 37 others. 2002. Functional organization of the yeast proteome by systematic analysis of protein complexes. *FASEB J.* 16: A523–A523. [12]

Ge, F., L.-S. Wang and J. Kim. 2005. The cobweb of life revealed by genome-scale estimates of horizontal gene transfer. *PLoS Biol.* 3: e316. [4]

Ge, H., A. J. Walhout and M. Vidal. 2003. Integrating 'omic' information: a bridge between genomics and systems biology. *Trends Genet.* 19: 551–560. [2]

Gerbod, D., V. P. Edgcomb, C. Noel, S. Vanacova, R. Wintjens, J. Tachezy, M. L. Sogin and E. Viscogliosi. 2001. Phylogenetic relationships of class II fumarase genes from trichomonad species. *Mol. Biol. Evol.* 18: 1574–1584. [4]

Gerdes, S. Y. and 21 others. 2003. Experimental determination and system level analysis of essential genes in *Escherichia coli* MG1655. *J. Bacteriol.* 185: 5673–5684. [7]

Gerrard, D. T. and D. A. Filatov. 2005. Positive and negative selection on Mammalian Y chromosomes. *Mol. Biol. Evol.* 22: 1423–1432. [10]

Gershenzon, N. I., E. N. Trifonov and I. P. Ioshikhes. 2006. The features of Drosophila core promoters revealed by statistical analysis. *BMC Genomics* 7: 161. [5]

Gerstein, M. 1997. A structural census of genomes: comparing bacterial, eukaryotic, and archaeal genomes in terms of protein structure. *J. Mol. Biol.* 274: 562–576. [6]

Giaever, G. and 71 others. 2002. Functional profiling of the *Saccharomyces cerevisiae* genome. *Nature* 418: 387–391. [7]

Gibson, G. 1996. Epistasis and pleiotropy as natural properties of transcriptional regulation. *Theor. Popul. Biol.* 49: 58–89. [5]

Gilad, Y., S. Rosenberg, M. Przeworski, D. Lancet and K. Skorecki. 2002. Evidence for positive selection and population structure at the human *MAO-A* gene. *Proc. Natl. Acad. Sci. USA* 99: 862–867. [13]

Gilad, Y., C. D. Bustamante, D. Lancet and S. Paabo. 2003. Natural selection on the olfactory receptor gene family in humans and chimpanzees. *Am. J. Hum. Genet.* 73: 489–501. [13]

Gilad, Y., S. A. Rifkin, P. Bertone, M. Gerstein and K. P. White. 2005. Multi-species microarrays reveal the effect of sequence divergence on gene expression profiles. *Genome Res.* 15: 674–680. [5]

Gilad, Y., A. Oshlack, G. K. Smyth, T. P. Speed and K. P. White. 2006. Expression profiling in primates reveals a rapid evolution of human transcription factors. *Nature* 440: 242–245. [13]

Gilbert, N. and W. A. Bickmore. 2006. The relationship between higher-order chromatin structure and transcription. *Biochem. Soc. Symp.* 73: 59–66. [5]

Gilbert, N., S. Lutz-Prigge and J. V. Moran. 2002. Genomic deletions created upon LINE-1 retrotransposition. *Cell* 110: 315–325. [13]

Gilbert, N., S. Boyle, H. Fiegler, K. Woodfine, N. P. Carter and W. A. Bickmor. 2004. Chromatin architecture of the human genome: Gene-rich domains are enriched in open chromatin fibers. *Cell* 118: 555–566. [5]

Gilbert, W. 1978. Why genes in pieces? *Nature* 271: 501. [6]

Gillespie, J. H. 1984. The molecular clock may be an episodic clock. *Proc. Natl. Acad. Sci. USA* 81: 8009–8013. [13]

Giot, L. and 48 others. 2003. A protein interaction map of *Drosophila melanogaster*. *Science* 302: 1727–1736. [6]

Girard A., R. Sachidanandam, G. J. Hannon and M. A. Carmell. 2006. A germline-specific class of small RNAs binds mammalian Piwi proteins. *Nature* 442: 199–202. [8]

Giraud, A., M. Radman, I. Matic and F. Taddei. 2001. The rise and fall of mutator bacteria. *Curr. Opin. Microbiol.* 4: 582–585. [4]

Gisiger, T. 2001. Scale invariance in biology: coincidence or footprint of a universal mechanism? *Biol. Rev. Camb. Philos. Soc.* 76: 161–209. [2]

Gitan, R. S., H. Shi, C. M. Chen, P. S. Yan and T. H. Huang. 2002. Methylation-specific oligonucleotide microarray: A new potential for high-throughput methylation analysis. *Genome Res.* 12: 158–164. [5]

Glasner, J. D. and 10 others. 2003. ASAP, a systematic annotation package for community analysis of genomes. *Nucleic Acids Res.* 31: 147–151. [7]

Glass, J. I. and 8 others. 2006. Essential genes of a minimal bacterium. *Proc. Natl. Acad. Sci. USA* 103: 425–430. [7]

Glazko, G. V., E. V. Koonin, I. B. Rogozin and S. A. Shabalina. 2003. A significant fraction of conserved noncoding DNA in human and mouse consists of predicted matrix attachment regions. *Trends Genet.* 19: 119–124. [5, 13]

Gleiss, P. M., P. F. Stadler, A. Wagner and D. A. Fell. 2001. Small cycles in small worlds. *Advances in Complex Systems* 4: 207–226. [12]

Goetze, S. and 9 others. 2007. The three-dimensional structure of human interphase chromosomes is related to the transcriptome map. *Mol. Cell. Biol.* 27: 4475–4487. [5]

Gogarten, J. P. and J. P. Townsend. 2005. Horizontal gene transfer, genome innovation and evolution. *Nat. Rev. Microbiol.* 3: 679–687. [4]

Gogarten, J. P., W. F. Doolittle and J. G. Lawrence. 2002. Prokaryotic evolution in light of gene transfer. *Mol Biol Evol.* 19: 2226–2238. [4]

Goldberg, A. L. 2003. Protein degradation and protection against misfolded or damaged proteins. *Nature* 426: 895–899. [2]

Goll, M. G. and T H. Bestor. 2005. Eukaryotic cytosine methyltransferases. *Annu. Rev. Biochem.* 74: 481–514. [8]

Gordo, I. and B. Charlesworth. 2000. The speed of Muller's ratchet with background selection, and the degeneration of Y chromosomes. *Genet. Res.* 78: 149–162. [10]

Gorelick, R. 2005. Environmentally alterable additive genetic effects. *Evol. Ecol. Res.* 7: 371–379. [5]

Goriely, A., G. A. McVean, A. M. van Pelt, A. W. O'Rourke, S. A. Wall, D. G. de Rooij and A. O. Wilkie. 2005. Gain-of-function amino acid substitutions drive positive selection of *FGFR2* mutations in human spermatogonia. *Proc. Natl. Acad. Sci. USA* 102: 6051–6056. [13]

Grantham, R., C. Gautier, M. Gouy, R. Mercier and A. Pavé. 1980. Codon catalog usage and the genome hypothesis. *Nucleic Acids Res.* 8: r49–62. [11]

Graveley, B. R. 2005. Mutually exclusive splicing of the insect Dscam pre-mRNA directed by competing intronic RNA secondary structures. *Cell* 123: 65–73. [5]

Graves, J. A. 2006. Sex chromosome specialization and degeneration in mammals. *Cell* 124: 90–114. [9]

Greshock, J. and 9 others. 2004. 1-Mb resolution array-based comparative genomic hybridization using a BAC clone set optimized for cancer gene analysis. *Genome Res.* 14: 179–187. [3]

Gressmann, H. and 9 others. 2005. Gain and loss of multiple genes during the evolution of *Helicobacter pylori*. *PLoS Genet.* 1: e43. [4]

Greve, B., S. Jensen, K. Brugger, W. Zillig and R. A. Garrett. 2004. Genomic comparison of archaeal conjugative plasmids from *Sulfolobus*. *Archaea* 1: 231–239. [4]

Grewal, S. I. and D. Moazed. 2003. Heterochromatin and epigenetic control of gene expression. *Science* 301: 798–802. [5]

Griffiths, A. J. F. 2002. *Modern Genetic Analysis : Integrating Genes and Genomes*. W. H. Freeman and Co., New York. [11]

Grishok, A. and 9 others. 2001. Genes and mechanisms related to RNA interference regulate expression of the small temporal RNAs that control *C. elegans* developmental timing. *Cell* 106: 23–34. [8]

Grishok, A., J. L. Sinskey and P. A. Sharp. 2005. Transcriptional silencing of a transgene by RNAi in the soma of *C. elegans*. *Genes Dev.* 19: 683–696. [8]

Grivna, S. T., E. Beyret, Z. Wang and H. Lin. 2006. A novel class of small RNAs in mouse spermatogenic cells. *Genes Dev.* 20: 1709–1714. [8]

Gu, Z. L., L. M. Steinmetz, X. Gu, C. Scharfe, R. W. Davis and W.-H. Li. 2003. Role of duplicate genes in genetic robustness against null mutations. *Nature* 421: 63–66. [7]

Gumucio, D. L., D. A. Shelton, W. J. Bailey, J. L. Slightom and M. Goodman. 1993. Phylogenetic footprinting reveals unexpected complexity in trans factor binding upstream from the epsilon-globin gene. *Proc. Natl. Acad. Sci. USA* 90: 6018–6022. [5]

Gupta, V. and 8 others. 2006. Global analysis of X-chromosome dosage compensation. *J. Biol.* 5: 1–23. [9, 10]

Haag-Liautard, C., M. Dorris, X. Maside, S. Macaskill, D. L. Halligan, B. Charlesworth and P. D. Keightley. 2007. Direct estimation of per nucleotide and genomic deleterious mutation rates in Drosophila. *Nature* 445: 82–85. [11]

Habermann, J. K. and 11 others. 2007. Stage-specific alterations of the genome, transcriptome, and proteome during colorectal carcinogenesis. *Genes Chromosomes Cancer* 46: 10–26. [2]

Hahn, M. W. and A. D. Kern. 2005. Comparative genomics of centrality and essentiality in three eukaryotic protein-interaction networks. *Mol. Biol. Evol.* 22: 803–806. [2, 7]

Hahn, M. W., J. E. Stajich and G. A. Wray. 2003. The effects of selection against spurious transcription factor binding sites. *Mol. Biol. Evol.* 20: 901–906. [13]

Hahn, M. W., G. C. Conant and A. Wagner. 2004a. Molecular evolution in large genetic networks: Does connectivity equal constraint? *J. Mol. Evol.* 58: 203–211. [7]

Hahn, M. W., M. V. Rockman, N. Soranzo, D. B. Goldstein and G. A. Wray. 2004b. Population genetic and phylogenetic evidence for positive selection on regulatory mutations at the factor VII locus in humans. *Genetics* 167: 867–877. [13]

Hajkova, P., S. Erhardt, N. Lane, T. Haaf, O. El-Maarri, W. Rex, J. Walter and M. A. Surani. 2002. Epigenetic reprogramming in mouse primordial germ cells. *Mech. Dev.* 117: 15–23. [5]

Haldane, J. B. S. 1933. The part played by recurrent mutation in evolution. *Amer. Nat.* 42: 5–19. [10]

Hall, C., S. Brachat and F. S. Dietrich. 2005. Contribution of horizontal gene transfer to the evolution of *Saccharomyces cerevisiae*. *Eukaryot. Cell* 4: 1102–1115. [4]

Hamilton, A., O. Voinnet, L. Chappell and D. Baulcombe. 2002. Two classes of short interfering RNA in RNA silencing. *EMBO J.* 21: 4671–4679. [8]

Hammond, S. M., E. Bernstein, D. Beach and G. J. Hannon. 2000. An RNA-directed nuclease mediates post-transcriptional gene silencing in Drosophila cells. *Nature* 404: 293–296. [8]

Hammond, S. M., S. Boettcher, A. A. Caudy, R. Kobayashi and G. J. Hanno. 2001. Argonaute2, a link between genetic and biochemical analyses of RNAi. *Science* 293: 1146–1150. [8]

Han, J.-D. and 10 others. 2004. Evidence for dynamically organized modularity in the yeast protein-protein interaction network. *Nature* 430: 88–93. [2]

Han, J.-D. J., D. Dupuy, N. Bertin, M. E. Cusick and M. Vidal. 2005. Effect of sampling on topology predictions of protein-protein interaction networks. *Nat. Biotechnol.* 23: 839–844. [7]

Handley, L., L. Berset-Brändli and N. Perrin. 2006. Disentangling reasons for low Y chromosome variation in the greater white-toothed shrew (*Crocidura russula*). *Genetics* 173: 935–942. [10]

Harding, R. M. and 10 others. 2000. Evidence for variable selective pressures at MC1R. *Am. J. Hum. Genet.* 66: 1351–1361. [13]

Hardison, R., J. L. Slightom, D. L. Gumucio, M. Goodman, N. Stojanovic and W. Miller. 1997. Locus control regions of mammalian beta-globin gene clusters: Combining phylogenetic analyses and experimental results to gain functional insights. *Gene* 205: 73–94. [5]

Hardison, R. C. and 17 others. 2003. Covariation in frequencies of substitution, deletion, transposition, and recombination during eutherian evolution. *Genome Res.* 13: 13–26. [13]

Hare, M. P. and S. R. Palumbi. 2003. High intron sequence conservation across three mammalian orders suggests functional constraints. *Mol. Biol. Evol.* 20: 969–978. [11]

Harmon, B. and J. Sedat. 2005. Cell-by-cell dissection of gene expression and chromosomal interactions reveals consequences of nuclear reorganization. *PLoS Biol.* 3: e67. [5]

Harper, J. T. and P. J. Keeling. 2004. Lateral gene transfer and the complex distribution of insertions in eukaryotic enolase. *Gene* 340: 227–235. [4]

Harris, E. E. and J. Hey. 2001. Human populations show reduced DNA sequence variation at the factor IX locus. *Curr. Biol.* 11: 774–778. [13]

Harrison, R., B. Papp, C. Pál, S. G. Oliver and D. Delneri. 2007. Plasticity of genetic interactions in metabolic networks of yeast. *Proc. Natl. Acad. Sci. USA* 104: 2307–2312. [7]

Hartner, J. C., C. Schmittwolf, A. Kispert, A. M. Muller, M. Higuchi and P. H. Seeburg. 2004. Liver disintegration in the mouse embryo caused by deficiency in the RNA-editing enzyme ADAR1. *J. Biol. Chem.* 279: 4894–4902. [8]

Hatey, F., G. Tosser-Klopp, C. Clouscard-Martinato, P. Mulsant and F. Gasser. 1998. Expressed sequence tags for genes: A review. *Genetics Sel. Evol.* 30: 521–541. [5]

Havilio, M., E. Y. Levanon, G. Lerman, M. Kupiec and E. Eisenberg. 2005. Evidence for abundant transcription of non-coding regions in the *Saccharomyces cerevisiae* genome. *BMC Genomics* 6: 93. [7]

Hayakawa, T., Y. Satta, P. Gagneux, A. Varki and N. Takahata. 2001. Alu-mediated inactivation of the human CMP- N-acetylneuraminic acid hydroxylase gene. *Proc. Natl. Acad. Sci. USA* 98: 11399–11404. [13]

He, X. and J. Zhang. 2005. Rapid subfunctionalization accompanied by prolonged and substantial neofunctionalization in duplicate gene evolution. *Genetics* 169: 1157–1164. [13]

He, X. and J. Zhang. 2006a. Higher duplicability of less important genes in yeast genomes. *Mol. Biol. Evol.* 23:144–151. [7]

He, X. and J. Zhang. 2006b. Transcriptional reprogramming and backup between duplicate genes: Is it a genomewide phenomenon? *Genetics* 172: 1363–1367. [7]

He, X. and J. Zhang. 2006c. Why do hubs tend to be essential in protein networks? *PLoS Genetics* 2: e88. [7]

Heard, E., P. Clerc and P. Avner. 1997. X-chromosome inactivation in mammals. *Annu. Rev. Genet.* 31: 571–610. [9]

Heintzman, N. D. and 14 others. 2007. Distinct and predictive chromatin signatures of transcriptional promoters and enhancers in the human genome. *Nat. Genet.* 39: 311–318. [5]

Hellman, A. and A. Chess. 2007. Gene body-specific methylation on the active X chromosome. *Science* 315: 1141–1143. [10]

Hellmann, I., S. Zollner, W. Enard, I. Ebersberger, B. Nickel and S. Pääbo. 2003. Selection on human genes as revealed by comparisons to chimpanzee cDNA. *Genome Res.* 13: 831–837. [11, 13]

Hellmann, I., K. Prufer, H. Ji, M. C. Zody, S. Paabo and S. E. Ptak. 2005. Why do human diversity levels vary at a megabase scale? *Genome Res.* 15: 1222–1231. [13]

Hendrix, R. W. 2002. Bacteriophages: Evolution of the majority. *Theor. Popul. Biol.* 61: 471–480. [4]

Heng, H. H. and 8 others. 2004. Chromatin loops are selectively anchored using scaffold/matrix-attachment regions. *J. Cell Sci.* 117: 999–1008. [5]

Henze, K., D. S. Horner, S. Suguri, D. V. Moore, L. B. Sanchez, M. Muller and T. M. Embley. 2001. Unique phylogenetic relationships of glucokinase and glucosephosphate isomerase of the amitochondriate eukaryotes *Giardia intestinalis*, *Spironucleus barkhanus* and *Trichomonas vaginalis*. *Gene* 281: 123–131. [4]

Herbeck, J. T. and D. P. Wall. 2005. Converging on a general model of protein evolution. *Trends Biotechnol.* 23: 485–487. [2]

Herman, H., M. Lu, M. Anggraini, A. Sikora, Y. Chang, B. J. Yoon and P. D. Soloway. 2003. Trans allele methylation and paramutation-like effects in mice. *Nat. Genet.* 34: 199–202. [5]

Herrgard, M. J., B. S. Lee, V. Portnoy and B. O. Palsson. 2006. Integrated analysis of regulatory and metabolic networks reveals novel regulatory mechanisms in *Saccharomyces cerevisiae*. *Genome Res.* 16: 627–635. [7]

Hickey, D. A. 1992. Evolutionary dynamics of transposable elements in prokaryotes and eukaryotes. *Genetica* 86: 269–274. [8]

Hinds, D. A., L. L. Stuve, G. B. Nilsen, E. Halperin, E. Eskin, D. G. Ballinger, K. A. Frazer and D. R. Cox. 2005. Whole-genome patterns of common DNA variation in three human populations. *Science* 307: 1072–1079. [13]

Hinds, D. A., A. P. Kloek, M. Jen, X. Chen and K. A. Frazer. 2006. Common deletions and SNPs are in linkage disequilibrium in the human genome. *Nat. Genet.* 38: 82–85. [13]

Hirsh, A. E. and H. B. Fraser. 2001. Protein dispensability and rate of evolution. *Nature* 411: 1046–1049. [2, 7]

Hirt, R. P., N. Harriman, A. V. Kajava and T. M. Embley. 2002. A novel potential surface protein in *Trichomonas vaginalis* contains a leucine-rich repeat shared by micro-organisms from all three domains of life. *Mol. Biochem. Parasitol.* 125: 195–199. [4]

Ho, Y. and 45 others. 2002. Systematic identification of protein complexes in *Saccharomyces cerevisiae* by mass spectrometry. *Nature* 415: 180–183. [12]

Hodgkin, J. 2002. Exploring the envelope. Systematic alteration in the sex-determination system of the nematode *Caenorhabditis elegans*. *Genetics* 162: 767–780. [9]

Hoffman, M. M. and E. Birney. 2007. Estimating the neutral rate of nucleotide substitution using introns. *Mol. Biol. Evol.* 24: 522–531. [11]

Hollick, J. B. and V. L. Chandler. 2001. Genetic factors required to maintain repression of a paramutagenic maize pl1 allele. *Genetics* 157: 369–378. [5]

Holliday, R. and J. E. Pugh. 1975. DNA modification mechanisms and gene activity during development. *Science* 187: 226–232. [5]

Honys, D. and D. Twell. 2004. Transcriptome analysis of haploid male gametophyte development in Arabidopsis. *Genome Bio.* 5: R85. [10]

Horner, D. S., P. G. Foster and T. M. Embley. 2000. Iron hydrogenases and the evolution of anaerobic eukaryotes. *Mol. Biol. Evol.* 17: 1695–1709. [4]

Hotopp, J. C. and 19 others. Widespread lateral gene transfer from intracellular bacteria to multicellular eukaryotes. *Science*, Aug. 30, 2007. (Epub ahead of print.) [4]

Howard-Ashby, M., S. C. Materna, C. T. Brown, Q. Tu, P. Oliveri, R. A. Cameron and E. H. Davidson. 2006. High regulatory gene use in sea urchin embryogenesis: Implications for bilaterian development and evolution. *Dev. Biol.* 300: 27–34. [2]

Huang, J., N. Mullapudi, T. Sicheritz-Ponten and J. C. Kissinger. 2004a. A first glimpse into the pattern and scale of gene transfer in Apicomplexa. *Int. J. Parasitol.* 34: 265–274. [4]

Huang, J., N. Mullapudi, C. A. Lancto, M. Scott, M. S. Abrahamsen and J. C. Kissinger. 2004b. Phylogenomic evidence supports past endosymbiosis, intracellular and horizontal gene transfer in *Cryptosporidium parvum*. *Genome Biol.* 5: R88. [4]

Huang, J., Y. Xu and J. P. Gogarten. 2005. The presence of a haloarchaeal type tyrosyl-tRNA synthetase marks the opisthokonts as monophyletic. *Mol. Biol. Evol.* 22: 2142–2146. [4]

Huang, Z. P., H. Zhou, H. L. He, C. L. Chen, D. Liang and L. H. Qu. 2005. Genome-wide analyses of two families of snoRNA genes from *Drosophila melanogaster*, demonstrating the extensive utilization of introns for coding of snoRNAs. *RNA* 11: 1303–1316. [11]

Huber, B. A. 2005. Sexual selection research on spiders: progress and biases. *Biol. Rev. Camb. Philos. Soc.* 80: 363–385. [9]

Hughes, A. L. and M. Yeager. 1998. Comparative evolutionary rates of introns and exons in murine rodents. *J. Mol. Evol.* 46: 497. [11]

Hughes, J. F., H. Skaletsky, T. Pyntikova, P. J. Minx, T. Graves, S. Rozen, R. K. Wilson and D. C. Page. 2005. Conservation of Y-linked genes during human evolution revealed by comparative sequencing in chimpanzee. *Nature* 437: 100. [10]

Huminiecki, L. and K. H. Wolfe. 2004. Divergence of spatial gene expression profiles following species-specific gene duplications in human and mouse. *Genome Res.* 14: 1870–1879. [13]

Hung, M. S., N. Karthikeyan, H. S. Huang, H. C. Koo, J. Kiger and C. J. Shen. 1999. *Drosophila* proteins related to vertebrate DNA (5-cytosine) methyltransferases. *Proc. Natl. Acad. Sci. USA* 96: 11940–11945. [5]

Hurst, L. D. and J. P. Randerson. 1999. An eXceptional chromosome. *Trends Genet.* 15: 383–385. [9]

Hurst, L. D. and N. G. C. Smith. 1999. Do essential genes evolve slowly? *Curr. Biol.* 9: 747–750. [2, 7]

Hurst, L. D., A. Atlan and B. O. Bengtsson. 1996. Genetic conflicts. *Q. Rev. Biol.* 71: 317–364. [13]

Hutchison, C. A., S. N. Peterson, S. R. Gill, R. T. Cline, O. White, C. M. Fraser, H. O. Smith and J. C. Venter. 1999. Global transposon mutagenesis and a minimal mycoplasma genome. *Science* 286: 2165–2169. [7]

Hutter, H. and 10 others. 2000. Conservation and novelty in the evolution of cell adhesion and extracellular matrix genes. *Science* 287: 989–94. [6]

Hutvagner, G., J. McLachlan, A. E. Pasquinelli, E. Balint, T. Tuschl and P. D. Zamore. 2001. A cellular function for the RNA-interference enzyme Dicer in the maturation of the let-7 small temporal RNA. *Science* 293: 834–838. [8]

Hwang, D. G. and P. Green. 2004. Bayesian Markov chain Monte Carlo sequence analysis reveals varying neutral substitution patterns in mammalian evolution. *Proc. Natl. Acad. Sci. USA* 101: 13994–14001. [11, 13]

Hynes, R. O. and Q. Zhao. 2000. The evolution of cell adhesion. *J. Cell Biol.* 150: F89–96. [6]

Iafrate, A. J., L. Feuk, M. N. Rivera, M. L. Listewnik, P. K. Donahoe, Y. Qi, S. W. Scherer and C. Lee. 2004. Detection of large-scale variation in the human genome. *Nat. Genet.* 36: 949–951. [13]

Ibarra, R. U., J. S. Edwards and B. O. Palsson. 2002. *Escherichia coli* K-12 undergoes adaptive evolution to achieve in silico predicted optimal growth. *Nature* 420: 186–189. [7]

International Chicken Genome Sequencing Consortium. 2004. Sequence and comparative analysis of the chicken genome provide unique perspectives on vertebrate evolution. *Nature* 432: 695–716. [3]

International Human Genome Sequencing Consortium. 2001. Initial sequencing and analysis of the human genome. *Nature* 409: 860–921. [13]

Intrieri, M. C. and M. Buiatti. 2001. The horizontal transfer of *Agrobacterium rhizogenes* genes and the evolution of the genus *Nicotiana. Mol. Phylogenet. Evol.* 20: 100–110. [4]

Itoh, Y. and 12 others. 2007. Dosage compensation is less effective in birds than in mammals. *J. Biol.* 6: 2. [10]

Ivakhno, S. 2007. From functional genomics to systems biology. *FEBS J.* 274: 2439–2448. [2]

Ivens, A. C. and 100 others. 2005. The genome of the kinetoplastid parasite, *Leishmania major. Science* 309: 436–442. [4]

Jablonka, E. and M. J. Lamb. 1989. The inheritance of acquired epigenetic variations. *J. Theor. Biol.* 139: 69–83. [5]

Jablonka, E. and M. J. Lamb. 2002. The changing concept of epigenetics. *Ann. NY Acad. Sci.* 981: 82–96. [5]

Jackson, J. P., A. M. Lindroth, X. Cao and S. E. Jacobsen. 2002. Control of CpNpG DNA methylation by the KRYPTONITE histone H3 methyltransferase. *Nature* 416: 556–560. [8]

Jacob, F. and J. Monod. 1961. Genetic regulatory mechanisms in the synthesis of proteins. *J. Mol. Biol.* 3: 318–356. [5]

Jacobs, M. A. and 14 others. 2003. Comprehensive transposon mutant library of *Pseudomonas aeruginosa. Proc. Natl. Acad. Sci. USA* 100: 14339–14344. [7]

Jaenisch, R. and A. Bird. 2003. Epigenetic regulation of gene expression: How the genome integrates intrinsic and environmental signals. *Nat. Genet.* 33: 245–254. [5]

Jain, R., M. C. Rivera and J. A. Lake. 1999. Horizontal gene transfer among genomes: The complexity hypothesis. *Proc. Natl. Acad. Sci. USA.* 96: 3801–3806.

Jeffreys, A. J. and R. Neumann. 2002. Reciprocal crossover asymmetry and meiotic drive in a human recombination hot spot. *Nat. Genet.* 31: 267–271. [13]

Jeffreys, A. J., A. Ritchie R. and Neumann. 2000. High resolution analysis of haplotype diversity and meiotic crossover in the human TAP2 recombination hotspot. *Hum. Mol. Genet.* 9: 725–733. [13]

Jeffreys, A. J., L. Kauppi and R. Neumann. 2001. Intensely punctate meiotic recombination in the class II region of the major histocompatibility complex. *Nat. Genet.* 29: 217–222. [13]

Jeffreys, A. J., R. Neumann, M. Panayi, S. Myers and P. Donnelly. 2005. Human recombination hot spots hidden in regions of strong marker association. *Nat. Genet.* 37, 601–606. [13]

Jenkins, C. and 9 others. 2002. Genes for the cytoskeletal protein tubulin in the bacterial genus Prosthecobacter. *Proc. Natl. Acad. Sci. USA* 99: 17049–17054. [4]

Jenuwein, T. and C. D. Allis. 2001. Translating the histone code. *Science* 293: 1074–1080. [5]

Jeong, H., B. Tombor, R. Albert, Z. N. Oltvai and A.-L. Barabasi. 2000. The large-scale organization of metabolic networks. *Nature* 407: 651–654. [7, 12]

Jeong, H., S. P. Mason, A.-L. Barabasi and Z. N. Oltvai. 2001. Lethality and centrality in protein networks. *Nature* 411: 41–42. [6, 7, 12]

Ji, Y. D., B. Zhang, S. F. Van Horn, P. Warren, G. Woodnutt, M. K. R. Burnham and M. Rosenberg. 2001. Identification of critical staphylococcal genes using conditional phenotypes generated by antisense RNA. *Science* 293: 2266–2269. [7]

Ji, Y., G. Woodnutt, M. Rosenberg and M. K. Burnham. 2002. Identification of essential genes in *Staphylococcus aureus* using inducible antisense RNA. *Methods Enzymol.* 358:123–128. [7]

Jiang, M., J. Ryu, M. Kiraly, K. Duke, V. Reinke and S. K. Kim. 2001. Genome-wide analysis of developmental and sex-regulated gene expression profiles in *Caenorhabditis elegans. Proc. Natl. Acad. Sci. USA* 98: 218–223. [9]

Jiang, N., Z. Bao, X. Zhang, S. Eddy and S. Wessler. 2004. Pack-MULE transposable elements mediate gene evolution in plants. *Nature* 431: 569–573. [3]

Jin, W., R. M. Riley, R. D. Wolfinger, K. P. White, G. Passador-Gurgel and G. Gibson. 2001. The contributions of sex, genotype and age to transcriptional variance in *Drosophila melanogaster. Nat. Genet.* 29: 389–395. [5]

Johnson, S. A. and T. Hunter. 2005. Kinomics: methods for deciphering the kinome. *Nat. Methods* 2: 17–25. [2]

Jones, C. D., A. W. Custer and D. J. Begun. 2005. Origin and evolution of a chimeric fusion gene in *Drosophila subobscura, D. madeirensis* and *D. guanche Genetics* 170: 207–219. [3]

Jopling, C. L., M. Yi, A. M. Lancaster, S. M. Lemon and P. Sarnow. 2005. Modulation of hepatitis C virus RNA abundance by a liver-specific MicroRNA. *Science* 309: 1577–1581. [8]

Jordan, I. K., I. B. Rogozin, Y. I. Wolf and E. V. Koonin. 2002. Essential genes are more evolutionarily conserved than are nonessential genes in bacteria. *Genome Res.* 12: 962–968. [2, 7]

Jordan, I. K., Y. I. Wolf and E. V. Koonin. 2003. No simple dependence between protein evolution rate and the number of protein-protein interactions: only the most prolific interactors tend to evolve slowly. *BMC Evol. Biol.* 3: 1. [2]

Jordan, I. K., L. Marino-Ramirez, Y. I. Wolf and E. V. Koonin. 2004. Conservation and coevolution in the scale-free human gene coexpression network. *Mol. Biol. Evol.* 21: 2058–2070. [2]

Jordan, I. K., L. Marino-Ramirez and E. V. Koonin. 2005. Evolutionary significance of gene expression divergence. *Gene* 345: 119–126. [2, 5]

Jorgenson, E. and 10 others. 2005. Ethnicity and human genetic linkage maps. *Am. J. Hum. Genet.* 76: 276–290. [13]

Kacser, H. and J. A. Burns. 1981. The molecular basis of dominance. *Genetics* 97: 639–666. [7]

Kafri, R., A. Bar-Even and Y. Pilpel. 2005. Transcription control reprogramming in genetic backup circuits. *Nat. Genet.* 37: 295–299. [7]

Kamath, R. S. and 12 others. 2003. Systematic functional analysis of the *Caenorhabditis elegans* genome using RNAi. *Nature* 421: 231–237. [1, 7]

Kaplan, N. L., R. R. Hudson and C. H. Langley. 1989. The "hitchhiking effect" revisited. *Genetics* 123: 887–899. [13]

Kato, M., A. Miura, J. Bender, S. E. Jacobsen and T. Kakutani, 2003. Role of CG and non-CG methylation in immobilization of transposons in Arabidopsis. *Curr. Biol.* 13: 421–426. [8]

Katz, L. and C. B. Burge. 2003. Widespread selection for local RNA secondary structure in coding regions of bacterial genes. *Genome Res.* 13: 2042–2051. [11]

Katze, M. G., Y. He and M. Gale, Jr. 2002. Viruses and interferon: A fight for supremacy. *Nat. Rev. Immunol.* 2: 675–687. [8]

Keeling, P. J. and Y. Inagaki. 2004. A class of eukaryotic GTPase with a punctate distribution suggesting multiple functional replacements of translation elongation factor 1alpha. *Proc. Natl. Acad. Sci. USA.* 101: 15380–15385. [4]

Keeling, P. J. and J. D. Palmer. 2001. Lateral transfer at the gene and subgenic levels in the evolution of eukaryotic enolase. *Proc. Natl. Acad. Sci. USA* 98: 10745–10750. [4]

Keightley, P. D. 1996. A metabolic basis for dominance and recessivity. *Genetics* 143: 621–625. [7]

Keightley, P. D. and A. Eyre-Walker. 2000. Deleterious mutations and the evolution of sex. *Science* 290: 331–333. [1, 11]

Keightley, P. D., G. V. Kryukov, S. Sunyaev, D. L. Halligan and D. J. Gaffney. 2005a. Evolutionary constraints in conserved nongenic sequences of mammals. *Genome Res.* 15: 1373–1378. [13]

Keightley, P. D., M. J. Lercher and A. Eyre-Walker. 2005b. Evidence for widespread degradation of gene control regions in hominid genomes. *PLoS Biol.* 3: e42. [5, 13]

Keller, E. F. 2005. Revisiting "scale-free" networks. *Bioessays* 27: 1060–1068. [2]

Kelly, W. G., C. E. Schaner, A. F. Dernburg, M. H. Lee, S. K. Kim, A. M. Villeneuve and V. Reinke. 2002. X-chromosome silencing in the germ line of *C. elegans*. *Development* 129: 479–492. [9]

Kessler, M. and M. B. Mathews. 1992. Premature termination and processing of human immunodeficiency virus type 1-promoted transcripts. *J. Virol.* 66: 4488–4496. [5]

Ketting, R. F. and R. H. Plasterk. 2000. A genetic link between co-suppression and RNA interference in *C. elegans*. *Nature* 404: 296–298. [8]

Ketting, R. F., T. H. Haverkamp, H. G. van Luenen and R. H. Plasterk. 1999. Mut-7 of *C. elegans*, required for transposon silencing and RNA interference, is a homolog of Werner syndrome helicase and RNaseD. *Cell* 99: 133–141. [8]

Ketting, R. F., S. E. Fischer, E. Bernstein, T. Sijen, G. J. Hannon and R. H. Plasterk. 2001. Dicer functions in RNA interference and in synthesis of small RNA involved in developmental timing in *C. elegans*. *Genes Dev.* 15: 2654–2659. [8]

Khaitovich, P. and 8 others. 2004. A neutral model of transcriptome evolution. *PLoS Biol.* 2: e132. [2, 5, 13]

Khaitovich, P. and 8 others. 2005. Parallel patterns of evolution in the genomes and transcriptomes of humans and chimpanzees. *Science* 309: 1850–1854. [5, 13]

Khaitovich, P., S. Paabo and G. Weiss. 2005. Toward a neutral evolutionary model of gene expression. *Genetics* 170: 929–939. [2]

Khil, P. P., N. A. Smirnova, P. J. Romanienko and R. D. Camerini-Otero. 2004. The mouse X chromosome is enriched for sex-biased genes not subject to selection by meiotic sex chromosome inactivation. *Nat. Genet.* 36: 642–646. [9]

Khil, P. P., B. Oliver and R. D. Camerini-Otero. 2005. X for intersection: retrotransposition both on and off the X chromosome is more frequent. *Trends Genet.* 21: 3–7. [9, 10]

Khvorova, A., A. Reynolds and S. D. Jayasena. 2003. Functional siRNAs and miRNAs exhibit strand bias. *Cell* 115: 209–216. [8]

Kidwell, M. G. 1985. Hybrid dysgenesis in Drosophila: Nature and inheritance of P element regulation. *Genetics* 111: 337–350. [8]

Kikuchi, K., W. Kai, A. Hosokawa, N. Mizuno, H. Suetake, K. Asahina and Y. Suzuki. 2007. The sex-determining locus in the Tiger Pufferfish, *Takifugu rubripes*. *Genetics* 175: 2039–2042. [10]

Kim, D. D., T. T. Kim, T. Walsh, Y. Kobayashi, T. C. Matise, S. Buyske and A. Gabriel. 2004. Widespread RNA editing of embedded alu elements in the human transcriptome. *Genome Res.* 14: 1719–1725. [8]

Kimchi-Sarfaty, C., J. M. Oh, I.-W. Kim, Z. E. Sauna, A. M. Calcagno, S. V. Ambudkar and M. M. Gottesman. 2007. A "silent" polymorphism in the MDR1 gene changes substrate specificity. *Science* 315: 525–528. [11]

Kimura, M. 1968. Evolutionary rate at the molecular level. *Nature* 217: 624–626. [11]

Kimura, M. 1977. Preponderance of synonymous changes as evidence for the neutral theory of molecular evolution. *Nature* 267: 275–276. [11]

Kimura, M. 1983. *The Neutral Theory of Molecular Evolution*. Cambridge University Press, Cambridge, U.K. [5, 11, 12, 13]

Kimura, M. 1991. The neutral theory of molecular evolution: a review of recent evidence. *Jpn. J. Genet.* 66: 367–386. [11]

Kimura, M. and T. Ohta. 1974. On some principles governing molecular evolution. *Proc. Natl. Acad. Sci. USA* 71: 2848–2852. [2, 7]

King, M. C. and A. C. Wilson. 1975. Evolution at two levels in humans and chimpanzees. *Science* 188: 107–116. [13]

Kirst, M., C. J. Basten, A. A. Myburg, Z. B. Zeng and R. R. Sederoff. 2005. Genetic architecture of transcript-level variation in differentiating xylem of a eucalyptus hybrid. *Genetics* 169: 2295–2303. [5]

Klotz, M. G., G. R. Klassen and P. C. Loewen. 1997. Phylogenetic relationships among prokaryotic and eukaryotic catalases. *Mol. Biol. Evol.* 14: 951–958. [4]

Knuth, K., H. Niesalla, C. J. Hueck and T. M. Fuchs. 2004. Large-scale identification of essential *Salmonella* genes by trapping lethal insertions. *Mol. Microbiol.* 51: 1729–1744. [7]

Kobayashi, K. and 98 others. 2003. Essential *Bacillus subtilis* genes. *Proc. Natl. Acad. Sci. USA* 100: 4678–4683. [7]

Kobe, B. and J. Deisenhofer. 1995. Proteins with leucine-rich repeats. *Curr. Opin. Struct. Biol.* 5: 409–416. [6]

Kolker, E. 2002. Editorial. *Omics* 6: 1. [2]

Kolodner, R. D., C. D. Putnam and K. Myung. 2002. Maintenance of genome stability in *Saccharomyces cerevisiae*. *Science* 297: 552–557. [7]

Kondo, M. and 10 others. 2006. Genomic organization of the sex-determining and adjacent regions of the sex chromosomes of medaka. *Genome Res.* 16: 815–826. [10]

Kondrashov, F. A., I. B. Rogozin, Y. I. Wolf and E. V. Koonin. 2002. Selection in the evolution of gene duplications. *Genome Biol.* 3: R8. [13]

Konecny, J., M. Schoniger, I. Hofacker, M. D. Weitze and G. L. Hofacker. 2000. Concurrent neutral evolution of mRNA secondary structures and encoded proteins. *J. Mol. Evol.* 50: 238–242. [11]

Kong, A. and 14 others. 2002. A high-resolution recombination map of the human genome. *Nat. Genet.* 31: 241–247. [13]

Kong, A. and 13 others. 2004. Recombination rate and reproductive success in humans. *Nat. Genet.* 36: 1203–1206. [13]

Kontstantinidis, K. and J. Tiedje. 2005. Genomic insights that advance the species definition for prokaryotes. *Proc. Natl. Acad. Sci. USA* 102: 2567–2572. [4]

Koonin, E. V. 2001. An apology for orthologs—or brave new memes. *Genome Biol.* 2: C1005. [2]

Koonin, E. V. and Y. I. Wolf. 2006. Evolutionary systems biology: links between gene evolution and function. *Curr. Opin. Biotechnol.* 17: 481–487. [2, 5]

Koonin, E. V., K. S. Makarova and L. Aravind. 2001. Horizontal gene transfer in prokaryotes: Quantification and classification. *Annu. Rev. Microbiol.* 55: 709–742. [4]

Koonin, E. V., Y. I. Wolf and G. P. Karev. 2002. The structure of the protein universe and genome evolution. *Nature* 420: 218–223. [2, 6]

Koonin, E. V. and 17 others. 2004. A comprehensive evolutionary classification of proteins encoded in complete eukaryotic genomes. *Genome Biol.* 5: R7. [2]

Koonin, E. V., Y. I. Wolf and G. P. Karev. 2006. *Power laws, scalefree networks and genome biology*. Landes Bioscience, Georgetown, TX. [2]

Kornegay, J. R., J. W. Schilling and A. C. Wilson. 1994. Molecular adaptation of a leaf-eating bird—stomach lysozyme of the hoatzin. *Mol. Biol. Evol.* 11: 921–928. [12]

Koski, L. B. and G. B. Golding. 2001. The closest BLAST hit is often not the nearest neighbor. *J. Mol. Evol.* 52: 540–542. [4]

Kouyos, R. D., O. K. Silander and S. Bonhoeffer. 2007. Epistasis between deleterious mutations and the evolution of recombination. *Trends Ecol. Evol.* 22: 308–315. [7]

Kouzminova, E. and E. U. Selker. 2001. *dim-2* encodes a DNA methyltransferase responsible for all known cytosine methylation in Neurospora. *EMBO J.* 20: 4309–4323. [8]

Krakauer, D. C. and J. B. Plotkin. 2002. Redundancy, antiredundancy, and the robustness of genomes. *Proc. Natl. Acad. Sci. USA* 99: 1405–1409. [7]

Krawczak, M., E. V. Ball and D. N. Cooper. 1998. Neighboring-nucleotide effects on the rates of germ-line single-base-pair substitution in human genes. *Am. J. Hum. Genet.* 63: 474–488. [11]

Krylov, D. M., Y. I. Wolf, I. B. Rogozin and E. V. Koonin. 2003. Gene loss, protein sequence divergence, gene dispensability, expression level, and interactivity are correlated in eukaryotic evolution. *Genome Res.* 13: 2229–2235. [2, 7]

Krzywinski, J., D. R. Nusskern, M. K. Kern and N. J. Besansky. 2004. Isolation and characterization of Y Chromosome sequences from the African malaria mosquito *Anopheles gambiae*. *Genetics* 166: 1291–1302. [10]

Krzywinski, J., M. A. Chrystal and N. J. Besansky. 2006. Gene finding on the Y: Fruitful strategy in *Drosophila* does not deliver in *Anopheles*. *Genetica* 126: 369–375. [10]

Kuepfer, L., U. Sauer and L. M. Blank. 2005. Metabolic functions of duplicate genes in *Saccharomyces cerevisiae*. *Genome Res.* 15: 1421–1430. [7]

Kunert, N., J. Marhold, J. Stanke, D. Stach and F. Lyko. 2003. A Dnmt2-like protein mediates DNA methylation in Drosophila. *Development* 130: 5083–5090. [5]

Kunin, V. and C. A. Ouzounis. 2003. The balance of driving forces during genome evolution in prokaryotes. *Genome Res.* 13: 1589–1594. [4]

Kunin, V., L. Goldovsky, N. Darzentas and C. A. Ouzounis. 2005. The net of life: Reconstructing the microbial phylogenetic network. *Genome Res.* 15: 954–949. [4]

Kurdistani, S. K., S. Tavazoie and M. Grunstein. 2004. Mapping global histone acetylation patterns to gene expression. *Cell* 117: 721–733. [5]

Kurland, C. G., B. Canback and O. G. Berg. 2003. Horizontal gene transfer: A critical view. *Proc. Natl. Acad. Sci. USA.* 100: 9658–9662. [4]

Lachaise, D., M. L. Cariou, J. R. David, F. Lemeunier and L. Tsacas. 1988. Historical biogeography of the *Drosophila melanogaster* species subgroup. *Evol. Biol.* 22: 159–225. [3]

Lachaise, D., M. Harry, M. Solignac, F. Lemeunier, V. Benassi and M. L. Cariou. 2000. Evolutionary novelties in island: *Drosophila santomea*, a new *melanogaster* sister species from Sao Tome. *Proc. R. Soc. Lond. Ser. B* 267: 1487–1495. [3]

Lacroix, B., T. Tzfira, A. Vainstein and V. Citovsky. 2006. A case of promiscuity: *Agrobacterium*'s endless hunt for new partners. *Trends Genet.* 22: 29–37.

Lahn, B. T. and D. C. Page. 1999. Four evolutionary strata on the human X chromosome. *Science* 286: 964–967. [10]

Lai, C. S., S. E. Fisher, J. A. Hurst, F. Vargha-Khadem and A. P. Monaco. 2001. A forkhead-domain gene is mutated in a severe speech and language disorder. *Nature* 413: 519–523. [13]

Lande, R. and S. Arnold. 1983. The measurement of selection on correlated characters. *Evolution* 37: 1210–1226. [5]

Landry, C. R., P. J. Wittkopp, C. H. Taubes, J. M. Ranz, A. G. Clark and D. L. Hartl. 2005. Compensatory cis-trans evolution and the dysregulation of gene expression in interspecific hybrids of *Drosophila*. *Genetics* 171: 1813–1822. [5]

Landry, C. R., J. Oh, D. L. Hartl and D. Cavalieri. 2006. Genome-wide scan reveals that genetic variation for transcriptional plasticity in yeast is biased towards multi-copy and dispensable genes. *Gene* 366: 343–351. [5]

Landry, C. R., B. Lemos, S. A. Rifkin, W. J. Dickinson and D. L. Hartl. 2007. Genetic properties influencing the evolvability of gene expression. *Science* 317:118–121. [5]

Landry, J. R., A. Rouhi, P. Medstrand and D. L. Mager. 2002. The Opitz syndrome gene Mid1 is transcribed from a human endogenous retroviral promoter. *Mol. Biol. Evol.* 19: 1934–1942. [13]

Laporte, V., D. A. Filatov, E. Kamau and D. Charlesworth. 2005. Indirect evidence from DNA sequence diversity for genetic degeneration of Y-chromosome in dioecious species of the plant Silene: the SlY4/SlX4 and DD44-X/DD44-Y gene pairs. *J. Evol. Biol.* 18: 337–347. [10]

Lau, N. C., L. P. Lim, E. G. Weinstein and D. P. Bartel. 2001. An abundant class of tiny RNAs with probable regulatory roles in *Caenorhabditis elegans*. *Science* 294: 858–862. [5]

Lau, N. C., A. G. Seto, J. Kim, S. Kuramochi-Miyagawa, T. Nakano, D. P. Bartel and R. E. Kingston. 2006. Characterization of the piRNA complex from rat testes. *Science* 313: 363–367. [8]

Lawrence, J. G. and J. R. Roth. 1996. Selfish operons: Horizontal transfer may drive the evolution of gene clusters. *Genetics* 143: 1843–1860. [4]

Lawson-Handley, L. J., H. Ceplitis and H. Ellegren. 2004. Evolutionary strata on the chicken Z chromosome: implications for sex chromosome evolution. *Genetics* 167: 367–376. [10]

Lecellier, C. H., P. Dunoyer, K. Arar, J. Lehmann-Che, S. Eyquem, C. Himber, A. Saib and O. Voinnet. 2005. A cellular microRNA mediates antiviral defense in human cells. *Science* 308: 557–560. [8]

Lee, R. C. and V. Ambros. 2001. An extensive class of small RNAs in *Caenorhabditis elegans*. *Science* 294: 862–864. [5]

Lee, T. I. and 20 others. 2002. Transcriptional regulatory networks in *Saccharomyces cerevisiae*. *Science* 298: 799–804. [5, 12]

Leighton, P. A., R. S. Ingram, J. Eggenschwiler, A. Efstratiadis and S. M. Tilghman. 1995. Disruption of imprinting caused by deletion of the H19 gene region in mice. *Nature* 375: 34–39. [7]

Leipe, D. D., E. V. Koonin and L. Aravind. 2004. STAND, a class of P-loop NTPases including animal and plant regulators of programmed cell death: Multiple, complex domain architectures, unusual phyletic patterns, and evolution by horizontal gene transfer. *J. Mol. Biol.* 343: 1–28. [4]

Lemos, B., C. D. Meiklejohn and D. L. Hartl. 2004. Regulatory evolution across the protein interaction network. *Nat. Genet.* 36: 1059–1060. [5]

Lemos, B., B. R. Bettencourt, C. D. Meiklejohn and D. L. Hartl. 2005a. Evolution of proteins and gene expression levels are coupled in *Drosophila* and are independently associated with mRNA abundance, protein length, and number of protein-protein interactions. *Mol. Biol. Evol.* 22: 1345–1354. [2, 5]

Lemos, B., C. D. Meiklejohn, M. Caceres and D. L. Hartl. 2005b. Rates of divergence in gene expression profiles of primates, mice, and flies: Stabilizing selection and variability among functional categories. *Evolution Int. J. Org. Evolution* 59: 126–137. [5]

Lengeler, K. B., D. S. Fox, J. A. Frase, A. Allen, K. Forrester, F. S. Dietrich and J. Heitman. 2002. Mating-type locus of *Cryptococcus neoformans*: a step in the evolution of sex chromosomes. *Eukaryot. Cell.* 1: 704–718. [10]

Leonhardt, H., A. W. Page, H. U. Weier and T. H. Bestor. 1992. A targeting sequence directs DNA methyltransferase to sites of DNA replication in mammalian nuclei. *Cell* 71: 865–873. [8]

Lerat, E., V. Daubin and N. A. Moran. 2003. From gene trees to organismal phylogeny in prokaryotes: The case of the gamma-Proteobacteria. *PLoS Biol.* 1: e19. [4]

Lerat, E., V. Daubin, H. Ochman and N. A. Moran. 2005. Evolutionary origins of genomic repertoires in bacteria. *PLoS Biol.* 3: e130. [4]

Lercher, M. J., A. O. Urrutia and L. D. Hurst. 2003. Evidence that the human X chromosome is enriched for male-specific but not female-specific genes. *Mol. Biol. Evol.* 20: 1113–1116. [9]

Lercher, M. J., J.-V. Chamary and L. D. Hurst. 2004. Genomic regionality in rates of evolution is not explained by clustering of genes of comparable expression profile. *Genome Res.* 14: 1002–1013. [7]

Lesage, G., A. M. Sdicu, P. Menard, J. Shapiro, S. Hussein and H. Bussey. 2004. Analysis of beta-1,3-glucan assembly in *Saccharomyces cerevisiae* using a synthetic interaction network and altered sensitivity to caspofungin. *Genetics* 167: 35–49. [7]

Levanon, E. Y. and 12 others. 2004. Systematic identification of abundant A-to-I editing sites in the human transcriptome. *Nat. Biotechnol.* 22: 1001–1005. [8]

Levitt, M. 2007. Growth of novel protein structural data. *Proc. Natl. Acad. Sci. USA* 104: 3183–3188. [6]

Lewinsky, R. H., T. G. Jensen, J. Moller, A. Stensballe, J. Olsen and J. T. Troelsen. 2005. T-13910 DNA variant associated with lactase persistence interacts with Oct-1 and stimulates lactase promoter activity in vitro. *Hum. Mol. Genet.* 14: 3945–3953. [13]

Li, B., M. Carey and J. L. Workman. 2007. The role of chromatin during transcription. *Cell* 128: 707–719. [5]

Li, E., T. H. Bestor and R. Jaenisch. 1992. Targeted mutation of the DNA methyltransferase gene results in embryonic lethality. *Cell* 69: 915–926. [8]

Li, F., T. Long, Y. Lu, Q. Ouyang and C. Tang. 2004. The yeast cell-cycle network is robustly designed. *Proc. Natl. Acad. Sci. USA* 101: 4781–4786. [7]

Li, S. and 47 others. 2004. A map of the interactome network of the metazoan *C. elegans. Science* 303: 540–543. [6]

Li, W.-H. 1997. *Molecular Evolution.* Sinauer Associates, Sunderland, MA. [12]

Li, W.-H. and D. Graur. 1991. *Fundamentals of Molecular Evolution.* Sinauer Associates, Sunderland, MA. [11]

Li, W.-H. and M. Tanimura. 1987. The molecular clock runs more slowly in man than in apes and monkeys. *Nature* 326: 93–96. [11]

Li, W.-H., M. Tanimura and P. M. Sharp. 1987. An evaluation of the molecular clock hypothesis using mammalian DNA sequences. *J. Mol. Evol.* 25: 330–342. [13]

Li, W.-H., S. Yi and K. Makova. 2002. Male-driven evolution. *Curr. Opin. Genet. Dev.* 12: 650–656. [13]

Li, W.-H., J. Yang and X. Gu. 2005. Expression divergence between duplicate genes. *Trends Genet.* 21: 602–607. [13]

Li, X., J. Liang, H. Yu, B. Su, C. Xiao, Y. Shang and W. Wang. 2007. Functional consequences of new exon acquisition in mammalian chromodomain Y-like (CDYL) genes. *Trends Genet.* 23:427–431. [3]

Liang, H. and W.-H. Li. 2007. Gene essentiality, gene duplicability and protein connectivity in human and mouse. *Trends Genet.* 23:375–378. [7]

Liao, B. Y. and J. Zhang. 2006. Low rates of expression profile divergence in highly expressed genes and tissue-specific genes during mammalian evolution. *Mol. Biol. Evol.* 23: 1119–1128. [5]

Liao, B. Y. and J. Zhang. 2007 Mouse duplicate genes are as essential as singletons. *Trends Genet.* 23:378–381. [7]

Liao, B. Y., N. M. Scott and J. Zhang. 2006. Impacts of gene essentiality, expression pattern, and gene compactness on the evolutionary rate of mammalian proteins. *Mol. Biol. Evol.* 23: 2072–2080. [7]

Liberati, N. T. and 8 others. 2006. An ordered, nonredundant library of *Pseudomonas aeruginosa* strain PA14 transposon insertion mutants. *Proc. Natl. Acad. Sci. USA* 103: 2833–2838. [7]

Lifschytz, E. and D. L. Lindsley. 1972. The role of X-chromosome inactivation during spermatogenesis. *Proc. Natl. Acad. Sci. USA* 69: 182–186. [9]

Lindell, D., M. B. Sullivan, Z. I. Johnson, A. C. Tolonen, F. Rohwer and S. W. Chisholm. 2004. Transfer of photosynthesis genes to and from *Prochlorococcus* viruses. *Proc. Natl. Acad. Sci. USA* 101: 11013–11018. [4]

Lindroth, A. M., X. Cao, J. P. Jackson, D. Zilberman, C. M. McCallum, S. Henikoff and S. E. Jacobsen. 2001. Requirement of CHROMOMETHYLASE3 for maintenance of CpXpG methylation. *Science* 292: 2077–2080. [8]

Lippman, Z. and 13 others. 2004. Role of transposable elements in heterochromatin and epigenetic control. *Nature* 430: 471–476. [5]

Lippman, Z., A. V. Gendrel, V. Colot and R. Martienssen. 2005. Profiling DNA methylation patterns using genomic tiling microarrays. *Nat. Methods* 2: 219–224. [5]

Liu, L. P., J. Q. Ni, Y. D. Shi, E. J. Oakeley and F. L. Sun. 2005. Sex-specific role of *Drosophila melanogaster* HP1 in regulating chromatin structure and gene transcription. *Nat. Genet.* 37: 1361–1366. [5]

Liu, Z. and 12 others. 2004. A primitive Y chromosome in Papaya marks incipient sex chromosome evolution. *Nature* 427: 348–352. [10]

Lloyd, D. G. 1974. Theoretical sex ratios of dioecious and gynodioecious angiosperms. *Heredity* 32: 11–34. [10]

Lloyd, D. G. 1975. The maintenance of gynodioecy and androdioecy in angiosperms. *Genetica* 45: 325–339. [10]

Lloyd, D. G. and C. J. Webb. 1986. The avoidance of interference between the presentation of pollen and stigmas in angiosperms. I. Dichogamy. *New Zeal. J. Bot.* 24: 135–162. [10]

Loftus, B. and 53 others. 2005. The genome of the protist parasite *Entamoeba histolytica*. *Nature* 433: 865–868. [4]

Loisel, D. A., M. V. Rockman, G. A. Wray, J. Altmann and S. C. Alberts. 2006. Ancient polymorphism and functional variation in the primate MHC-DQA1 5′ cis-regulatory region. *Proc. Natl. Acad. Sci. USA* 103: 16331–16336. [5]

Long, M. and C. H. Langley. 1993. Natural selection and the origin of *jingwei*, a chimeric processed functional gene in *Drosophila*. *Science* 260: 91–95. [3]

Long, M., E. Betran, K. Thornton and W. Wang. 2003. The origin of new genes: glimpses from the young and old. *Nat. Rev. Genet.* 4: 865–875. [3]

Loppin, B., D. Lepetit, S. Dorus, P. Couble and T. L. Karr. 2005. Origin and neofunctionalization of a *Drosophila* paternal effect gene essential for zygote viability. *Curr. Biol.* 15: 87–93. [3, 7]

Louie, E., J. Ott and J. Majewski. 2003. Nucleotide frequency variation across human genes. *Genome Res.* 13: 2594–2601. [11]

Lu, J. and C. I. Wu. 2005. Weak selection revealed by the whole-genome comparison of the X chromosome and autosomes of human and chimpanzee. *Proc. Natl. Acad. Sci. USA* 102: 4063–4067. [11]

Lunter, G., C. P. Ponting and J. Hein. 2006. Genome-wide identification of human functional DNA using a neutral indel model. *PLoS Comput. Biol.* 2: e5. [13]

Luscombe, N., J. Qian, Z. Zhang, T. Johnson and M. Gerstein. 2002. The dominance of the population by a selected few: power-law behaviour applies to a wide variety of genomic properties. *Genome Biol.* 3: R40. [2]

Lyko, F., B. H. Ramsahoye and R. Jaenisch. 2000. DNA methylation in *Drosophila melanogaster*. *Nature* 408: 538–540. [5]

Lynch, M. 1988. The rate of polygenic mutation. *Genet. Res.* 51: 137–148. [5]

Lynch, M. and J. S. Conery. 2000. The evolutionary fate and consequences of duplicate genes. *Science* 290: 1151–1155. [12, 13]

Lynch, M. and J. S. Conery. 2003. The origins of genome complexity. *Science* 302: 1401–1404. [1]

Ma, H. and 8 others. 2004. High-density linkage mapping revealed suppression of recombination at the sex determination locus in papaya. *Genetics* 166: 419–436. [10]

Maas, S. and A. Rich. 2000. Changing genetic information through RNA editing. *Bioessays* 22: 790–802. [5]

MacLeod, D., R. L. Charlebois, F. Doolittle and E. Bapteste. 2005. Deduction of probable events of lateral gene transfer through comparison of phylogenetic trees by recursive consolidation and rearrangement. *BMC Evol. Biol.* 5: 27. [4]

Mahadevan, R. and B. O. Palsson. 2005. Properties of metabolic networks: Structure versus function. *Biophys. J.* 88: L07–09. [7]

Mahner, M. and M. Kary. 1997. What exactly are genomes, genotypes and phenotypes? And what about phenomes? *J. Theor. Biol.* 186: 55–63. [2]

Makova, K. D., M. Ramsay, T. Jenkins and W.-H. Li. 2001. Human DNA sequence variation in a 6.6-kb region containing the melanocortin 1 receptor promoter. *Genetics* 158: 1253–1268. [13]

Malone, J. H., D. L. Hawkins and P. Michalak. 2006. Sex-biased gene expression in a ZW sex determination system. *J. Mol. Evol.* 63: 427–436. [9]

Mangan, S. and U. Alon. 2003. Structure and function of the feed-forward loop network motif. *Proc. Natl. Acad. Sci. USA* 100: 11980–11985. [12]

Mangan, S., A. Zaslaver and U. Alon. 2003. The coherent feedforward loop serves as a sign-sensitive delay element in transcription networks. *J. Mol. Biol.* 334: 197–204. [12]

Mangeat, B., P. Turelli, G. Caron, M. Friedli, L. Perrin and D. Trono 2003. Broad antiretroviral defence by human APOBEC3G through lethal editing of nascent reverse transcripts. *Nature* 424: 99–103. [8]

Mank, J. E., D. E. L. Promislow and J. C. Avise. 2006. Evolution of alternative sex-determining mechanisms in teleost fishes *Biol. J. Linn. Soc.* 87: 83–93. [10]

Margulies, E. H. and 10 others. 2005. An initial strategy for the systematic identification of functional elements in the human genome by low-redundancy comparative sequencing. *Proc. Natl. Acad. Sci. USA* 102: 4795–4800. [13]

Mariani, R. and 8 others. 2003. Species-specific exclusion of APOBEC3G from HIV-1 virions by Vif. *Cell* 114: 21–31. [8]

Mariotti, B., R. Navajas-Perez, R. Lozano, J. S. Parker, R. d. l. Herran, C. R. Rejon, R. M. R, M. Garrido-Ramos and M. Jamilena. 2006. Cloning and characterization of dispersed repetitive DNA derived from microdissected sex chromosomes of *Rumex acetosa*. *Genome* 49: 114–121. [10]

Marques, A. C., I. Dupanloup, N. Vinckenbosch, A. Reymond and H. Kaessmann. 2005. Emergence of young human genes after a burst of retroposition in primates. *PLoS Biol.* 3: E357. [3, 13]

Martens, J. A., L. Laprade and F. Winston. 2004. Intergenic transcription is required to repress the *Saccharomyces cerevisiae* SER3 gene. *Nature* 429: 571–574. [7]

Martin, A. P. and S. R. Palumbi. 1993. Body size, metabolic rate, generation time, and molecular clock. *Proc. Natl. Acad. Sci. USA* 90: 4087–91. [13]

Martin, W. and E. V. Koonin. 2006. Introns and the origin of nucleus-cytosol compartmentalization. *Nature* 440: 41–45. [6]

Martin, W. and 9 others. 2002. Evolutionary analysis of Arabidopsis, cyanobacterial, and chloroplast genomes reveals plastid phylogeny and thousands of cyanobacterial genes in the nucleus. *Proc. Natl. Acad. Sci. USA* 99: 12246–12251. [4]

Martone, R. and 11 others. 2003. Distribution of NF-kappaB-binding sites across human chromosome 22. *Proc. Natl. Acad. Sci. USA* 100: 12247–12252. [5]

Matassi, G., P. M. Sharp and C. Gautier. 1999. Chromosomal location effects on gene sequence evolution in mammals. *Curr. Biol.* 9: 786–791. [11]

Matsubara, K., H. Tarui, M. Toriba, K. Yamada, C. Nishida-Umehara, K. Agata and Y. Matsuda. 2006. Evidence for different origin of sex chromosomes in snakes, birds, and mammals and stepwise differentiation of snake sex chromosomes. *Proc. Natl. Acad. Sci. U S A* 103: 18190–18195. [10]

Matsunaga, S., E. Isono, E. Kejnovsky, B. Vyskot, S. Kawano and D. Charlesworth. 2003. Duplicative transfer of a MADS box gene to a plant Y chromosome. *Mol. Biol. Evol.* 20: 1062–1069. [10]

Matthysse, A. G., K. Deschet, M. Williams, M. Marry, A. R. White and W. C. Smith. 2004. A functional cellulose synthase from ascidian epidermis. *Proc. Natl. Acad. Sci. USA* 101: 986–991. [4]

Mattick, J. S. 2007. A new paradigm for developmental biology. *J. Exp. Biol.* 210: 1526–1547. [1]

Maynard Smith, J. and J. Haigh. 1974. The hitch-hiking effect of a favourable gene. *Genet. Res.* 23: 23–35. [13]

McCarroll, S. A. and 10 others. 2006. Common deletion polymorphisms in the human genome. *Nat. Genet.* 38: 86–92. [13]

McClintock, B. 1946. Maize Genetics. *Carnegie Inst. Washington Year Book* 45: 176–186. [8]

McDaniel, S. F., J. H. Willis and A. J. Shaw. 2007. A linkage map reveals a complex basis for segregation distortion in an interpopulation cross in the moss *Ceratodon purpureus*. *Genetics* 176: 2489–2500. [10]

McDonald, J. F., M. A. Matzke and A. J. Matzke. 2005. Host defenses to transposable elements and the evolution of genomic imprinting. *Cytogenet. Genome Res.* 110: 242–249. [5]

McDonald, J. H. and M. Kreitman. 1991. Adaptive protein evolution at the *Adh* locus in *Drosophila*. *Nature* 351: 652–654. [3, 13]

McGregor, A. P., V. Orgogozo, I. Delon, J. Zanet, D. G. Srinivasan, I. Payne and D. L. Stern. 2007. Morphological evolution through multiple *cis*-regulatory mutations at a single gene. *Nature* 448: 587–591. [1]

McQueen, H. A., D. McBride, G. Miele, A. P. Bird and M. Clinton. 2001. Dosage compensation in birds. *Current Biology* 11: 253–257. [10]

McVean, G. A., S. R. Myers, S. Hunt, P. Deloukas, D. R. Bentley and P. Donnelly. 2004. The fine-scale structure of recombination rate variation in the human genome. *Science* 304: 581–584. [13]

Medina, M. 2005. Genomes, phylogeny, and evolutionary systems biology. *Proc. Natl. Acad. Sci. USA* 102 Suppl. 1: 6630–6635. [2]

Meibom, K. L., M. Blokesch, N. A. Dolganov, C. Y. Wu and G. K. Schoolnik. 2005. Chitin induces natural competence in *Vibrio cholerae*. *Science* 310: 1824–1827. [4]

Meiklejohn, C. D., J. Parsch, J. M. Ranz and D. L. Hartl. 2003. Rapid evolution of male-biased gene expression in Drosophila. *Proc. Natl. Acad. Sci. USA* 100: 9894–9899. [5]

Mekel-Bobrov, N., S. L. Gilbert, P. D. Evans, E. J. Vallender, J. R. Anderson, R. R. Hudson, S. A. Tishkoff and B. T. Lahn. 2005. Ongoing adaptive evolution of *ASPM*, a brain size determinant in *Homo sapiens*. *Science* 309: 1720–1722. [13]

Merino, E. and C. Yanofsky. 2005. Transcription attenuation: A highly conserved regulatory strategy used by bacteria. *Trends Genet.* 21: 260–264. [5]

Mette, M. F., W. Aufsatz, J. van der Winden, M. A. Matzke and A. J. Matzke. 2000. Transcriptional silencing and promoter methylation triggered by double-stranded RNA. *EMBO J.* 19: 5194–5201. [8]

Meunier, J. and L. Duret. 2004. Recombination drives the evolution of GC-content in the human genome. *Mol. Biol. Evol.* 21: 984–990. [13]

Meunier, J., A. Khelifi, V. Navratil and L. Duret. 2005. Homology-dependent methylation in primate repetitive DNA. *Proc. Natl. Acad. Sci. USA* 102: 5471–5476. [5]

Meyer, I. M. and I. Miklos. 2005. Statistical evidence for conserved, local secondary structure in the coding regions of eukaryotic mRNAs and pre-mRNAs. *Nucleic Acids Res.* 33: 6338–6348. [11]

Mi, S. and 11 others. 2000. Syncytin is a captive retroviral envelope protein involved in human placental morphogenesis. *Nature* 403: 785–789. [13]

Mikkelsen, T. S. and 66 others. 2005. The Chimpanzee Sequencing and Analysis Consortium. Initial sequence of the chimpanzee genome and comparison with the human genome. *Nature* 437: 69–87. [11, 13]

Mikkelsen, T. and 63 others. 2007. Genome of the marsupial *Monodelphis domestica* reveals innovation in non-coding sequences. *Nature* 447: 167–177. [10]

Milo, R., S. Shen-Orr, S. Itzkovitz, N. Kashtan, D. Chklovskii and U. Alon. 2002. Network motifs: Simple building blocks of complex networks. *Science* 298: 824–827. [12]

Mirkin, B. G., T. I. Fenner, M. Y. Galperin and E. V. Koonin. 2003. Algorithms for computing parsimonious evolutionary scenarios for genome evolution, the last universal common ancestor and dominance of horizontal gene transfer in the evolution of prokaryotes. *BMC Evol Biol.* 3: 2. [4]

Mita, K. and 11 others. 2003. The construction of an EST database for Bombyx mori and its application. *Proc. Natl. Acad. Sci. USA* 100: 14121–14126. [4]

Miura, I., H. Ohtani, M. Nakamura, Y. Ichikawa and K. Saitoh. 1998. The origin and differentiation of the heteromorphic sex chromosomes Z, W, X, and Y in the frog *Rana rugosa*, inferred from the sequences of a sex-linked gene, ADP/ATP translocase. *Mol. Biol. Evol.* 15: 1612–1619. [10]

Miyata, T. and T. Yasunaga. 1980. Molecular evolution of mRNA: a method for estimating evolutionary rates of synonymous and amino acid substitutions from homologous nucleotide sequences and its application. *J. Mol. Evol.* 16: 23–36. [11]

Mochizuki, K. and M. A. Gorovsky. 2004. Small RNAs in genome rearrangement in Tetrahymena. *Curr. Opin. Genet. Dev.* 14: 181–187. [8]

Mochizuki, K., N. A. Fine, T. Fujisawa and M. A. Gorovsky. 2002. Analysis of a piwi-related gene implicates small RNAs in genome rearrangement in tetrahymena. *Cell* 110: 689–699. [8]

Mockler, T. C., S. Chan, A. Sundaresan, H. Chen, S. E. Jacobsen and J. R. Ecker. 2005. Applications of DNA tiling arrays for whole-genome analysis. *Genomics* 85: 1–15. [5]

Mohd-Zain, Z. and 10 others. 2004. Transferable antibiotic resistance elements in *Haemophilus influenzae* share a common evolutionary origin with a diverse family of syntenic genomic islands. *J. Bacteriol.* 186: 8114–8122. [4]

Molin, S. and T. Tolker-Nielsen. 2003. Gene transfer occurs with enhanced efficiency in biofilms and induces enhanced stabilisation of the biofilm structure. *Curr. Opin. Biotechnol.* 14: 255–261. [4]

Mongodin, E. F. and 18 others. 2005a The genome of *Salinibacter ruber*: Convergence and gene exchange among hyperhalophilic bacteria and archaea. *Proc. Natl. Acad. Sci. USA* 102: 18147–18152. [4]

Mongodin, E. F., I. R. Hance, R. T. Deboy, S. R. Gill, S. Daugherty, R. Huber, C. M. Fraser, K. Stetter and K. E. Nelson. 2005b. Gene transfer and genome plasticity in *Thermotoga maritima*, a model hyperthermophilic species. *J. Bacteriol.* 187: 4935–4944. [4]

Montell, H., A.-K. Fridolfsson and H. Ellegren. 2001. Contrasting levels of nucleotide diversity on the avian Z and W sex chromosomes. *Mol. Biol. Evol.* 18: 2010–2016. [10]

Montgomery, S. B. and 8 others. 2006. ORegAnno: An open access database and curation system for literature-derived promoters, transcription factor binding sites and regulatory variation. *Bioinformatics* 22: 637–640. [5]

Moore, R. C. and M. D. Purugganan. 2003. The early stages of duplicate gene evolution. *Proc. Natl. Acad. Sci. USA* 100: 15682–15687. [3]

Morley, M., C. M. Molony, T. M. Weber, J. L. Devlin, K. G. Ewens, R. G. Spielman and V. C. Cheung. 2004. Genetic analysis of genome-wide variation in human gene expression. *Nature* 430: 743–747. [5]

Moses, A. M., D. Y. Chiang, M. Kellis, E. S. Lander and M. B. Eisen. 2003. Position specific variation in the rate of evolution in transcription factor binding sites. *BMC Evol. Biol.* 3: 19. [5]

Motamedi, M. R., A. Verdel, S. U. Colmenares, S. A. Gerber, S. P. Gygi and D. Moazed. 2004. Two RNAi complexes, RITS and RDRC, physically interact and localize to noncoding centromeric RNAs. *Cell* 119: 789–802. [8]

Mouse Genome Sequencing Consortium. 2002. Initial sequencing and comparative analysis of the mouse genome. *Nature* 420: 520–562. [10]

Mueller, M., L. Martens and R. Apweiler. 2007. Annotating the human proteome: Beyond establishing a parts list. *Biochim. Biphys. Acta* 1774: 175–191. [2]

Muhlrad, D. and R. Parker. 1994. Premature translational termination triggers mRNA decapping. *Nature* 370: 578–581. [5]

Mukherjee, S., M. F. Berger, G. Jona, X. S. Wang, D. Muzzey, M. Snyder, R. A. Young and M. L. Bulyk. 2004. Rapid analysis of the DNA-binding specificities of transcription factors with DNA microarrays. *Nat. Genet.* 36: 1331–1339. [5]

Muller, F., M. A. Demeny and L. Tora. 2007. New problems in RNA polymerase II transcription initiation: Matching the diversity of core promoters with a variety of promoter recognition factors. *J. Biol. Chem.* 282: 14685–14689. [5]

Muller, H. J. 1914. A gene for the fourth chromosome of Drosophila. *J. Exp. Zool.* 17: 325–336. [10]

Murphy, W. J., S. Sun, Z.-Q. Chen, J. Pecon-Slattery and S. J. O'Brien. 1999. Extensive conservation of sex chromosome organization between cat and human revealed by parallel radiation hybrid mapping. *Genome Res.* 9: 1223–1230. [10]

Mutch, D. M., L. Fauconnot, M. Grigorov and L. B. Fay. 2006. Putting the 'Ome' in lipid metabolism. *Biotechnol. Annu. Rev.* 12: 67–84. [2]

Myers, S., L. Bottolo, C. Freeman, G. McVean and P. Donnelly. 2005. A fine-scale map of recombination rates and hotspots across the human genome. *Science* 310: 321–324. [13]

Nachman, M. W. and S. L. Crowell. 2000a. Contrasting evolutionary histories of two introns of the duchenne muscular dystrophy gene, Dmd, in humans. *Genetics* 155: 1855–1864. [13]

Nachman, M. W. and S. L. Crowell. 2000b. Estimate of the mutation rate per nucleotide in humans. *Genetics* 156: 297–304. [11]

Nagai, H., J. C. Kagan, X. Zhu, R. A. Kahn and C. R. Roy. 2002. A bacterial guanine nucleotide exchange factor activates ARF on Legionella phagosomes. *Science* 295: 679–682. [4]

Nagylaki, T. 1983. Evolution of a finite population under gene conversion. *Proc. Natl. Acad. Sci. USA* 80: 6278–6281. [13]

Nakamura, Y., T. Itoh, H. Matsuda and T. Gojobori. 2004. Biased biological functions of horizontally transferred genes in prokaryotic genomes. *Nat. Genet.* 36: 760–766. [4]

Nakashima, K., L. Yamada, Y. Satou, J. Azuma and N. Satoh. 2004. The evolutionary origin of animal cellulose synthase. *Dev. Genes Evol.* 214: 81–88. [4]

Napoli, C., C. Lemieux and R. Jorgensen. 1990. Introduction of a chimeric chalcone synthase gene into petunia results in reversible co-suppression of homologous genes in trans. *Plant Cell* 2: 279–289. [8]

Nardone, J., D. U. Lee, K. M. Ansel and A. Rao. 2004. Bioinformatics for the 'bench biologist': How to find regulatory regions in genomic DNA. *Nat. Immunol.* 5: 768–774. [5]

Nathans, J., D. Thomas and D. S. Hogness. 1986. Molecular genetics of human color vision: the genes encoding blue, green, and red pigments. *Science* 232: 193–202. [13]

Nei, M. 1969. Linkage modification and sex difference in recombination. *Genetics* 63: 681–699. [10]

Nei, M. 1970. Accumulation of nonfunctional genes on sheltered chromosomes. *Amer. Nat.* 104: 311–322. [10]

Nei, M. 2005. Selectionism and neutralism in molecular evolution. *Mol. Biol. Evol.* 22: 2318–2342. [11]

Nei, M. and S. Kumar. 2000. *Molecular Evolution and Phylogenetics.* Oxford University Press, Oxford, UK. [11]

Nei, M., Zhang, J. and S. Yokoyama. 1997. Color vision of ancestral organisms of higher primates. *Mol. Biol. Evol.* 14: 611–618. [13]

Nekrutenko, A. and W.-H. Li. 2001. Transposable elements are found in a large number of human protein-coding genes. *Trends Genet.* 17: 619–621. [13]

Nekrutenko, A., W. Y. Chung and W.-H. Li. 2003. An evolutionary approach reveals a high protein-coding capacity of the human genome. *Trends Genet.* 19: 306–310. [11]

Nelson, K. E. and 29 others. 1999. Evidence for lateral gene transfer between Archaea and bacteria from genome sequence of *Thermotoga maritima*. *Nature* 399: 323–329. [4]

Nesbø, C. L., K. E. Nelson and W. F. Doolittle. 2002. Suppressive subtractive hybridization detects extensive genomic diversity in *Thermotoga maritima*. *J. Bacteriol.* 184: 4475–4488. [4]

Nguyen, D. K. and C. M. Disteche. 2006. Dosage compensation of the active X chromosome in mammals. *Nat. Genet.* 38: 47–53. [9, 10]

Nicolas, M. and 9 others. 2005. A gradual process of recombination restriction in the evolutionary history of the sex chromosomes in dioecious plants. *PLoS Biology* 3: 47–56. [10]

Nielsen, R. 2001. Statistical tests of selective neutrality in the age of genomics. *Heredity* 86: 641–647. [13]

Nielsen, R. 2004. Population genetic analysis of ascertained SNP data. *Hum. Genomics* 1: 218–224. [13]

Nielsen, R. 2005. Molecular signatures of natural selection. *Annu. Rev. Genet.* 39: 197–218. [13]

Nielsen, R. and J. Signorovitch. 2003. Correcting for ascertainment biases when analyzing SNP data: applications to the estimation of linkage disequilibrium. *Theor. Popul. Biol.* 63: 245–255. [13]

Nielsen, R. and Z. Yang. 1998. Likelihood models for detecting positively selected amino acid sites and applications to the HIV-1 envelope gene. *Genetics* 148: 929–936. [11]

Nielsen, R., M. J. Hubisz and A. G. Clark. 2004. Reconstituting the frequency spectrum of ascertained single-nucleotide polymorphism data. *Genetics* 168: 2373–2382. [13]

Nielsen, R. and 12 others. 2005a. A scan for positively selected genes in the genomes of humans and chimpanzees. *PLoS Biol.* 3: e170. [5, 11, 13]

Nielsen, R., S. Williamson, Y. Kim, M. J. Hubisz, A. G. Clark and C. Bustamante. 2005b. Genomic scans for selective sweeps using SNP data. *Genome Res.* 15: 1566–1575. [13]

Nisole, S., C. Lynch, J. P. Stoye and M. W. Yap. 2004. A Trim5-cyclophilin A fusion protein found in owl monkey kidney cells can restrict HIV-1. *Proc. Natl. Acad. Sci. USA* 101: 13324–13328. [3]

Nixon, J. E., A. Wang, J. Field, H. G. Morrison, A. G. McArthur, M. L. Sogin, B. J. Loftus and J. Samuelson. 2002. Evidence for lateral transfer of genes encoding ferredoxins, nitroreductases, NADH oxidase, and alcohol dehydrogenase 3 from anaerobic prokaryotes to Giardia lamblia and Entamoeba histolytica. *Eukaryot. Cell* 1: 181–190. [4]

Nixon, J. E., J. Field, A. G. McArthur, M. L. Sogin, N. Yarlett, B. J. Loftus and J. Samuelson. 2003. Iron-dependent hydrogenases of Entamoeba histolytica and Giardia lamblia: Activity of the recombinant entamoebic enzyme and evidence for lateral gene transfer. *Biol. Bull.* 204: 1–9. [4]

Nobrega, M. A., I. Ovcharenko, V. Afzal and E. M. Rubin. 2003. Scanning human gene deserts for long-range enhancers. *Science* 302: 413. [13]

Nobrega, M. A., Y. Zhu, I. Plajzer-Frick, V. Afzal and E. M. Rubin. 2004. Megabase deletions of gene deserts result in viable mice. *Nature* 431: 988–993. [13]

Noma, K., T. Sugiyama, H. Cam, A. Verdel, M. Zofall, S. Jia, D. Moazed and S. I. Grewal. 2004. RITS acts in cis to promote RNA interference-mediated transcriptional and post-transcriptional silencing. *Nat. Genet.* 36: 1174–1180. [8]

Nowak, M. A., M. C. Boerlijst, J. Cooke and J. M. Smith. 1997. Evolution of genetic redundancy. *Nature* 388: 167–171. [7]

Nozawa, M., T. Aotsuka and K. Tamura, K. 2005. A novel chimeric gene, siren, with retroposed promoter sequence in the *Drosophila bipectinata* complex. *Genetics* 171: 1719–1727. [3]

Nurminsky, D. I., M. V. Nurminskaya, D. De Aguiar and D. L. Hartl. 1998. Selective sweep of a newly evolved sperm-specific gene in Drosophila. *Nature* 396: 572–575. [3]

Nuzhdin, S. V., M. L. Wayne, K. L. Harmon and L. M. McIntyre. 2004. Common pattern of evolution of gene expression level and protein sequence in Drosophila. *Mol. Biol. Evol.* 21: 1308–1317. [5]

Ny, T., F. Elgh and B. Lund. 1984. The structure of human tissue-type plasminogen activator gene: correlation of intron and exon structures to functional and structural domains. *Proc. Natl. Acad. Sci. USA* 81: 5355–5359. [6]

Oakley, T. H., Z. Gu, E. Abouheif, N. H. Patel and W.-H. Li. 2005. Comparative methods for the analysis of gene-expression evolution: An example using yeast functional genomic data. *Mol. Biol. Evol.* 22: 40–50. [5]

Ochman, H., S. Elwyn and N. A. Moran. 1999. Calibrating bacterial evolution. *Proc. Natl. Acad. Sci. USA* 96: 12638–12643. [11]

Ochman, H., J. G. Lawrence and E. A. Groisman. 2000. Lateral gene transfer and the nature of bacterial innovation. *Nature* 405: 299–304. [4]

Oeffner, F. and 10 others. 2000. Significant association between a silent polymorphism in the neuromedin B gene and body weight in German children and adolescents. *Acta Diabetol.* 37: 93–101. [11]

Ohno, S. 1970. *Evolution by Gene Duplication.* Springer, Berlin. [3]

Ohno, S. 1972. *Evolution by Gene Duplication.* Springer, New York. [6]

Okano, M., D. W. Bell, D. A. Haber and E. Li. 1999. DNA methyltransferases Dnmt3a and Dnmt3b are essential for de novo methylation and mammalian development. *Cell* 99: 247–257. [8]

Oleksiak, M. F., G. A. Churchill and D. L. Crawford. 2002. Variation in gene expression within and among natural populations. *Nat. Gen.* 32: 261–266. [5]

Oliver, B. and M. Parisi. 2004. Battle of the Xs. *Bioessays* 26: 543–548. [9]

Omelchenko, M. V., K. S. Makarova, Y. I. Wolf, I. B. Rogozin and E. V. Koonin. 2003. Evolution of mosaic operons by horizontal gene transfer and gene displacement *in situ*. *Genome Biol.* 4: R55.

Omelchenko, M. V., Y. I. Wolf, E. K. Gaidamakova, V. Y. Matrosova, A. Vasilenko, M. Zhai, M. J. Daly, E. V. Koonin and K. S. Makarova. 2005. Comparative genomics of *Thermus thermophilus* and *Deinococcus radiodurans*: Divergent routes of adaptation to thermophily and radiation resistance. *BMC Evol. Biol.* 5: 57. [4]

Ono, K., H. Suga, N. Iwabe, K. Kuma and T. Miyata. 1999. Multiple protein tyrosine phosphatases in sponges and explosive gene duplication in the early evolution of animals before the parazoan-eumetazoan split. *J. Mol. Evol.* 48: 654–662. [6]

Ophir, R. and D. Graur. 1997. Patterns and rates of indel evolution in processed pseudogenes from humans and murids. *Gene* 205: 191–202. [7]

Orgel, L. E. and F. H. Crick. 1980. Selfish DNA: The ultimate parasite. *Nature* 284: 604–607. [8]

Orian, A. and 13 others. 2003. Genomic binding by the Drosophila Myc, Max, Mad/Mnt transcription factor network. *Genes Dev.* 179: 1101–1114. [5]

Orr, H. A. 1991. A test of Fisher's theory of dominance. *Proc. Natl. Acad. Sci. USA* 88: 11413–11415. [7]

Orr, H. A. and Y. Kim. 1998. An adaptive hypothesis for the evolution of the Y chromosome. *Genetics* 150: 1693–1698. [10]

Osier, M. V. and 13 others. 2002. A global perspective on genetic variation at the *ADH* genes reveals unusual patterns of linkage disequilibrium and diversity. *Am. J. Hum. Genet.* 71: 84–99. [13]

Otto, S. P. and J. Whitton. 2000. Polyploid incidence and evolution. *Annu. Rev. Genet.* 34: 401–437. [3]

Ou, H. Y. and 10 others. 2007. MobilomeFINDER: web-based tools for in silico and experimental discovery of bacterial genomic islands. *Nucleic Acids Res.* 35: W97–W104. [2]

Pagel, M. 2006. Editorial Foreword. Evolutionary Bioinformatics 2: 56–57. [1]

Pagel, M. and R. A. Johnstone. 1992. Variation across species in the size of the nuclear genome supports the junk-DNA explanation for the C-value paradox. *Proc. Biol. Sci.* 249: 119–124. [8]

Pagel, M., A. Meade and D. Scott. 2007. Assembly rules for protein interaction networks. *BMC Evol. Biol.* 7 (Suppl. 1): S16. [1, 12]

Pál, C. and L. D. Hurst. 2000. The evolution of gene number: are heritable and non-heritable errors equally important? *Heredity* 84: 393–400. [7]

Pál, C., B. Papp and L. D. Hurst. 2001. Highly expressed genes in yeast evolve slowly. *Genetics* 158: 927–931. [2]

Pál, C., B. Papp and L. D. Hurst. 2003. Genomic function: Rate of evolution and gene dispensability. *Nature* 421: 496–497; discussion, 497–498. [2, 7]

Pál, C., B. Papp and M. J. Lercher. 2005. Adaptive evolution of bacterial metabolic networks by horizontal gene transfer. *Nat. Genet.* 37: 1372–1375. [4]

Pál, C., B. Papp and M. J. Lercher. 2006a. An integrated view of protein evolution. *Nat. Rev. Genet.* 7: 337–348. [7]

Pál, C., B. Papp, M. J. Lercher, P. Csermely, S. G. Oliver and L. D. Hurst. 2006b. Chance and necessity in the evolution of minimal metabolic networks. *Nature* 440: 667–670. [7]

Palauqui, J. C. and S. Balzergue. 1999. Activation of systemic acquired silencing by localised introduction of DNA. *Curr. Biol.* 9: 59–66. [8]

Palauqui, J. C., T. Elmayan, J. M. Pollien and H. Vaucheret. 1997. Systemic acquired silencing: Transgene-specific post-transcriptional silencing is transmitted by grafting from silenced stocks to non-silenced scions. *EMBO J.* 16: 4738–4745. [8]

Pal-Bhadra, M., U. Bhadra and J. A. Birchler. 1997. Cosuppression in Drosophila: Gene silencing of Alcohol dehydrogenase by white-Adh transgenes is Polycomb dependent. *Cell* 90: 479–490. [8]

Pal-Bhadra, M., U. Bhadra and J. A. Birchler. 1999. Cosuppression of nonhomologous transgenes in Drosophila involves mutually related endogenous sequences. *Cell* 99: 35–46. [8]

Pal-Bhadra, M., U. Bhadra and J. A. Birchler. 2002. RNAi related mechanisms affect both transcriptional and posttranscriptional transgene silencing in Drosophila. *Mol. Cell* 9: 315–327. [8]

Papp, B., C. Pál and L. D. Hurst. 2003a. Dosage sensitivity and the evolution of gene families in yeast. *Nature* 424: 194–197. [7]

Papp, B., C. Pál and L. D. Hurst. 2003b. Evolution of *cis*-regulatory elements in duplicated genes of yeast. *Trends Genet.* 19: 417–422. [7, 12]

Papp, B., C. Pál and L. D. Hurst. 2004. Metabolic network analysis of the causes and evolution of enzyme dispensability in yeast. *Nature* 429: 661–664. [7]

Parish, D., P. Vise, H. Wichman, J. Bull and R. Baker. 2002. Distribution of LINEs and other repetitive elements in the karyotype of the bat *Carollia*: implications for X-chromosome inactivation. *Cyto. Genome Res.* 96: 191–197. [10]

Parisi, M., R. Nuttal, D. Naiman, G. Bouffard, J. Malley, J. Andrews, S. Eastman and B. Oliver. 2003. Paucity of genes on the Drosophila X chromosome showing male-biased expression. *Science* 299: 697–700. [5, 9]

Parisi, M. and 11 others. 2004. A survey of ovary-, testis-, and soma-biased gene expression in *Drosophila melanogaster* adults. *Genome Biol.* 5: R40. [9]

Park, D., S. Lee, D. Bolser, M. Schroeder, M. Lappe, D. Oh and J. Bhak. 2005. Comparative interactomics analysis of protein family interaction networks using PSIMAP (protein structural interactome map). *Bioinformatics* 21: 3234–3240. [6]

Park, J. M., J. F. Manen and G. M. Schneeweiss. 2007. Horizontal gene transfer of a plastid gene in the non-photosynthetic flowering plants *Orobanche* and *Phelipanche* (Orobanchaceae). *Mol. Phylogenet. Evol.* 43: 974–985. [3]

Parker, J. S. 1990. Sex-chromosome and sex differentiation in flowering plants. *Chromosomes Today* 10: 187–198. [10]

Parker, J. S., S. M. Roe and D. Barford, D. 2004. Crystal structure of a PIWI protein suggests mechanisms for siRNA recognition and slicer activity. *EMBO J.* 23: 4727–4737. [8]

Parmley, J. L., J. V. Chamary and L. D. Hurst. 2006. Evidence for purifying selection against synonymous mutations in mammalian exonic splicing enhancers. *Mol. Biol. Evol.* 23: 301–309. [11]

Pastinen, T. and T. J. Hudson. 2004. Cis-acting regulatory variation in the human genome. *Science* 306: 647–650. [5]

Patthy, L. 1985. Evolution of the proteases of blood coagulation and fibrinolysis by assembly from modules. *Cell* 41: 657–663. [6]

Patthy, L. 1987. Intron-dependent evolution: preferred types of exons and introns. *FEBS Lett.* 214: 1–7. [6]

Patthy, L. 1991. Modular exchange principles in proteins. *Curr. Opin. Struct. Biol.* 1: 351–361. [6]

Patthy, L. 1994. Exons and introns. *Curr. Opin. Struct. Biol.* 4: 383–392. [6]

Patthy, L. 1996. Exon shuffling and other ways of module exchange. *Matrix Biology* 15: 301–310. [6]

Patthy, L. 1999. Genome evolution and the evolution of exon-shuffling—a review. *Gene* 238: 103–114. [6]

Patthy, L. 2003. Modular assembly of genes and the evolution of new functions. *Genetica* 118: 217–231. [6]

Pecon-Slattery, J., L. Sanner-Wachter and S. J. O'Brien. 2000. Novel gene conversion between X-Y homologues located in the nonrecombining region of the Y chromosome in Felidae (Mammalia). *Proc. Natl. Acad. Sci. USA* 97: 5307–5312. [10]

Pedulla, M. L. and 19 others. 2003. Origins of highly mosaic mycobacteriophage genomes. *Cell* 113: 171–182. [4]

Peichel, C. L. and 9 others. 2004. The master sex-determination locus in threespine sticklebacks is on a nascent Y chromosome. *Curr. Biol.* 14: 1416–1424. [10]

Pennisi, E. 2005. How will big pictures emerge from a sea of biological data? *Science* 309: 94. [2]

Pereira-Leal, J. B., B. Audit, J. M. Peregrin-Alvarez and C. A. Ouzounis. 2005. An exponential core in the heart of the yeast protein interaction network. *Mol. Biol. Evol.* 22: 421–425. [7]

Pérez Jurado, L. A., R. Peoples, P. Kaplan, B. C. Hamel and U. Francke. 1996. Molecular definition of the chromosome 7 deletion in Williams syndrome and parent-of-origin effects on growth. *Am. J. Hum. Genet.* 59: 781–792. [13]

Perfeito, L., L. Fernandes, C. Mota and I. Gordo. 2007. Adaptive mutations in bacteria: high rate and small effects. *Science* 317: 813–815. [1]

Persico, M., A. Ceol, C. Gavrila, R. Hoffmann, A. Florio and G. Cesareni. 2005. HomoMINT: an inferred human network based on orthology mapping of protein interactions discovered in model organisms. *BMC Bioinformatics* 6 Suppl. 4: S21. [6]

Petsko, G. A. 2001. Homologuephobia. *Genome Biol.* 2: C1002. [2]

Pfeiffer, T., O. S. Soyer and S. Bonhoeffer. 2005. The evolution of connectivity in metabolic networks. *PLoS Biology* 3: e228. [7]

Pickersgill, H., B. Kalverda, E. de Wit, W. Talhout, M. Fornerod and B. van Steensel. 2006. Characterization of the *Drosophila melanogaster* genome at the nuclear lamina. *Nat. Genet.* 38: 1005–1014. [5]

Pinkel, D. and 13 others. 1998. High resolution analysis of DNA copy number variation using comparative genomic hybridization to microarrays. *Nat. Genet.* 20: 207–211. [3]

Prasad, N. G., S. Bedhomme, T. Day and A. K. Chippindale. 2007. An evolutionary cost of separate genders revealed by male-limited evolution. *Amer. Nat.* 169: 29–37. [10]

Price, M. N., K. H. Huang, A. P. Arkin and E. J. Alm. 2005. Operon formation is driven by co-regulation and not by horizontal gene transfer. *Genome Res.* 15: 809–819. [4]

Prince, V. E. and F. B. Pickett. 2002. Splitting pairs: the diverging fates of duplicated genes. *Nat. Rev. Genet.* 3: 827–837. [13]

Proulx, S. R. and P. C. Phillips. 2005. The opportunity for canalization and the evolution of genetic networks. *Am. Nat.* 165: 147–162. [7]

Proulx, S. R., D. E. L. Promislow and P. C. Phillips. 2005. Network thinking in ecology and evolution. *Trends Ecol. Evol.* 20: 345–353. [5]

Provost, P. and 9 others. 2002. Dicer is required for chromosome segregation and gene silencing in fission yeast cells. *Proc. Natl. Acad. Sci. USA* 99: 16648–16653. [8]

Przeworski, M. 2002. The signature of positive selection at randomly chosen loci. *Genetics* 160: 1179–1189. [13]

Ptak, S. and 8 others. 2005. Fine-scale recombination patterns differ between chimpanzees and humans. *Nat. Genet.* 37: 429–434. [13]

Ptak, S. E., A. D. Roeder, M. Stephens, Y. Gilad, S. Paabo and M. Przeworski. 2004. Absence of the TAP2 human recombination hotspot in chimpanzees. *PLoS Biol.* 2: E155. [13]

Ptashne, M. and A. Gann. 2002. *Genes & Signals.* Cold Spring Harbor Laboratory Press, Cold Spring Harbor, New York. [5]

Qian, Q. and P. J. Keeling, P. J. 2001. Diplonemid glyceraldehyde-3-phosphate dehydrogenase (GAPDH) and prokaryote-to-eukaryote lateral gene transfer. *Protist* 152: 193–201. [4]

Quackenbush, J. 2002. Microarray data normalization and transformation. *Nat. Genet.* 32 Suppl: 496–501. [5]

Raes, J. and Y. Van de Peer. 2005. Functional divergence of proteins through frameshift mutations. *Trends Genet.* 21: 428–431. [3]

Ragan, M. A., T. J. Harlow and R. B. Beiko. 2006. Do different surrogate methods detect lateral genetic transfer events of different relative ages? *Trends Microbiol.* 14: 4–8. [4]

Ranz, J. M., C. I. Castillo-Davis, C. D. Meiklejohn and D. L. Hartl. 2003. Sex-dependent gene expression and evolution of the Drosophila transcriptome. *Science* 300: 1742–1745. [5, 9]

Rassoulzadegan, M., V. Grandjean, P. Gounon, S. Vincent, I. Gillot and F. Cuzin. 2006. RNA-mediated non-mendelian inheritance of an epigenetic change in the mouse. *Nature* 441: 469–474. [5]

Razin, A. and A. D. Riggs. 1980. DNA methylation and gene function. *Science* 210: 604–610. [11]

Redon, R. and 42 others. 2006. Global variation in copy number in the human genome. *Nature* 444: 444–454. [1]

Reed, F. A. and S. A. Tishkoff. 2006. Positive selection can create false hotspots of recombination. *Genetics* 172: 2011–2014. [13]

Reich, D. E., S. F. Schaffner, M. J. Daly, G. McVean, J. C. Mullikin, J. M. Higgins, D. J. Richter, E. S. Lander and D. Altshuler. 2002. Human genome sequence variation and the influence of gene history, mutation and recombination. *Nat. Genet.* 32: 135–142. [13]

Reinhart, B. J. and D. P. Bartel. 2002. Small RNAs correspond to centromere heterochromatic repeats. *Science* 297: 1831. [8]

Reinke, V., I. S. Gil, S. Ward and K. Kazmer. 2004. Genome-wide germline-enriched and sex-biased expression profiles in *Caenorhabditis elegans*. *Development* 131: 311–323. [9]

Remold, S. K. and R. E. Lenski. 2004. Pervasive joint influence of epistasis and plasticity on mutational effects in *Escherichia coli*. *Nat. Genet.* 36: 423–426. [7]

Repping, S. and 10 others. 2006. High mutation rates have driven extensive structural polymorphism among human Y chromosomes. *Nature Genet.* 38: 463–467. [10]

Rhounim, L., J. L. Rossignol and G. Faugeron. 1992. Epimutation of repeated genes in *Ascobolus immersus*. *EMBO J.* 11: 4451–4457. [8]

Rice, W. R. 1984. Sex chromosomes and the evolution of sexual dimorphism. *Evolution* 38: 735–742. [9]

Rice, W. R. 1987a. Genetic hitch-hiking and the evolution of reduced genetic activity of the Y sex chromosome. *Genetics* 116: 161–167. [10]

Rice, W. R. 1987b. The accumulation of sexually antagonistic genes as a selective agent promoting the evolution of reduced recombination between primitive sex-chromosomes. *Evolution* 41: 911–914. [10]

Richards, T. A., J. B. Dacks, S. A. Campbell, J. L. Blanchard, P. G. Foster, R. McLeod and C. W. Roberts. 2006. Evolutionary origins of the eukaryotic shikimate pathway: Gene fusions, horizontal gene transfer, and endosymbiotic replacements. *Eukaryot. Cell* 5: 1517–1531. [4]

Richardson, A. O. and J. D. Palmer. 2007. Horizontal gene transfer in plants. *J. Exp. Bot.* 58: 1–9. [3]

Riddle, N. C. and E. J. Richards 2002. The control of natural variation in cytosine methylation in Arabidopsis. *Genetics* 162: 355–363. [5]

Rifkin, S. A., J. Kim and K. P. White. 2003. Evolution of gene expression in the *Drosophila melanogaster* subgroup. *Nat. Genet.* 33: 138–144. [5]

Rifkin, S. A., D. Houle, J. Kim and K. P. White. 2005. A mutation accumulation assay reveals a broad capacity for rapid evolution of gene expression. *Nature* 438: 220–223. [5]

Ringner, M. and M. Krogh. 2005. Folding free energies of 5'-UTRs impact post-transcriptional regulation on a genomic scale in yeast. *PLoS Comput. Biol.* 1: e72. [5]

Rinn, J. L. and M. Snyder. 2005. Sexual dimorphism in mammalian gene expression. *Trends Genet.* 21: 298–305. [9]

Rivas, F. V., N. H. Tolia, J. J. Song, J. P. Aragon, J. Liu, G. J. Hannon and L. Joshua-Tor. 2005. Purified Argonaute2 and an siRNA form recombinant human RISC. *Nat. Struct. Mol. Biol.* 12: 340–349. [8]

Roberts, R. J. 1985. Restriction and modification enzymes and their recognition sequences. *Nucleic Acids Res.* 13 Suppl: r165–200. [8]

Robinson, W. P., J. Waslynka, F. Bernasconi, M. Wang, S. Clark, D. Kotzot and A. Schinzel. 1996. Delineation of 7q11.2 deletions associated with Williams-Beuren syndrome and mapping of a repetitive sequence to within and to either side of the common deletion. *Genomics* 34: 17–23. [13]

Rocha, E. P. C. and A. Danchin. 2004. An analysis of determinants of amino acids substitution rates in bacterial proteins. *Mol. Biol. Evol.* 21: 108–116. [2, 7]

Rockman, M. V. and G. A. Wray. 2002. Abundant raw material for cis-regulatory evolution in humans. *Mol. Biol. Evol.* 19: 1991–2004. [5, 13]

Rockman, M. V., M. W. Hahn, N. Soranzo, D. B. Goldstein and G. A. Wray. 2003. Positive selection on a human-specific transcription factor binding site regulating IL4 expression. *Curr. Biol.* 13: 2118–2123. [13]

Rockman, M. V., M. W. Hahn, N. Soranzo, D. A. Loisel, D. B. Goldstein and G. A. Wray. 2004. Positive selection on *MMP3* regulation has

shaped heart disease risk. *Curr. Biol.* 14: 1531–1539. [13]

Rockman, M. V., M. W. Hahn, N. Soranzo, F. Zimprich, D. B. Goldstein and G. A. Wray. 2005. Ancient and recent positive selection transformed opioid cis-regulation in humans. *PLoS Biol.* 3: e387. [13]

Rogers, D. W., M. Carr and A. Pomiankowski. 2003. Male genes: X-pelled or X-cluded? *Bioessays* 25: 739–741. [1, 9]

Rollins, R. A., F. Haghighi, J. R. Edwards, R. Das, M. Q. Zhang, J. Ju and T. H. Bestor. 2006. Large-scale structure of genomic methylation patterns. *Genome Res.* 16:157–163. [8]

Romano, N. and G. Macino. 1992. Quelling: Transient inactivation of gene expression in *Neurospora crassa* by transformation with homologous sequences. *Mol. Microbiol.* 6: 3343–3353. [8]

Ross, J. 1995. mRNA stability in mammalian cells. *Microbiol. Rev.* 59: 423–450. [5]

Ross, M. T. and 281 others. 2005. The DNA sequence of the human X chromosome. *Nature* 434: 325–337. [10]

Rozen, S., H. Skaletsky, J. D. Marszalek, P. J. Minx, H. S. Cordum, R. H. Waterston, R. K. Wilson and D. C. Page. 2003. Abundant gene conversion between arms of palindromes in human and ape Y chromosomes. *Nature* 423: 873–876. [10]

Rual, J. F. and 37 others. 2005. Towards a proteome-scale map of the human protein-protein interaction network. *Nature* 437: 1173–1178. [6]

Rzhetsky, A. and S. M. Gomez. 2001. Birth of scale-free molecular networks and the number of distinct DNA and protein domains per genome. *Bioinformatics* 17: 988–996. [12]

Sabeti, P. C. and 16 others. 2002. Detecting recent positive selection in the human genome from haplotype structure. *Nature* 419: 832–837. [13]

Sabo, P. J. and 23 others. 2006. Genome-scale mapping of DNase I sensitivity in vivo using tiling DNA microarrays. *Nat. Methods* 3: 511–518. [5]

Sachidanandam, R. and 40 others. 2001. A map of human genome sequence variation containing 1.42 million single nucleotide polymorphisms. *Nature* 409: 928–933.

Sahara, K. and 10 others. 2003. W-derived BAC probes as a new tool for identification of the W chromosome and its aberrations in *Bombyx mori*. *Chromosoma* 112: 48–55. [10]

Saito, K., K. M. Nishida, T. Mori, Y. Kawamura, K. Miyoshi, T. Nagami, H. Siomi and M. C. Siomi. 2006. Specific association of Piwi with rasiRNAs derived from retrotransposon and heterochromatic regions in the *Drosophila* genome, *Genes Dev.* 20: 2214–2222. [8]

Salama, N. R., B. Shepherd and S. Falkow. 2004. Global transposon mutagenesis and essential gene analysis of *Helicobacter pylori. J. Bacteriol.* 186: 7926–7935. [7]

Saleh, M. and 15 others. 2004. Differential modulation of endotoxin responsiveness by human caspase-12 polymorphisms. *Nature* 429: 75–79. [13]

Sandelin, A., P. Carninci, B. Lenhard, J. Ponjavic, Y. Hayashizaki and D. A. Hume. 2007. Mammalian RNA polymerase II core promoters: insights from genome-wide studies. *Nat. Rev. Genet.* 8: 424–436. [5]

Sanjuan, R. and S. F. Elena. 2006. Epistasis correlates to genomic complexity. *Proc. Natl. Acad. Sci. USA* 103: 14402–14405. [7]

Santoro, R. 2005. The silence of the ribosomal RNA genes. *Cell. Mol. Life Sci.* 62: 2067–2079. [5]

Sapp, J. 2005. (ed.) *Microbial Phylogeny and Evolution: Concepts and Controversies.* Oxford University Press, Oxford. [4]

Sassetti, C. M., D. H. Boyd and E. J. Rubin. 2003. Genes required for mycobacterial growth defined by high density mutagenesis. *Mol. Microbiol.* 48: 77–84. [7]

Sauer, S., B. M. Lange, J. Gobom, L. Nyarsik, H. Seitz and H. Lehrach. 2005. Miniaturization in functional genomics and proteomics. *Nat. Rev. Genet.* 6: 465–476. [3]

Saunders, M. A., M. F. Hammer and M. W. Nachman. 2002. Nucleotide variability at G6pd and the signature of malarial selection in humans. *Genetics* 162: 1849–1861. [13]

Savageau, M. A. 1976. *Biochemical Systems Analysis: A Study of Function and Design in Molecular Biology.* Addison-Wesley, Reading, MA. [12]

Sayah, D. M., E. Sokolskaja, L. Berthoux and J. Luban. 2004. Cyclophilin A retrotransposition into TRIM5 explains owl monkey resistance to HIV-1. *Nature* 430: 569–673. [3]

Schadt, E. E. and 13 others. 2003. Genetics of gene expression surveyed in maize, mouse and man. *Nature* 422: 297–302. [5]

Schilling, C. H., J. S. Edwards and B. O. Palsson. 1999. Toward metabolic phenomics: Analysis of genomic data using flux balances. *Biotechnology Progr.* 15: 288–295. [12]

Schlieper, D., M. A. Oliva, J. M. Andreu and J. Lowe. 2005. Structure of bacterial tubulin BtubA/B: Evidence for horizontal gene transfer. *Proc. Natl. Acad. Sci. USA* 102: 9170–9175. [4]

Scholl, E. H., J. L. Thorne, J. P. McCarter and D. M. Bird. 2003. Horizontally transferred genes in plant-parasitic nematodes: A high-throughput genomic approach. *Genome Biol.* 4: R39. [4]

Schorderet, D. F. and S. M. Gartler. 1992. Analysis of CpG suppression in methylated and nonmethylated species. *Proc. Natl. Acad. Sci. USA* 89: 957–961. [8]

Schübeler, D., M. C. Lorincz, D. M. Cimbora, A. Telling, Y. Q. Feng, E. E. Bouhassira and M. Groudine. 2000. Genomic targeting of methylat-ed DNA: Influence of methylation on transcription, replication, chromatin structure, and histone acetylation. *Mol. Cell. Biol.* 20: 9103–9112. [8]

Schübeler, D. and 11 others. 2004. The histone modification pattern of active genes revealed through genome-wide chromatin analysis of a higher eukaryote. *Genes Dev.* 18: 1263–1271. [5]

Schutt, C. and R. Nothiger. 2000. Structure, function and evolution of sex-determining systems in Dipteran insects. *Development* 127: 667–677. [10]

Schwartz, A. and 8 others. 1998. Reconstructing hominid Y evolution: X-homologous block, created by X-Y transposition, was disrupted by Yp inversion through LINE-LINE recombination. *Hum. Molec. Genet.* 7: 1–11. [10]

Schwarz, D. S., G. Hutvagner, T. Du, Z. Xu, N. Aronin and P. D. Zamore. 2003. Asymmetry in the assembly of the RNAi enzyme complex. *Cell* 115: 199–208. [8]

Scott, R. J. and M. Spielman. 2004. Epigenetics: Imprinting in plants and mammals—the same but different? *Curr. Biol.* 14: R201–203. [5]

Sebat, J. and 20 others. 2004. Large-scale copy number polymorphism in the human genome. *Science* 305: 525–528. [13]

Segre, D., D. Vitkup and G. Church. 2002. Analysis of optimality in natural and perturbed metabolic networks. *Proc. Natl. Acad. Sci. USA* 99:15112–15117. [12]

Semon, M. and L. Duret. 2004. Evidence that functional transcription units cover at least half of the human genome. *Trends Genet.* 20: 229–232. [7]

Serre, D., R. Nadon and T. J. Hudson. 2005. Large-scale recombination rate patterns are conserved among human populations. *Genome Res.* 15: 1547–1552. [13]

Seymour, R. M. and A. Pomiankowski. 2005. ESS gene expression of X-linked imprinted genes subject to sexual selection. *J. Theor. Biol.* 241: 81–93. [9]

Sharp, A. J. and 13 others. 2005. Segmental duplications and copy-number variation in the human genome. *Am. J. Hum. Genet.* 77: 78–88. [13]

Shen, J. C., W. M. Rideout, 3rd and P. A. Jones. 1994. The rate of hydrolytic deamination of 5-methylcytosine in double-stranded DNA. *Nucleic Acids Res.* 22: 972–976. [13]

Shen-Orr, S. S., R. Milo, S. Mangan and U. Alon. 2002. Network motifs in the transcriptional regulation network of *Escherichia coli. Nat. Genet.* 31: 64–68. [5, 12]

Shibata, F., M. Hizume and Y. Kuroki. 2000. Differentiation and the polymorphic nature of the Y chromosomes revealed by repetitive sequences in the dioecious plant, *Rumex acetosa. Chromosome Res.* 8: 229–236. [10]

Shields, D. C., P. M. Sharp, D. G. Higgins and F. Wright. 1988. Silent sites in Drosophila genes are not neutral—evidence of selection among synonymous codons. *Molec. Biol. Evol.* 5: 704–716. [11]

Shiu, P. K., N. B. Raju, D. Zickler and R. L. Metzenberg. 2001. Meiotic silencing by unpaired DNA. *Cell* 107: 905–916. [8]

Sicheritz-Ponten, T. and S. G. Andersson. 2001. A phylogenomic approach to microbial evolution. *Nucleic Acids Res.* 29: 545–552. [2, 4]

Siegal, M. L., D. E. Promislow and A. Bergman. 2007. Functional and evolutionary inference in gene networks: Does topology matter? *Genetica* 129: 83–103. [2]

Siepel, A. and 15 others. 2005. Evolutionarily conserved elements in vertebrate, insect, worm, and yeast genomes. *Genome Res.* 15: 1034–1050. [13]

Siller, S. 2001. Sexual selection and the maintenance of sex. *Nature* 411: 689–692. [1]

Skaletsky, H. and 39 others. 2003. The male-specific region of the human Y chromosome is a mosaic of discrete sequence classes. *Nature* 423: 825–837. [9, 10]

Sliwa, P. and R. Korona. 2005. Loss of dispensable genes is not adaptive in yeast. *Proc. Natl. Acad. Sci. USA* 102: 17670–17674. [7]

Smale, S. T. and J. T. Kadonaga. 2003. The RNA polymerase II core promoter. *Annu. Rev. Biochem.* 72: 449–479. [5]

Smit, A. F. and A. D. Riggs. 1996. Tiggers and DNA transposon fossils in the human genome. *Proc. Natl. Acad. Sci. USA* 93: 1443–1448. [8]

Smith, N. G. C. and A. Eyre-Walker. 2002. Adaptive protein evolution in *Drosophila*. *Nature* 415: 1022–1024. [1]

Smith, N. G., M. T. Webster, H. and Ellegren. 2002. Deterministic mutation rate variation in the human genome. *Genome Res.* 12: 1350–1356. [13]

Snel, B., M. A. Huynen and B. E. Dutilh. 2005. Genome trees and the nature of genome evolution. *Annu. Rev. Microbiol.* 59: 191–209. [4]

Sole, R. V., R. Pastor-Satorras, E. D. Smith and T. Kepler. 2002. A model of large-scale proteome evolution. *Advances in Complex Systems* 5: 43–54. [12]

Sorek, R., G. Ast and D. Graur. 2002. Alu-containing exons are alternatively spliced. *Genome Res.* 12: 1060–1067. [13]

Sørensen, S. J., M. Bailey, L. H. Hansen, N. Kroer and S. Wuertz. 2005. Studying plasmid horizontal transfer *in situ*: A critical review. *Nat. Rev. Microbiol.* 3: 700–710. [4]

Spilianakis, C. G. and R. A. Flavell. 2006. Molecular biology. Managing associations between different chromosomes. *Science* 312: 207–208. [5]

Spilianakis, C. G., M. D. Lalioti, T. Town, G. R. Lee and R. A. Flavell. 2005. Interchromosomal associations between alternatively expressed loci. *Nature* 435: 637–645. [5]

Springael, D. and E. M. Top. 2004. Horizontal gene transfer and microbial adaptation to xenobiotics: New types of mobile genetic elements and lessons from ecological studies. *Trends Microbiol.* 12: 53–58. [4]

Stam, M. and O. Mittelsten Scheid. 2005. Paramutation: An encounter leaving a lasting impression. *Trends Plant Sci.* 10: 283–290. [5]

Stankiewicz, P. and J. R. Lupski. 2002. Genome architecture, rearrangements and genomic disorders. *Trends Genet.* 18: 74–82. [13]

Stanojevic, D., S. Small and M. Levine. 1991. Regulation of a segmentation stripe by overlapping activators and repressors in the Drosophila embryo. *Science* 254: 1385–1387. [5]

Stechmann, A., M. Baumgartner, J. D. Silberman and A. J. Roger. 2006. The glycolytic pathway of *Trimastix pyriformis* is an evolutionary mosaic. *BMC Evol. Biol.* 6: 101. [4]

Stedman, H. H. and 9 others. 2004. Myosin gene mutation correlates with anatomical changes in the human lineage. *Nature* 428: 415–418. [13]

Steele, R. E., S. E. Hampson, N. A. Stover, D. F. Kibler and H. R. Bode. 2004. Probable horizontal transfer of a gene between a protist and a cnidarian. *Curr. Biol.* 14: R298–299. [4]

Stefansson, H. and 26 others. 2005. A common inversion under selection in Europeans. *Nat. Genet.* 37: 129–137. [13]

Stehlik, I. and S. C. H. Barrett. 2005. Mechanisms governing sex-ratio variation in dioecious *Rumex nivalis*. *Evolution* 59: 814–825. [10]

Steinmetz, L. M. and 10 others. 2002. Systematic screen for human disease genes in yeast. *Nat. Genet.* 31: 400–404. [1, 7]

Stelzl, U. and 22 others. 2005. A human protein-protein interaction network: a resource for annotating the proteome. *Cell* 122: 957–968. [6]

Stewart, C. B., J. W. Schilling and A. C. Wilson. 1987. Adaptive evolution in the stomach lysozymes of foregut fermenters. *Nature* 330: 401–404. [12]

Stone, A. C., R. C. Griffiths, S. L. Zegura and M. F. Hammer. 2002. High levels of Y-chromosome nucleotide diversity in the genus *Pan*. *Proc. Natl. Acad. Sci. USA* 99: 43–48. [10]

Storchova, R., Divina, P., 2006. Nonrandom representation of sex-biased genes on chicken Z chromosome. *J. Mol. Evol.* 63: 676–681. [9]

Streelman, J. T. and T. D. Kocher. 2000. From phenotype to genotype. *Evol. Dev.* 2: 166–173. [2]

Striepen, B., A. J. Pruijssers, J. Huang, C. Li, M. J. Gubbels, N. N. Umejiego, L. Hedstrom and J. C. Kissinger. 2004. Gene transfer in the evolution of parasite nucleotide biosynthesis. *Proc. Natl. Acad. Sci. USA* 101: 3154–3159. [4]

Striepen, B., M. W. White, C. Li, M. N. Guerini, S. B. Malik, J. M. Logsdon, Jr., C. Liu and M. S. Abrahamsen. 2002. Genetic complementation in apicomplexan parasites. *Proc. Natl. Acad. Sci. USA* 99: 6304–6309. [4]

Stuart, J. M., E. Segal, D. Koller and S. K. Kim. 2003. A gene-coexpression network for global discovery of conserved genetic modules. *Science* 302: 249–255. [5]

Stumpf, M. P., W. P. Kelly, T. Thorne and C. Wiuf. 2007. Evolution at the system level: The natural history of protein interaction networks. *Trends Ecol. Evol.* 22: 366–373. [2]

Subramanian, S. and S. Kumar. 2003. Neutral substitutions occur at a faster rate in exons than in non-coding DNA in primate genomes. *Genome Res.* 13: 838–844. [11]

Subramanian, S. and S. Kumar. 2004. Gene expression intensity shapes evolutionary rates of the proteins encoded by the vertebrate genome. *Genetics* 168: 373–381. [7]

Subramanian, S. and S. Kumar. 2006. Higher intensity of purifying selection on >90% of the human genes revealed by the intrinsic replacement mutation rates. *Mol. Biol. Evol.* 23: 2283–2287. [11]

Sucena, E., I. Delon, I. Jones, F. Payre and D. L. Stern. 2003. Regulatory evolution of shavenbaby/ovo underlies multiple cases of morphological parallelism. *Nature* 424: 935–938. [5]

Sudarsan N., J. E. Barrick and R. R. Breaker. 2003. Metabolite-binding RNA domains are present in the genes of eukaryotes. *RNA* 9: 644–647. [5]

Suga, H., M. Koyanagi, D. Hoshiyama, K. Ono, N. Iwabe, K. Kuma and T. Miyata. 1999. Extensive gene duplication in the early evolution of animals before the parazoan-eumetazoan split demonstrated by G proteins and protein tyrosine kinases from sponge and hydra. *J. Mol. Evol.* 48: 646–653. [6]

Sugino, R. P. and H. Innan. 2006. Selection for more of the same product as a force to enhance concerted evolution of duplicated genes. *Trends Genet.* 22: 642–644. [7]

Suguri, S., K. Henze, L. B. Sanchez, D. V. Moore and M. Muller. 2001. Archaebacterial relationships of the phosphoenolpyruvate carboxykinase gene reveal mosaicism of Giardia intestinalis core metabolism. *J. Eukaryot. Microbiol.* 48: 493–497. [4]

Sundin, G. W. 2007. Genomic insights into the contribution of phytopathogenic bacterial plasmids to the evolutionary history of their hosts. *Annu. Rev. Phytopathol.* 45: 129–151. [2]

Suzuki, N., N. Okai, H. Nonaka, Y. Tsuge, M. Inui and H. Yukawa. 2006. High-throughput transposon mutagenesis of *Corynebacterium glutamicum* and construction of a single-gene disruptant mutant library. *Appl. Environ. Microbiol.* 72: 3750–3755. [7]

Suzuki, Y. and M. Nei. 2001. Reliabilities of parsimony-based and likelihood-based methods for detecting positive selection at single amino acid sites. *Mol. Biol. Evol.* 18: 2179–2185. [11]

Suzuki, Y., T. Gojobori and M. Nei. 2001. ADAPTSITE: detecting natural selection at single amino acid sites. *Bioinformatics* 17: 660–661. [11]

Swanson, W. J., A. G. Clark, H. M. Waldrip-Dail, M. F. Wolfner and C. F. Aquadro. 2001. Evolutionary EST analysis identifies rapidly evolving male reproductive proteins in *Drosophila*. *Proc. Natl. Acad. Sci. USA* 98: 7375–7379. [13]

Swanson, W. J., R. Nielsen and Q. Yang. 2003. Pervasive adaptive evolution in mammalian fertilization proteins. *Mol. Biol. Evol.* 20: 18–20. [13]

Symer, D. E., C. Connelly, S. T. Szak, E. M. Caputo, G. J. Cost, G. Parmigiani and J. D. Boeke. 2002. Human l1 retrotransposition is associated with genetic instability in vivo. *Cell* 110: 327–338. [13]

Szabo, Z., S. A. Levi-Minzi, A. M. Christiano, C. Struminger, M. Stoneking, M. A. Batzer and C. D. Boyd. 1999. Sequential loss of two neighboring exons of the tropoelastin gene during primate evolution. *J. Mol. Evol.* 49: 664–671. [13]

Tajima, F. 1989. Statistical method for testing the neutral mutation hypothesis by DNA polymorphism. *Genetics* 123: 585–595. [13]

Takahashi, A., S. C. Tsaur, J. A. Coyne and C. I. Wu. 2001. The nucleotide changes governing cuticular hydrocarbon variation and their evolution in *Drosophila melanogaster*. *Proc. Natl. Acad. Sci. USA* 98: 3920–3925. [5]

Tamaru, H. and E. U. Selker. 2001. A histone H3 methyltransferase controls DNA methylation in *Neurospora crassa*. *Nature* 414: 277–283. [8]

Tamura, K. and S. Kumar. 2002. Evolutionary distance estimation under heterogeneous substitution pattern among lineages. *Mol. Biol. Evol.* 19: 1727–1736. [11]

Tamura, K., S. Subramanian and S. Kumar. 2004. Temporal patterns of fruit fly (Drosophila) evolution revealed by mutation clocks. *Mol. Biol. Evol.* 21: 36–44. [11]

Tatusov, R. L., E. V. Koonin and D. J. Lipman. 1997. A genomic perspective on protein families. *Science* 278: 631–637. [6]

Teichmann, S. A. and M. M. Babu. 2004. Gene regulatory network growth by duplication. *Nat. Genet.* 36: 492–496. [5]

Teichmann, S. A. and G. Mitchison. 1999. Is there a phylogenetic signal in prokaryote proteins? *J. Mol. Evol.* 49: 98–107. [4]

Temporini, E. D. and H. D. VanEtten. 2004. An analysis of the phylogenetic distribution of the pea pathogenicity genes of *Nectria haematococca* MPVI supports the hypothesis of their origin by horizontal transfer and uncovers a potentially new pathogen of garden pea: *Neocosmospora boniensis*. *Curr. Genet.* 46: 29–36. [4]

Tettelin, H. and 45 others. 2005. Genome analysis of multiple pathogenic isolates of *Streptococcus agalactiae*: Implications for the microbial "pangenome". *Proc. Natl. Acad. Sci USA*. 102: 13950–13955. [4]

Thatcher, J. W., J. M. Shaw and W. J. Dickinson. 1998. Marginal fitness contributions of nonessential genes in yeast. *Proc. Natl. Acad. Sci. USA* 95: 253–257. [7]

The International HapMap Consortium. 2005. A haplotype map of the human genome. *Nature* 437: 1299–1320. [13]

The International SNP MAP Working Group. 2001. A map of human genome sequence variation containing 1.42 million single nucleotide polymorphisms. *Nature* 409: 928–933. [13]

The Yeast Genome Directory. 1997. *Nature* 387 (6632 Suppl.): 1–105. [1]

Thomas, C. M. and K. M. Nielsen. 2005. Mechanisms of, and barriers to, horizontal gene transfer between bacteria. *Nature Reviews Microbiology* 3: 711–721. [4]

Thomas, M. C. and C. M. Chiang. 2006. The general transcription machinery and general cofactors. *Crit. Rev. Biochem. Mol. Biol.* 41: 105–178. [5]

Thomas, R. and R. D'Ari. 1990. *Biological Feedback.* CRC Press, Boca Raton, FL. [12]

Thompson, E. E., H. Kuttab-Boulos, D. Witonsky, L. Yang, B. A. Roe and A. Di Rienzo. 2004. *CYP3A* variation and the evolution of salt-sensitivity variants. *Am. J. Hum. Genet.* 75: 1059–1069. [13]

Thompson, J. R. and 8 others. 2005. Genotypic diversity within a natural coastal bacterioplankton population. *Science* 307: 1311–1313. [4]

Thornburg, B. G., V. Gotea and W. Makalowski. 2006. Transposable elements as a significant source of transcription regulating signals. *Gene* 365:104–110. [13]

Thornton, K. and M. Long. 2002. Rapid divergence of gene duplicates on the *Drosophila melanogaster* X chromosome. *Mol. Biol. Evol.* 19: 918–925. [3]

Thornton, K., D. Bachtrog and P. Andolfatto. 2006. X chromosomes and autosomes evolve at similar rates in Drosophila: No evidence for faster-X protein evolution. *Genome Res.* 16: 498–504. [9]

Tijsterman, M., K. L. Okihara, K. Thijssen and R. H. Plasterk. 2002. PPW-1, a PAZ/PIWI protein required for efficient germline RNAi, is defective in a natural isolate of *C. elegans. Curr. Biol.* 12: 1535–1540. [8]

Timmis, G. C., M. A. Ayliffe, C. Y. Haung and W. Martin. 2004. Endosymbiotic gene transfer: organelle genomes forge eukaryotic chromosomes. *Nat. Rev. Genet.* 5: 123–135. [4]

Tirosh, I., A. Weinberger, M. Carmi and N. Barkai. 2006. A genetic signature of interspecies variations in gene expression. *Nat. Genet.* 38: 830–834. [5]

Tishkoff, S. A. and 16 others. 2001. Haplotype diversity and linkage disequilibrium at human *G6PD*: recent origin of alleles that confer malarial resistance. *Science* 293: 455–462. [13]

Tjian, R. 1996. The biochemistry of transcription in Eukaryotes: A paradigm for multisubunit regulatory complexes. *Phil. Trans. Biol. Sci.* 351: 491–499. [5]

Tobe, T. and 11 others. 2007. An extensive repertoire of type III secretion effectors in *Escherichia coli* O157 and the role of lambdoid phages in their dissemination. *Proc. Natl. Acad. Sci. USA* 103: 14941–14946. [4]

Toda, T., S. Cameron, P. Sass, M. Zoller and M. Wigler. 1987. Three different genes in *S. cerevisiae* encode the catalytic subunits of the cAMP-dependent protein kinase. *Cell* 50: 277–287. [7]

Toma, D. P., K. P. White, J. Hirsch and R. J. Greenspan. 2002. Identification of genes involved in *Drosophila melanogaster* geotaxis, a complex behavioral trait. *Nat. Genet.* 31: 349–353. [5]

Toomajian, C. and M. Kreitman. 2002. Sequence variation and haplotype structure at the human *HFE* locus. *Genetics* 161: 1609–1623. [13]

Tordai, H., A. Nagy, K. Farkas, L. Banyai and L. Patthy. 2005. Modules, multidomain proteins and organismic complexity. *FEBS J.* 272: 5064–5078. [6]

Torgerson, D. G. and R. S. Singh. 2003. Sex-linked mammalian sperm proteins evolve faster than autosomal ones. *Mol. Biol. Evol.* 20: 1705–1709. [13]

Torgerson, D. G. and R. S. Singh. 2006. Enhanced adaptive evolution of sperm-expressed genes on the mammalian X chromosome. *Heredity* 96: 39–44. [13]

Toruner, G. A., D. L. Streck, M. N. Schwalb and J. J. Dermody. 2007. An oligonucleotide based array-CGH system for detection of genome wide copy number changes including subtelomeric regions for genetic evaluation of mental retardation. *Am. J. Med. Genet.* 143: 824–829. [3]

Townsend, J. P., D. Cavalieri and D. L. Hartl. 2003a. Population genetic variation in genome-wide gene expression. *Mol. Biol Evol.* 20: 955–963. [5]

Townsend, J. P., K. M. Nielsen, D. S. Fisher and D. L. Hartl. 2003b. Horizontal acquisition of divergent chromosomal DNA in bacteria: effects of mutator phenotypes. *Genetics* 164: 13–21. [4]

Traut, W. and U. Willhoeft. 1990. A jumping sex determining factor in the fly *Megaselia scalaris. Chromosoma* 99: 407–412. [10]

Traut, W., K. Sahara, T. D. Otto and F. Marec. 1999. Molecular differentiation of sex chromosomes probed by comparative genomic hybridization. *Chromosoma* 108: 173–180. [10]

Tsaparas, P., L. Mariño-Ramírez, O. Bodenreider, E. V. Koonin and I. K. Jordan. 2006. Global similarity and local divergence in human and mouse gene co-expression networks. *BMC Biol.* 6: 70.[2]

Tucker, B. J. and R. R. Breaker. 2005. Riboswitches as versatile gene control elements. *Curr. Opin. Struct. Biol.* 15: 342–348. [5]

Turelli, P., S. Vianin and D. Trono. 2004. The innate antiretroviral factor APOBEC3G does not affect human LINE-1 retrotransposition in a cell culture assay. *J. Biol. Chem.* 279: 43371–43373. [8]

Turner, B. M. 2007. Defining an epigenetic code. *Nat. Cell Biol.* 9: 2–6. [5]

Tuschl, T., P. D. Zamore, R. Lehmann, D. P. Bartel and P. A. Sharp. 1999. Targeted mRNA degradation by double-stranded RNA in vitro. *Genes Dev.* 13: 3191–3197. [8]

Tuzun, E. and 11 others. 2005. Fine-scale structural variation of the human genome. *Nat. Genet.* 37: 727–732. [13]

Uetz, P. and 19 others. 2000. A comprehensive analysis of protein-protein interactions in *Saccharomyces cerevisiae*. *Nature* 403: 623–627. [12]

Ullu, E. and C. Tschudi.1984. Alu sequences are processed 7SL RNA genes. *Nature* 312: 171–172. [8]

Usakin, L. A., G. L. Kogan, A. I. Kalmykova and V. A. Gvozdev. 2005. An alien promoter capture as a primary step of the evolution of testes-expressed repeats in the *Drosophila melanogaster* genome. *Mol. Biol. Evol.* 22: 1555–1560. [3]

Vallender, E. J. and B. T. Lahn. 2004. How mammalian sex chromosomes acquired their peculiar gene content. *Bioessays* 26: 159–169. [9]

van den Berg, M. A., P. de Jong-Gubbels, C. J. Kortland, J. P. van Dijken, J. T. Pronk and H. Y. Steensma. 1996. The two acetyl-coenzyme A synthetases of *Saccharomyces cerevisiae* differ with respect to kinetic properties and transcriptional regulation. *J. Biol. Chem.* 271: 28953–28959. [7]

van der Giezen, M., S. Cox and J. Tovar. 2004. The iron-sulfur cluster assembly genes iscS and iscU of Entamoeba histolytica were acquired by horizontal gene transfer. *BMC Evol. Biol.* 4: 7. [4]

van Noort, V., B. Snel and M. A. Huynen. 2004. The yeast coexpression network has a small-world, scale-free architecture and can be explained by a simple model. *EMBO Rep.* 5: 280–284. [12]

van Steensel, B. and S. Henikoff. 2000. Identification of in vivo DNA targets of chromatin proteins using tethered dam methyltransferase. *Nat. Biotechnol.* 18: 424–428. [5]

van Steensel, B. and S. Henikoff. 2003. Epigenomic profiling using microarrays. *Biotechniques* 35: 346–350, 352–354, 356–357. [5]

Varki, A. and T. K. Altheide. 2005. Comparing the human and chimpanzee genomes: searching for needles in a haystack. *Genome Res.* 15: 1746–1758. [13]

Varma, A. and B. O. Palsson. 1993. Metabolic capabilities of *Escherichia coli*: Synthesis of biosynthetic precursors and cofactors. *J. Theor. Biol.* 165: 477–502. [12]

Verdel, A., S. Jia, S. Gerber, T. Sugiyama, S. Gygi, S. I. Grewal and D. Moazed. 2004. RNAi-mediated targeting of heterochromatin by the RITS complex. *Science* 303: 672–676. [8]

Vicoso, B. and B. Charlesworth. 2006. Evolution on the X chromosome: Unusual patterns and processes. *Nat. Rev. Genet.* 7: 645–653. [9]

Vinckenbosch, N., I. Dupanloup and H. Kaessmann. 2006. Evolutionary fate of retroposed gene copies in the human genome. *Proc. Natl. Acad. Sci. USA* 103: 3220–3225. [13]

Vitkup, D., P. Kharchenko and A. Wagner. 2006. Influence of metabolic network structure and function on enzyme evolution. *Genome Biol.* 7: R39. [2, 12]

Voight, B. F., S. Kudaravalli, X. Wen and J. K. Pritchard. 2006. A map of recent positive selection in the human genome. *PLoS Biol.* 4: e72. [13]

Volpe, T. A., C. Kidner, I. M. Hall, G. Teng, S. I. Grewal and R. A. Martienssen. 2002. Regulation of heterochromatic silencing and histone H3 lysine-9 methylation by RNAi. *Science* 297: 1833–1837. [8]

Volpe, T., V. Schramke, G. L. Hamilton, S. A. White, G. Teng, R. A. Martienssen and R. C. Allshire. 2003. RNA interference is required for normal centromere function in fission yeast. *Chromosome Res.* 11: 137–146. [8]

von Dassow, G., E. Meir, E. M. Munro and G. M. Odell. 2000. The segment polarity network is a robust developmental module. *Nature* 406: 188–192. [7]

von Mering, C., R. Krause, B. Snel, M. Cornell, S. G. Oliver, S. Fields and P. Bork. 2002. Comparative assessment of large-scale data sets of protein-protein interactions. *Nature* 417: 399–403. [7]

Wagner, A. 1996. Does evolutionary plasticity evolve? *Evolution* 50: 1008–1023. [5]

Wagner, A. 1999. Redundant gene functions and natural selection. *J. Evol. Biol.* 12: 1–16. [7]

Wagner, A. 2000. Robustness against mutations in genetic networks of yeast. *Nat. Genet.* 24: 355–361. [7]

Wagner, A. 2001. The yeast protein interaction network evolves rapidly and contains few redundant duplicate genes. *Mol. Biol. Evol.* 18: 1283–1292. [12]

Wagner, A. 2003. How the global structure of protein interaction networks evolves. *Proc. Roy. Soc. Lond. Ser. B.* 270: 457–466. [12]

Wagner, A. 2005a. Distributed robustness versus redundancy as causes of mutational robustness. *Bioessays* 27: 176–188. [7]

Wagner, A. 2005b. Circuit topology and the evolution of robustness in two-gene circadian oscillators. *Proc. Natl. Acad. Sci. USA* 102: 11775–11780. [7]

Wagner, A. 2005c. *Robustness and Evolvability in Living Systems*. Princeton University Press, Princeton, NJ. [12]

Wagner, A. and D. Fell. 2001. The small world inside large metabolic networks. *Proc. Roy. Soc. Lond. Ser. B.* 280: 1803–1810. [12]

Wagner, A. and J. Wright. 2005. Compactness and cycles in signal transduction and transcriptional regulation networks: A signature of natural selection? *Advances in Complex Systems* 7: 419–432. [12]

Wagner, A. and J. Wright. 2007. Alternative pathways and mutational robustness in molecular networks. *BioSystems* 88: 163–172. [12]

Wagner, A. and P. F. Stadler. 1999. Viral RNA and evolved mutational robustness. *J. Exp. Zool.* 285: 119–127. [7]

Wagstaff, B. J. and D. J. Begun. 2005. Comparative genomics of accessory gland protein genes in *Drosophila melanogaster* and *D. pseudoobscura*. *Mol. Biol. Evol.* 22: 818–832. [9]

Wall, D. P., A. E. Hirsh, H. B. Fraser, J. Kumm, G. Giaever, M. B. Eisen and M. W. Feldman. 2005. Functional genomic analysis of the rates of protein evolution. *Proc. Natl. Acad. Sci. USA* 102: 5483–5488. [2, 7]

Wallace, M. R., L. B. Andersen, A. M. Saulino, P. E. Gregory, T. W. Glover and F. S. Collins. 1991. A de novo Alu insertion results in neurofibromatosis type 1. *Nature* 353: 864–866. [13]

Waller, R. F, C. H. Slamovits and P. J. Keeling. 2006. Lateral gene transfer of a multigene region from cyanobacteria to dinoflagellates resulting in a novel plastid-targeted fusion protein. *Mol. Biol. Evol.* 23: 1437–1443. [4]

Walsh, C. P., J. R. Chaillet and T. H. Bestor. 1998. Transcription of IAP endogenous retroviruses is constrained by cytosine methylation. *Nat. Genet.* 20: 116–117. [8]

Wang, L. and 10 others. 2002. Redundant pathways for negative feedback regulation of bile acid production. *Dev. Cell* 2: 721–731. [7]

Wang, P. J., J. R. McCarrey, F. Yang and D. C. Page. 2001. An abundance of X-linked genes expressed in spermatogonia. *Nat. Genet.* 27: 422–426. [9]

Wang, Q., M. Miyakoda, W. Yang, J. Khillan, D. L. Stachura, M. J. Weiss and K. Nishikura. 2004. Stress-induced apoptosis associated with null mutation of ADAR1 RNA editing deaminase gene. *J. Biol. Chem.* 279: 4952–4961. [8]

Wang, W., F. G. Brunet, E. Nero and M. Long. 2002. Origin of *sphinx*, a young chimeric RNA gene in *Drosophila melanogaster*. *Proc. Natl. Acad. Sci. USA* 99: 4448–4453. [3]

Wang, W., H. Yu and M. Long. 2004. Duplication-degeneration as a mechanism of gene fission and the origin of new genes in *Drosophila* species. *Nat. Genet.* 36: 523–527. [3]

Wang, W. and 16 others. 2005. Origin and evolution of new exons in rodents. *Genome Res.* 15: 1258–1264. [3]

Wang, W. and 13 others. 2006. High rate of chimeric gene origination by retroposition in plant genomes. *Plant Cell* 18: 1791–1802. [3]

Wang, Y., C. L. Liu, J. D. Storey, R. J. Tibshirani, D. Herschlag and P. O. Brown. 2002. Precision and functional specificity in mRNA decay. *Proc. Natl. Acad. Sci. USA* 99: 5860–5865. [5]

Waterston, R. H. and 222 others. 2002. Initial sequencing and comparative analysis of the mouse genome. *Nature* 420: 520–562. [13]

Weaver, I. C. and 8 others. 2004. Epigenetic programming by maternal behavior. *Nat. Neurosci.* 7: 847–854. [5]

Weaver, I. C. and 8 others. 2005. Reversal of maternal programming of stress responses in adult offspring through methyl supplementation: Altering epigenetic marking later in life. *J. Neurosci* 25: 11045–11054. [5]

Webster, M. T., R. S. Wells and J. B. Clegg. 2002. Analysis of variation in the human α-globin gene cluster using a novel DHPLC technique. *Mutat. Res.* 501: 99–103. [13]

Webster, M. T., N. G. Smith and H. Ellegren. 2003. Compositional evolution of noncoding DNA in the human and chimpanzee genomes. *Mol. Biol. Evol.* 20: 278–286. [13]

Weinmann, A. S. and P. J. Farnham. 2002. Identification of unknown target genes of human transcription factors using chromatin immuno-precipitation. *Methods* 26: 37–47. [5]

Weir, B. S., L. R. Cardon, A. D. Anderson, D. M. Nielsen and W. G. Hill. 2005. Measures of human population structure show heterogeneity among genomic regions. *Genome Res.* 15: 1468–1476. [13]

Welch, R. A and 19 others. 2002. Extensive mosaic structure revealed by the complete genome sequence of uropathogenic *Escherichia coli*. *Proc. Natl. Acad. Sci. USA.* 99: 17020–17024. [4]

Wenzl, P., L. Wong, K. Kwang-won and R. A. Jefferson. 2005. A functional screen identifies lateral transfer of beta-glucuronidase (gus) from bacteria to fungi. *Mol. Biol. Evol.* 22: 308–316. [4]

Westergaard, M. 1958. The mechanism of sex determination in dioecious plants. *Adv. Genet.* 9: 217–281. [10]

Wigby, S. and T. Chapman. 2004. Sperm competition. *Curr. Biol.* 14: R100–102. [13]

Wilby, A. S. and J. S. Parker. 1986. Continuous variation in Y-chromosome structure of *Rumex acetosa*. *Heredity* 57: 247–254. [10]

Wilke, C. O. and D. A. Drummond. 2006. Population genetics of translational robustness. *Genetics* 173: 473–481. [2]

Wilke, C. O., J. L. Wang, C. Ofria, R. E. Lenski and C. Adami. 2001. Evolution of digital organisms at high mutation rates leads to survival of the flattest. *Nature* 412: 331–333. [7]

Wilkins, J. F. and D. Haig. 2003. What good is genomic imprinting: The function of parent-specific gene expression. *Nat. Rev. Genet.* 4: 359–368. [5]

Willhoeft, U. and W. Traut. 1990. Molecular differentiation of the homomorphic sex chromosomes in *Megaselia scalaris* (Diptera) detected by random DNA probes. *Chromosoma* 99: 237–242. [10]

Williamson, S., A. Fledel-Alon and C. D. Bustamante. 2004. Population genetics of polymorphism and divergence for diploid selection models with arbitrary dominance. *Genetics* 168: 463–475. [13]

Williamson, S. H., R. Hernandez, A. Fledel-Alon, L. Zhu, R. Nielsen and C. D. Bustamante. 2005. Simultaneous inference of selection and population growth from patterns of variation in the human genome. *Proc. Natl. Acad. Sci. USA* 102: 7882–7887. [13]

Wilson, A. C., S. S. Carlson and T. J. White. 1977. Biochemical evolution. *Annu. Rev. Biochem.* 46: 573–639. [2, 5, 7]

Wilson, L., Y. H. Ching, M. Farias, S. A. Hartford, G. Howell, H. Shao, M. Bucan and J. C. Schimenti. 2005. Random mutagenesis of proximal mouse chromosome 5 uncovers predominantly embryonic lethal mutations. *Genome Res.* 15:1095–1105. [1, 7]

Winkler, W. C. and R. R. Breaker. 2005. Regulation of bacterial gene expression by riboswitches. *Annu. Rev. Microbiol.* 59: 487–517. [5]

Winkler, W., A. Nahvi and R. R. Breaker. 2002. Thiamine derivatives bind messenger RNAs directly to regulate bacterial gene expression. *Nature* 419: 952–956. [5]

Winckler, W. and 10 others. 2005. Comparison of fine-scale recombination rates in humans and chimpanzees. *Science* 308: 107–111. [13]

Winston, W. M., C. Molodowitch and C. P. Hunter. 2002. Systemic RNAi in *C. elegans* requires the putative transmembrane protein SID-1. *Science* 295: 2456–2459. [8]

Wittkopp, P. J., B. K. Haerum and A. G. Clark. 2004. Evolutionary changes in *cis* and *trans* gene regulation. *Nature* 430: 85–88. [5]

Woese, C. R. 1987. Bacterial evolution. *Microbiol. Rev.* 51: 221–271. [4]

Wolf, Y. I. 2006. Coping with the quantitative genomics 'elephant': the correlation between the gene dispensability and evolution rate. *Trends Genet.* 22: 354–357. [2]

Wolf, Y. I., I. B, Rogozin, N. V. Grishin, R. L. Tatusov and E. V. Koonin. 2001. Genome trees constructed using five different approaches suggest new major bacterial clades. *BMC Evol. Biol.* 1: 8. [4]

Wolf, Y. I., L. Carmel and E. V. Koonin. 2006a. Correlations between quantitative measures of genome evolution, expression and function. In *Discovering Biomolecular Mechanisms with Computational Biology* (ed. F. Eisenhaber), pp. 133–144. Landes Bioscience and Springer Science, Georgetown, TX/New York. [2]

Wolf, Y. I., L. Carmel and E. V. Koonin. 2006b. Unifying measures of gene function and evolution. *Proc. Biol. Sci.* 273: 1507–1515. [2]

Wolfe, K. 2004. Evolutionary genomics: yeasts accelerate beyond BLAST. *Curr. Biol.* 14: R392–394. [7]

Wolfe, K. H. and D. C. Shields. 1997. Molecular evidence for an ancient duplication of the entire yeast genome. *Nature* 387: 708–713. [12]

Wolfe, K. H., P. M. Sharp and W.-H. Li. 1989. Mutation rates differ among regions of the mammalian genome. *Nature* 337: 283–285. [11]

Wolffe, A. P. and M. A. Matzke. 1999. Epigenetics: Regulation through repression. *Science* 286: 481–486. [5]

Won, H. and S. S. Renner. 2003. Horizontal gene transfer from flowering plants to Gnetum. *Proc. Natl. Acad. Sci. USA* 100: 10824–10829. [4]

Wong, S. L. and F. P. Roth. 2005. Transcriptional compensation for gene loss plays a minor role in maintaining genetic robustness in *Saccharomyces cerevisiae*. *Genetics* 171: 829–833. [7]

Wong, W. S. and R. Nielsen. 2004. Detecting selection in noncoding regions of nucleotide sequences. *Genetics* 167: 949–958. [13]

Woods, C. G., J. Bond and W. Enard. 2005. Autosomal recessive primary microcephaly (MCPH): a review of clinical, molecular, and evolutionary findings. *Am. J. Hum. Genet.* 76: 717–728. [13]

Woolfe, A. and 15 others. 2005. Highly conserved noncoding sequences are associated with vertebrate development. *PLoS Biol.* 3: e7. [13]

Wray, G. A. 2007. The evolutionary significance of *cis*-regulatory mutations. *Nat. Rev. Genet.* 8: 206–216. [1, 5]

Wray, G. A., M. W. Hahn, E. Abouheif, J. P. Balhoff, M. Pizer, M. V. Rockman and L. A. Romano. 2003. The evolution of transcriptional regulation in eukaryotes. *Mol. Biol. Evol.* 20: 1377–1419. [5, 13]

Wright, G. D. 2007. The antibiotic resistome: The nexus of chemical and genetic diversity. *Nat. Rev. Microbiol.* 5: 175–186. [2]

Wright, S. 1931. Evolution in Mendelian populations. *Genetics* 16: 97–126. [11]

Wright, S. 1934. Physiological and evolutionary theories of dominance. *Am. Nat.* 68: 25–53. [7]

WTCCC (The Wellcome Trust Case Control Consortium) 2007. Genome-wide association study of 14,000 cases of seven common diseases and 3,000 shared controls *Nature* 447: 661–678. [1]

Wu, C. I. and E. Y. Xu. 2003. Sexual antagonism and X inactivation—the SAXI hypothesis. *Trends Genet.* 19: 243–247. [9]

Wuchty, S. 2001. Scale-free behavior in protein domain networks. *Mol. Biol. Evol.* 18: 1694–1702. [6, 12]

Wuchty, S. 2002. Interaction and domain networks of yeast. *Proteomics* 2: 1715–1723. [12]

Wuchty, S. and E. Almaas. 2005. Evolutionary cores of domain co-occurrence networks. *BMC Evol. Biol.* 5: 24. [6]

Wyckoff, G. J., W. Wang and C. I. Wu. 2000. Rapid evolution of male reproductive genes in the descent of man. *Nature* 403: 304–309. [13]

Wyckoff, G. J., C. M. Malcom, E. J. Vallender and B. T. Lahn. 2005. A highly unexpected strong correlation between fixation probability of nonsynonymous mutations and mutation rate. *Trends Genet.* 21: 381–385. [5]

Xu, L., H. Chen, X. Hu, R. Zhang, Z. Zhang and Z. W. Luo. 2006. Average gene length is highly conserved in prokaryotes and eukaryotes and diverges only between the two kingdoms. *Mol. Biol.* 23: 1107–1108. [1]

Xue, Y. and 13 others. 2006. Spread of an inactive form of caspase-12 in humans is due to recent positive selection. *Am. J. Hum. Genet.* 78: 659–670. [13]

Yamamoto, M., T. Wakatsuki, A. Hada and A. Ryo. 2001. Use of serial analysis of gene expression (SAGE) technology. *J. Immunol. Methods* 250: 45–66. [5]

Yamato, K. T. and 33 others. 2007. Gene organization of the liverwort Y chromosome reveals distinct sex chromosome evolution in a haploid system. *Proc. Natl. Acad. Sci. USA* 104: 6472–6477. [10]

Yampolsky, L. Y., F. A. Kondrashov and A. S. Kondrashov. 2005. Distribution of the strength of selection against amino acid replacements in human proteins. *Hum. Mol. Genet.* 14: 3191–3201. [11]

Yang, F. and 9 others. 2003. Reciprocal chromosome painting among human, aardvark, and elephant (superorder Afrotheria) reveals the likely eutherian ancestral karyotype. *Proc. Natl. Acad. Sci. USA* 100: 1062–1066. [10]

Yang, F., H. Skaletsky and P. J. Wang. 2007. Ub14b, an X-derived retrogene, is specifically expressed in post-meiotic germ cells in mammals. *Gene Expr. Patterns* 7: 131–136. [3]

Yang, J., Z. L. Gu and W.-H. Li. 2003. Rate of protein evolution versus fitness effect of gene deletion. *Mol. Biol. Evol.* 20: 772–774. [7]

Yang, Z. 2006. *Computational Molecular Evolution*. Oxford University Press, Oxford, UK. [11]

Yang, Z. and J. P. Bielawski. 2000. Statistical methods for detecting molecular adaptation. *Trends Ecol. Evol.* 15: 496–503. [11]

Ye, Y. and A. Godzik. 2004. Comparative analysis of protein domain organization. *Genome Res.* 14: 343–353. [6]

Yi, S. and B. Charlesworth. 2003. Unusual pattern of single nucleotide polymorphism at the exuperantia2 locus of *Drosophila pseudoobscura*. *Genet Res.* 82: 101–106. [3]

Yi, S., D. L. Ellsworth and W.-H. Li. 2002. Slow molecular clocks in Old World monkeys, apes, and humans. *Mol. Biol. Evol.* 19: 2191–2198. [13]

Yoder, J. A., N. S. Soman, G. L. Verdine and T. H. Bestor. 1997a. DNA (cytosine-5)-methyltransferases in mouse cells and tissues. Studies with a mechanism-based probe. *J. Mol. Biol.* 270: 385–395. [8]

Yoder, J. A., C. P. Walsh and T. H. Bestor. 1997b. Cytosine methylation and the ecology of intragenomic parasites. *Trends Genet.* 13: 335–340. [8]

You, L. and J. Yin. 2002. Dependence of epistasis on environment and mutation severity as revealed by in silico mutagenesis of phage T7. *Genetics* 160: 1273–1281. [7]

Yu, A. and 10 others. 2001. Comparison of human genetic and sequence-based physical maps. *Nature* 409: 951–953. [13]

Yu, H., D. Greenbaum, H. Xin Lu, X. Zhu and M. Gerstein. 2004. Genomic analysis of essentiality within protein networks. *Trends Genet.* 20: 227–231. [7]

Yu, H., H. Jiang, Q. Zhou, J. Yang, Y. Cun, B. Su, C. Xiao and W. Wang. 2006. Origination and evolution of a human-specific transmembrane protein gene, *clorf37-dup*. *Hum. Mol. Genet.* 15: 1870–1875. [3]

Yu, Q. and 15 others. 2007. Chromosomal location and gene paucity of the male specific region on papaya Y chromosome. *Mol. Genet. Gen.* 278: 177–185. [10]

Yvert, G., R. B. Brem, J. Whittle, J. M. Akey, E. Foss, E. N. Smith, R. Mackelprang and L. Kruglyak. 2003. Trans-acting regulatory variation in *Saccharomyces cerevisiae* and the role of transcription factors. *Nat. Genet.* 35: 57–64. [5]

Zalacain, M. and 19 others. 2003. A global approach to identify novel broad-spectrum antibacterial targets among proteins of unknown function. *J. Mol. Microbiol. Biotechnol.* 6: 109–126. [7]

Zeidner, G., J. P. Bielawski, M. Shmoish, D. J. Scanlan, G. Sabehi and O. Beja. 2005. Potential photosynthesis gene recombination between *Prochlorococcus* and *Synechococcus* via viral intermediates. *Environ. Microbiol.* 7: 1505–1513. [4]

Zhang, H., B. Yang, R. J. Pomerantz, C. Zhang, S. C. Arunachalam and L. Gao. 2003. The cytidine deaminase CEM15 induces hypermutation in newly synthesized HIV-1 DNA. *Nature* 424: 94–98. [8]

Zhang, J. 2003. Evolution by gene duplication: an update. *Trends Ecol. Evol.* 18: 292–298. [13]

Zhang, J. Z. and X. L. He. 2005. Significant impact of protein dispensability on the instantaneous rate of protein evolution. *Mol. Biol. Evol.* 22: 1147–1155. [2, 7]

Zhang, J., D. M. Webb and O. Podlaha. 2002a. Accelerated protein evolution and origins of human-specific features: Foxp2 as an example. *Genetics* 162: 1825–1835. [13]

Zhang, J., Y. P. Zhang, H. F. Rosenberg. 2002b. Adaptive evolution of a duplicated pancreatic ribonuclease gene in a leaf-eating monkey. *Nat. Genet.* 30: 411–415. [3]

Zhang, J., A. M. Dean, F. Brunet and M. Long. 2004. Evolving protein functional diversity in new genes of Drosophila. *Proc. Natl. Acad. Sci. USA* 101: 16246–16250. [3]

Zhang, J., R. Nielsen and Z. Yang. 2005. Evaluation of an improved branch-site likelihood method for detecting positive selection at the molecular level. *Mol. Biol. Evol.* 22: 2472–2479. [11]

Zhang, L. Q. and W.-H. Li. 2004. Mammalian housekeeping genes evolve more slowly than tissue-specific genes. *Mol. Biol. Evol.* 21: 236–239. [7]

Zhang, X. Y., O. Clarenz, S. Cokus, Y. V. Bernatavichute, M. Pellegrini, J. Goodrich and S.

E. Jacobsen. 2007. Whole-genome analysis of histone H3 lysine 27 trimethylation in Arabidopsis. *PLoS Biol.* 5: 1026–1035. [5]

Zhang, Y. and B. Oliver. 2007. Dosage compensation goes global. *Curr. Opin. Genet. Devel.* 17: 113–20. [9]

Zhang, Y., Y. Wu, Y. Liu and B. Han. 2005. Computational identification of 69 retroposons in *Arabidopsis*. *Plant Physiol.* 138: 935–948. [3]

Zhang, Z. and G. G. Carmichael. 2001. The fate of dsRNA in the nucleus: A p54(nrb)-containing complex mediates the nuclear retention of promiscuously A-to-I edited RNAs. *Cell* 106: 465–475. [8]

Zhang, Z. and M. Gerstein. 2004. Large-scale analysis of pseudogenes in the human genome. *Curr. Opin. Genet. Dev.* 14: 328–335. [13]

Zhang, Z., N. Carriero and M. Gerstein. 2004. Comparative analysis of processed pseudogenes in the mouse and human genomes. *Trends Genet.* 20: 62–67. [13]

Zhao, C., U. S. Jung, P. Garrett-Engele, T. Roe, M. S. Cyert and D. E. Levin. 1998. Temperature-induced expression of yeast FKS2 is under the dual control of protein kinase C and calcineurin. *Mol. Cell. Biol.* 18: 1013–1022. [7]

Zhaxybayeva, O., J. P. Gogarten, R. L. Charlebois, W. F. Doolittle and R. T. Papke. 2006. Phylogenetic analyses of cyanobacterial genomes: quantification of horizontal gene transfer events. *Genome Res.* 16: 1099–1108. [4]

Zhu, X., M. Gerstein and M. Snyder. 2007. Getting connected: Analysis and principles of biological networks. *Genes Dev.* 21: 1010–1024. [2]

Zilberman, D. and S. Henikoff. 2004. Silencing of transposons in plant genomes: Kick them when they're down. *Genome Biol.* 5: 249. [8]

Zuckerkandl, E. 1976. Evolutionary processes and evolutionary noise at the molecular level. I. Functional density in proteins. *J. Mol. Evol.* 7: 167–183. [2]

Zuckerkandl, E. and L. Pauling. 1965. Molecules as documents of evolutionary history. *J. Theor. Biol.* 8: 357–366. [13]

Index

About the book

Editor: Andrew D. Sinauer

Project Editor: Kathaleen Emerson

Production Manager: Christopher Small

Copy Editor: Laura Green

Book Design: Joan Gemme

Book Layout and Production: Joan Gemme

Illustrations: The Format Group, LLC

Cover Illustration and Design: Joanne Delphia

Subject Index: Sharon Hughes